Boris N. Zakhariev · Allina A. Suzko

Direct and Inverse Problems

Boris N. Zakhariev · Allina A. Suzko

Direct and Inverse Problems

Potentials in Quantum Scattering

Translated from the Russian by G. Pontecorvo

With 42 Figures

Springer-Verlag Berlin Heidelberg NewYork
London Paris Tokyo Hong Kong Barcelona

Professor Dr. Boris N. Zakhariev

Laboratory of Theoretical Physics
Joint Institute for Nuclear Research
Dubna, Moscow Region, USSR

Dr. Allina A. Suzko

Luikov Heat and Mass Transfer Institute
BSSR Academy of Science
ul. Brovki 15
220728 Minsk, USSR

Translator:

Dr. Gil B. Pontecorvo

Joint Institute for Nuclear Research
141980 Dubna
Moscow, USSR

Library of Congress Cataloging-in-Publication Data

Zakhariev, B. N. (Boris Nikolaevich)
[Potentsialy i kvantovoe rasseianie. English]
Direct and inverse problems, potentials in quantum
scattering / Boris N. Zakhariev, Allina A. Suzko; translated
from the Russian by G. Pontecorvo. p. cm.
Translation of: Potentsialy i kvantovoe rasseianie. Includes bibliographical references.
1. Scattering (Physics) 2. Potential scattering. 3. Inverse problems (Differential equations)
4. Quantum theory. I. Suz'ko, A. A. (Allina Alekseevna) II. Title.
ISBN-13: 978-3-540-52484-7 e-ISBN-13: 978-3-642-95615-7
DOI: 10.1007/ 978-3-642-95615-7
QC794.6.S3Z3513 1990 539.7'58–dc20 90-9726

Typesetting: Macmillan India Ltd., Bangalore, India.
2157/3020-543210 – Printed on acid-free paper

Preface to the English Edition

Rapid progress in quantum theory brings us new important results which are often not immediately clear to all who need them. But fortunately, this is also followed by simplifications and unifications of our previous concepts. The inverse problem method ("The most beautiful idea of the XX-th century" – Zakharov et al., 1980) has just both these aspects.

It is rather astonishing that it took 50 years after the foundation of quantum mechanics for the creation of the "pictures" showing the direct connection of observables with interactions. Recently, illustrations of this type began to appear in the literature (e.g., how potentials are deformed with the shift of one energy level or change of some resonance reduced width). Although they are transparent to those studying the quantum world and can be included within the necessary elements of quantum literacy, they are still largely unknown even to many specialists. For the first time, the most interesting of these pictures enriching our quantum intuition are collected here and placed at your disposal.

The readers of this monograph have the advantage of getting the latest information which became available after the publication of the Russian edition. It has been incorporated here in the simplest presentation possible. For example, new sections concerning exactly solvable models, including the multi-channel, multi-dimensional ones and with time dependent potentials have been added. The first attempts in solving the three-body inverse problem are also mentioned.

The authors consider it an honour that their book is published by Springer Verlag making it accessible to a wide contingent of physicists. Many soviet scientists made key contributions to inverse scattering theory (Gelfand, Levitan, Marchenko, Krein, Faddeev, Zahkarov, Henkin, Novikov et al.) and many new references to the Russian literature can be found in this book. The list of references includes the results published up to the beginning of 1990.

Unfortunately, one of the authors (A.S.) couldn't take part in the revision of the book for the English edition. Although the new material is presented in the same style (the text was written by B.Z.), the lack of useful discussions and criticism by A.S. makes me solely responsible for any shortcomings in the added sections.

Moscow, May 1990 B. N. Zakhariev

Preface

All problems pertaining to nonrelativistic quantum mechanics fall into one of the two groups:

The direct problem ($V \to S$), where known interaction potentials V are used to derive the scattering data, matrix S and for spectral data, and the inverse problem ($S \to V$), where the potentials, i.e., the forces acting in the systems, are determined from S which is obtained experimentally.

When studying new micro-objects one must deal with either unknown interactions between their constituent elementary particles (quarks, nucleons, etc.) or with unknown effective potentials between complexes formed by them. Therefore, in principle, the need for solutions of S V problems is not less than for solutions of the $V \to S$ problem.

At present, however, our understanding of the direct problem is for more complete. For a long period, quantum mechanics developed without the contributions of the inverse problem formulation derived in the USSR by Gelfand and Levitan (1951), Marchenko (1955), and Krein (1951). Since that time methods of reconstructing potentials from scattering data have continued to develop, but they still lag significantly behind the methods applied for solution of the direct problem.

The number of instances in which the inverse problem can be applied, in nuclear and atomic physics, optics, acoustics, and geophysics continues to increase. The large "soliton boom" in the investigation of nonlinear equations (Zakharov et al., 1980), as well as other studies has also contributed to this growth.

In our opinion, because they are of a general theoretical importance, elements of the inverse problem should be presented to students at an early stage, along with the first notions of the Schrödinger equation. Our experience in teaching quantum mechanics has revealed that the fundamentals of quantum-mechanical laws are best understood when the material is presented from two parallel points of view. A vivid example of this is shown by the exact solutions to various quantum models presented in Chap. 2 in their explicit analytic form. Many of these are far easier to find by solving the inverse problem, although they may simultaneously serve as solutions to the direct problem.

Besides yielding exactly solvable models, the inverse formulation of quantum mechanical problems allows a simplicity of presentation by applying the algebraic approach based on the finite-difference approximation to the Schrödinger equation (Zakhariev et al., 1977). The first chapter of this book begins with a discussion of this approximation.

Very few books on the quantum inverse problem exist. The first book was published by Agranovich and Marchenko (1960). After a long interval in which only chapters of several excellent texts or review articles were devoted to the inverse problem (Alfare and Regge, 1965; Marchenko, 1972, 1977; Newton, 1982, Faddeev, 1963, 1976), the monograph by Chadan and Sabatier (1977) was published. For specialists in theoretical physics, to whom the book was addressed, this turned out to be an important event. At last, a concise publication with a systematic presentation of $S \rightarrow V$ problems was at their disposal. The inflow of new data, and the incorporation of older results omitted by Chadan and Sabatier motivated the writing of this text.

At present, after a long period of purely mathematical investigations, recently reviewed by Levitan (1984), the development of the inverse problem has reached a stage where most efforts are concentrated on developing practical solution algorithms. The existing difficulties are related to the so called ill-posedness of the problems. Thus, under the assumption that a solution exists its uniqueness and intensitivity to small errors in the initial parameters may be violated. Regularization of the solutions of the inverse problem remains an art to a large extent. Nevertheless, more and more frequent publications in which the interaction potential is successfully derived from spectral parameters and scattering data are appearing. In addition, examples of algorithms for solving one-dimensional, single-channel, multi-dimensional, multi-channel and multi-particle equations of the direct and inverse problems, are provided.

Most results in this book are either new, presented for the first time, or older ones given a new interpretation. Such is the case for the approximate reconstruction of potentials, where the accepted deviations can be controlled. A review of the various versions of the Levinson theorem, and effects related to the penetrability of barriers for tunneling by simple and composite particles, are given. Additional topics include the finite-element methods and step-wise potentials in the direct and inverse problems, as well as a number of other issues.

Various types of potentials are discussed such as those that are transparent at all energies, "confining" potentials, phase- and spectral-equivalent potentials, singular ones, those dependent on energy and velocity, potentials permitting exact solutions of the Schrödinger and the Gelfand-Levitan-Marchenko equations, potentials that are local or nonlocal with respect to some of the variables, short- and long-range, spherically and hyperspherically symmetric potentials (as well as potentials with violated symmetries), one-, two-, and multi-particle potentials, matrix potentials, and others.

This book consists of two parts: Part I is devoted to one-dimensional and one-channel systems, Part II deals with multi-channel, multi-dimensionals, and multi-particle systems.

New concepts are generally introduced by using simple examples. They are then applied, without proof, to more general cases.

To aid the reader in achieving a more thorough understanding of this subject, we have provided comments on the literature at the end of each chapter.

The writing of this book has been made possible by collaborations during various periods with I.B. Amirkhanov, P. Beregy, V.P. Zhigunov, N.A. Kostov, V.N. Mel'nikov, P.Yu. Nikishov, S.A. Niyazgulov, V.N. Pivovarchik, E.B. Plekhanov,

B. V. Rudyak, and Ya. A. Smorodinsky, who long ago triggered the curiosity of one of us (B. N. Z.) about inverse problems. We remember our collaboration as a time of happy discoveries and revelations (no matter how significant they turned out to be in an absolute sense). Support was also given by O. G. Martynenko and V. L. Kolpaschikov. We wish to convey our gratitude to all individuals involved.

Moscow/Minsk, July 1990 B. N. Zakhariev, A. A. Suzko

Table of Contents

Part I: One-dimensional, One-channel Systems

Part II. Multi-channel, Multi-dimensional, Multi-particle Problems

Part I. One-dimensional, One-channel Systems

Chapter 1
The Principal Equations of Scattering Theory

1.1 General Remarks

The four chapters of Part I are devoted to problems related to the one-dimensional motion of a single particle in an external field. The main concepts and equations are introduced in Chap. 1 within the framework of various formulations of the direct and inverse problems. Both the general aspects and special features of the respective formalisms are emphasized.

Specific solutions of the equations are dealt with in Chaps. 2 and 3. Cases involving the explicit form of known exact solutions are considered in Chap. 2, while approximate methods for computing wave functions and potentials with the aid of computers are discussed in Chap. 3. Chapter 4 ends Part I with a discussion of the various forms of the Levinson theorem. This postulate establishes relationships between parameters of discrete and continuous spectra for all kinds of interactions.

For a first demonstration of how a wave function is derived from a given potential V and, alternatively, how V is reconstructed from spectral data, the simplest "discrete model of quantum mechanics" involving a difference analogue of the Schrödinger equation is extremely convenient. With the discrete space variable x_n $n = 0, \pm 1, \pm 2 \ldots$ instead of a continuous variable x, all operations in quantum mechanics become purely algebraic and more straightforward. In Sect. 1.2 the equations of the inverse problem are derived from the equations of motion and the completeness relations for the eigenstates within the utmost elementary approach. In Sect. 1.3 we present the procedure for obtaining the Gelfand–Levitan–Marchenko equations by using as examples their algebraic analogues. In Sect. 1.4 various forms of direct and inverse problems are considered. It should be noted that certain formulas are assigned the same numbers in different parts of the book because their meaning is best understood by comparison. The letter "a" indicates finite-difference analogues of formulas corresponding to continuous space variables.

1.2 Elements of the Direct and Inverse Problems

The laws of quantum motion are concentrated in the Schrödinger equations. In order to determine the properties of any system it should be sufficient to solve the Schrödinger equation with corresponding boundary conditions.

Usually students learn quantum mechanics via a study of the one-dimensional motion of a particle in an external field V. In this case the Schrödinger equation is just an ordinary differential equation,

$$-\frac{\hbar^2}{2m}\psi''(x) + V(x)\psi(x) = E\psi(x)\,, \tag{1.2.1}$$

where $\psi(x)$ is the wave function determining the probability $|\psi(x)|^2$ of finding a particle of mass m at point x; E is the total energy of the particle. Naturally, ψ also depends on an energy variable, but we shall write this argument of ψ in its explicit form only if necessary.

1.2.1 The Simplest Difference Schrödinger Equation

In spite of its simplicity, (1.2.1) can be analytically solved only in certain special cases. One usually deals with the solutions of discrete, for example, finite-difference, analogues of (1.2.1), where energy or spatial variables have been discretized. Such an approach not only simplifies the mathematics, it can also be surprisingly revealing about laws governing the microcosm and yield a deeper understanding of already known phenomena.

Let us now consider the formulation and the most simple solutions of the direct and inverse problems with the aid of difference equations. For clarity we shall make use of the most simple, the Euler, version of the finite-difference Schrödinger equation which, although being less accurate than other approximations of (1.2.1), is especially close in structure to its differential prototype:

$$-\frac{\hbar^2}{2m}\left\{\frac{\psi(n+1) - 2\psi(n) + \psi(n-1)}{\Delta^2}\right\} + V(n)\psi(n) = E\psi(n)\,, \tag{1.2.1a}$$

where $\psi(n)$ represents the value of the wave function at the point $x_n = n\Delta$; Δ is the constant finite-difference step; the expression in the braces is the second finite-difference derivative. Equation (1.2.1a) represents a set of recursion relations, each of which relates the values of ψ at three neighboring points x_n, $x_n \pm 1$.

Equations (1.2.1,1a) possess many common properties. They will be pointed out as the need arises. For a unique solution, (1.2.1,1a) must be supplemented with boundary conditions. Since both the differential and difference operators in (1.2.1,1a) are of second order, two boundary conditions are necessary. The general solutions of (1.2.1,1a) are constructed in the form of linear combinations of two independent partial solutions with arbitrary constant coefficients.

1.2.2 Potential Wells of Infinite Depth

As the first example for illustrating the direct and inverse problems we shall consider the motion of a particle in a potential well with infinite walls at $x \leqslant 0$ and $x \geqslant a$, where a represents the width of the well. There are two arguments in favor of that choice. First, we shall have to deal with a *purely discrete spectrum* of eigenstates, which, together with the discrete nature of the coordinate in (1.2.1a), will further simplify solution of the inverse problem [will render finite the number of parameters used for the reconstruction of $V(x)$]. Second, one can readily pass from this case to potentials of a finite range, equal to zero for $x \geqslant a$ when a continuous spectrum exists along with the discrete spectrum. This is achieved by applying the R-matrix scattering theory, in which states of the continuum spectrum are parametrized by using a discrete set of numbers: the bound state levels of an infinitely deep well simply become the points of R-matrix resonances when the infinite potential wall at $x \geqslant a$ is removed. (Actually, this occurs only when certain boundary conditions are imposed on the functions of the R-resonance states.)

As boundary conditions in the case of an infinite well we shall require the eigenfunctions to be zero at the points $x = 0$ and $x = a$:

$$\psi(0) = \psi(a) = 0 \, . \tag{1.2.2}$$

Now let us partition the interval $[0, a]$ with the aid of N points at equal distances from each other, i.e., $\Delta = a/(N + 1)$. Using the discrete variable of the difference problem one can rewrite conditions (1.2.2):

$$\psi(0) = \psi[(N + 1)\Delta] = 0 \, . \tag{1.2.2a}$$

Equalities (1.2.2a) separate from the infinite chain of equations (1.2.1a), where $-\infty < n < \infty$, a *finite* closed set of N homogeneous algebraic equations which can be represented in matrix form shown on p. 4.

The three-diagonal matrix (1.2.3a) is equivalent to the combination of (1.2.1a,2a). Unlike the problem (1.2.1,2) which possesses an infinite number of eigensolutions, only N bound states with $E = E_\lambda$, where $\lambda = 1, \ldots, N$, correspond to (1.2.3a). The values E_λ are derived from the condition that the determinant of the matrix $H - E$ be equal to zero. We thus have at our disposal a closed quantum-mechanical model that is not only discrete, but also finite-dimensional (where we refer to the dimensionality of the Hilbert space of wave functions). The fact that N is finite allows complete and exact solution of the direct and inverse problems with the aid of a finite number of algebraic operations.

The finite-difference method has not been chosen here for the purpose of introducing the modern methods of precise numerical solution of quantum equations. Rather, it is more important for us to clarify the main points of the procedure for constructing ψ from a known $V(x)$ and for reconstruction of $V(x)$ from the boundary values of ψ. We note, in passing, that the discrete space variable may not only serve as a convenient approximation, but may also inherently contain the physical essence of a problem, as in the case of elastic

$$\begin{vmatrix} \frac{\hbar^2}{\Delta^2 m}+V(1) & -\frac{\hbar^2}{2\Delta^2 m} & 0 & & & \\ -\frac{\hbar^2}{2\Delta^2 m} & \frac{\hbar^2}{\Delta^2 m}+V(2) & -\frac{\hbar^2}{2\Delta^2 m} & 0 & & \\ 0 & -\frac{\hbar^2}{2\Delta^2 m} & & & & \\ & 0 & & & 0 & \\ & & & & -\frac{\hbar^2}{2\Delta^2 m} & 0 \\ & & 0 & -\frac{\hbar^2}{2\Delta^2 m} & \frac{\hbar^2}{\Delta^2 m}+V(N-1) & -\frac{\hbar^2}{2\Delta^2 m} \\ & & & 0 & -\frac{\hbar^2}{2\Delta^2 m} & \frac{\hbar^2}{\Delta^2 m}+V(N) \end{vmatrix}$$

$$\times \begin{pmatrix} \psi_{(1)} \\ \psi_{(2)} \\ \cdot \\ \cdot \\ \cdot \\ \psi_{(N-1)} \\ \psi_{(N)} \end{pmatrix} = E \begin{pmatrix} \psi_{(1)} \\ \psi_{(2)} \\ \cdot \\ \cdot \\ \cdot \\ \psi_{(N-1)} \\ \psi_{(N)} \end{pmatrix} \tag{1.2.3a}$$

waves in crystals. To show how close the inverse problem can be made to approach the direct one, we shall recall the familiar method of constructing $\psi(x)$ when the potential $V(x)$ is given.

1.2.3 The Direct Problem

Numerical solution of (1.2.1a) is more convenient when the values of ψ are known at two adjacent points near one of the edges of the interval $[0, a]$, while (1.2.3a) corresponds to values of ψ (1.2.2a) being given at different ends of the interval $[0, a]$. Therefore, instead of ψ, we shall search for so-called regular solutions φ for which the boundary conditions are the following:

$$\varphi(0) = 0\,, \quad \varphi'(0) = 1 \quad \text{for (1.2.1)}\,; \tag{1.2.4}$$

$$\varphi(0) = 0\,, \quad \varphi(1) = \Delta \quad \text{for (1.2.1a)}\,. \tag{1.2.4a}$$

The φ are called regular because the corresponding radial functions $\varphi(r)/r$ of the three-dimensional problem remain finite as $r \to 0$. Equalities (1.2.4a) correspond to the requirement that the difference derivative be equal to unity at the origin, just like the conventional derivative in (1.2.4),

$$[\varphi(1) - \varphi(0)]/\Delta = 1 .$$

The sought solutions $\psi(x)$ are expressed through $\varphi(x)$ that are subjected to *inhomogeneous* boundary conditions and are determined at all energies. At $E = E_\lambda$, where ψ also exists, both solutions can only differ by a constant factor:

$$\psi(E_\lambda, n) = c_\lambda \varphi(E_\lambda, n) , \tag{1.2.5}$$

since one of the boundary conditions is common for both of them:

$$\psi(0) = \varphi(0) = 0 .$$

The solution of the second-order equation (1.2.1a) is uniquely determined by values at two points. At $n = 0$ functions φ and ψ coincide, at $n = 1$ $\varphi(1) = \Delta$, while $\psi(E_\lambda, 1)$ are certain numbers Δc_λ. Hence, if we multiply $\varphi(n)$ by $c_\lambda = \psi(E_\lambda, 1)/\Delta$, we will obtain a function coinciding with ψ at two points, $n = 0$ and $n = 1$ and, consequently, at all other points.

In the direct problem the numerical coefficients c_λ are determined after $\varphi(E_\lambda, n)$ is derived from the normalization conditions:

$$\sum_{n=1}^{N} \psi^2(E_\lambda, n)\Delta = \sum_{n=1}^{N} c_\lambda^2 \varphi^2(E_\lambda, n)\Delta = 1 ;$$

$$c_\lambda^{-2} = \Delta \sum_{n=1}^{N} \varphi^2(E_\lambda, n) . \tag{1.2.6}$$

We have drawn special attention to the normalization coefficients c_λ, since they play an important role in the inverse problem. Together with the set E_λ they form the initial spectral information for determination of the potential $V(n)$. In the R-matrix scattering theory the amplitudes γ_λ of reduced widths of R-resonances serve as such normalizing coefficients. Unlike c_λ, the amplitudes γ_λ are related to the eigenfunctions at the other edge of the interval $[0, a]$, where the interaction region touches the external region of free motion and where the functions being matched fix the parameters of the scattered waves (Zakhariev et al 1977) considers the R-matrix in greater detail.

Usually, determination of c_λ from experiment is quite complicated, but in the case of quark-antiquark systems $q\bar{q}$ with confining potentials c_λ can be calculated directly from the probabilities of the $q\bar{q}$ lepton decays occurring in the vicinity of $x = 0$ (Thaker 1978; Thaker 1980).

Thus, having defined $\varphi(n)$ at two adjacent points near the edge of the interval of finite-difference "integration" we find in succession the function φ at point x_2 from (1.2.1a). Then $\varphi(1)$ and $\varphi(2)$ give φ at the next point x_3, and so on, throughout the whole length of the interval $[0, a]$. From (1.2.6) we determine c_λ, which yields the desired eigenfunctions $\psi(E_\lambda, n) = c_\lambda \varphi(E_\lambda, n)$. The direct problem is, thus, solved. Now we shall proceed with the reconstruction of the potentials $V(n)$ from the set of spectral parameters $\{E_\lambda, c_\lambda\}$.

1.2.4 The Inverse Problem

The inverse problem can be solved within the finite-difference approach with the aid of (1.2.1a) in a way that closely resembles the procedure just described. This solution serves as an example for rendering the more general Gelfand–Levitan–Marchenko method, considered in the next section, more comprehensible.

The constants c_λ fix the values of ψ at the point $n=1$ for $E=E_\lambda$, $\psi(E_\lambda, 1) = c_\lambda \Delta$, i.e., ψ is known at two neighboring points, at $n=0$ in accordance with (1.2.2a) and at $n=1$. However, this is insufficient for solving the inverse problem. Its peculiarity lies in that now one must not only determine $\psi(E_\lambda, n = 2, 3, \ldots)$, as in the direct problem, but also $V(n = 1, 2, \ldots)$. Equations (1.2.1a) or (1.2.3a) are now insufficient: there are half as many equations as the number of unknowns. Thus, in (1.2.1a) when $n=1$, $E=E_\lambda$, or in the first line of (1.2.3a) two unknown values occur, $\psi(E_\lambda, 2)$ and $V(1)$. Shifting the arguments of the functions in (1.2.1a) by a single step or passing to the second line in (1.2.3a) yields a new algebraic equation involving two new unknowns. To get rid of the extra unknown, further relations between the desired quantities are necessary. The role of these may be attributed to the relations of completeness for the eigenfunctions (see their derivation below):

$$\sum_{\lambda=1}^{N} \psi(E_\lambda, n)\psi(E_\lambda, m) = \delta_{mn}/\Delta \ . \tag{1.2.7}$$

Actually, this also represents the conditions of orthonormalization with respect to the energy variable, instead of the space variable. As usual:

$$\sum_{n=1}^{N} \psi(E_\lambda, n)\psi(E_\mu, n)\Delta = \delta_{\lambda,\mu} \ . \tag{1.2.8}$$

We shall take advantage of the orthogonality of the functions ψ (1.2.7) at different points of the interval in order to discard the unknown $\psi(E_\lambda, 2)$ in (1.2.1a) with $n=1$, $E=E_\lambda$. To this end we multiply (1.2.1a) by $\psi(E_\lambda, 1)$ and take the sum over λ. As a result, we obtain an equation with a single unknown, $V(1)$:

$$V(1) = \Delta \sum_{\lambda=1}^{N} E_\lambda \psi^2(E_\lambda, 1) - \frac{\hbar^2}{m\Delta^2} \ , \tag{1.2.9}$$

where $\psi(E_\lambda, 1) = c_\lambda \Delta$ according to (1.2.4a,5).

The first hurdle in solving the inverse problem has been overcome—an expression has been obtained for the potential at point $n=1$ in terms of the set of spectral parameters $\{E_\lambda, c_\lambda\}$. Having determined $V(1)$, we shall now find from (1.2.3a) for $n=1$, the values of the function at the next point

$$\psi(E_\lambda, 2) = \left\{\frac{2m}{\hbar^2}\Delta^2[V(1) - E_\lambda] + 2\right\}\psi(E_\lambda, 1) \ . \tag{1.2.10}$$

Repeating the operations used in determining $V(1)$ and $\psi(E_\lambda, 2)$ we find $V(2)$, $\psi(E_\lambda, 3)$ and so on. Thus, solution of the inverse problem reduces to alternating application of the formulas

$$V(n) = \Delta \sum_{\lambda=1}^{N} E_\lambda \psi^2(E_\lambda, n) - \frac{\hbar^2}{m\Delta^2};$$

$$\psi(E_\lambda, n+1) = \left\{\frac{2m}{\hbar^2}\Delta^2[V(n) - E_\lambda] + 2\right\}\psi(E_\lambda, n) - \psi(E_\lambda, n-1). \tag{1.2.11}$$

Unlike the direct problem, solution of the inverse problem involves states from the *entire* spectrum of the system, owing to the use of the relations of completeness (1.2.7).

An essential part in the reconstruction of V besides the condition of completeness, is the fact that the potential is *local* since its matrix as given in (1.2.3a) is diagonal. If the interaction were nonlocal, instead of $V(n)$ in (1.2.1a) there would occur at the single point n the sum $\sum_{n'} V(n, n')\psi(n')$ and recursion relations such as (1.2.11) would not be appropriate for determining $V(n, n')$. One can consider the *three-diagonal* potential matrix which is used for velocity-dependent forces in the finite-element method as a deviation from locality. This is also true when instead of (1.2.1a), the Schrödinger–Störmer equation is applied, *Zakhariev* et al (1979).

1.2.5 Scattering by a Potential of Finite Range

In the R-matrix scattering theory minimal changes are required in the case of an interaction of the finite range $a = (N+1)/\Delta$ in problems involving the continuous spectrum. We shall demonstrate how a discrete set of resonance constants $\{E_\lambda, \gamma_\lambda\}$, similar to the set $\{E_\lambda, c_\lambda\}$ used above for bound states, parametrizes the scattering data.

Outside the region of action of the potential the solutions of (1.2.1,1a) have the form of free waves. There exists a correspondence between the functions $\exp(\pm ikx)$, $\sin kx$, $\cos kx$ and $\exp(\pm i\theta n)$, $\sin\theta n$, $\cos\theta n$, which satisfy the differential Schrödinger equation (1.2.1) with $V(x) \equiv 0$ and its finite-difference analogue (1.2.1a) with $V(n) \equiv 0$. The quantity θ, taking the role of the wave number in discrete-coordinate problems, is related to the energy E by:

$$\cos\theta = 1 - E\Delta^2 m/\hbar^2 .$$

This can be compared with usual relationship between wave number and energy $k = \sqrt{2mE}/\hbar$. To the radial part of the physical wave function $\psi(k, r)$ with the asymptotics for $l = 0$

$$\psi(k, r) \approx \frac{i}{2}[\exp(-ikr) - s(k)\exp(ikr)] \quad \text{for } r \geqslant a \tag{1.2.12}$$

corresponds the following solution:

$$\psi(\theta, n) = \exp(-i\theta n) - s_{fd}(\theta)\exp(i\theta n) \quad \text{for } n > N, \tag{1.2.12a}$$

where s_{fd} is the finite-difference scattering function.

There are also discrete analogues of the hypergeometric functions which form a wide class of exact solutions for often-used potentials (Suslov 1989).

The function of free motion (1.2.12) must be joined smoothly with the wave function ψ at $r \leqslant a$ on the boundary of the interaction region. Therefore, giving the scattering S-function is equivalent to fixing the logarithmic derivative $\psi'/\psi|_{r=a} \equiv \mathscr{R}^{-1}(E)$ at the edge of the interaction region. In the finite-difference case

$$\mathscr{R}^{-1}(E) \equiv [\psi(N+1) - \psi(N)]a/\Delta\psi(N) .$$

Since ψ can be expanded in this region into a discrete set of eigenfunctions

$$u(E_\lambda, n) \equiv u_\lambda(n)$$

of the Schrödinger operator with the homogeneous boundary conditions $u_\lambda(0) = u_\lambda(N+1) = 0$, the energy dependences $R(E)$ and $s(E)$ are parametrized by a discrete set of parameters. In the finite-difference problem the number of such parameters

$$E_\lambda, \gamma_\lambda \equiv u_\lambda(N)\hbar/\sqrt{2am}$$

also becomes finite and equal to $2N$. This parametrization assumes an especially compact form for the function

$$R(E) = \mathscr{R}(E)/[1 + \mathscr{R}(E)a/\Delta] ,$$

which is known as the R-matrix:

$$R(E) = \sum_{\lambda}^{N} \gamma_\lambda^2/(E_\lambda - E) , \tag{1.2.13}$$

where E_λ represent the positions of the R-resonances, and γ_λ^2 are their reduced widths. Although in the one-dimensional one-channel case the R-matrix becomes a scalar function, in the literature the term "matrix" is conventionally used in all problems. In special case $R = \mathscr{R}$. In the general case $u_\lambda(N+1) = u_\lambda(N)(1 + \Delta B/a)$, which corresponds to $u'_\lambda(r)/u_\lambda(r)|_{r=a} = B$; $R = \mathscr{R}/(1 - \mathscr{R}B)$; and at $B = 0$, $R = \mathscr{R}$; in our case $B = -a/\Delta$.

A simple derivation of this formula will be given below. The chosen auxiliary functions $u_\lambda(n)$ coincide with the bound-state functions $\psi(E_\lambda, n)$ of infinite potential well except that in the present case at the point $n = N + 1$ we have $V(N+1) = 0$ instead of an infinite potential wall. Nevertheless we require that $u_\lambda(N+1) = 0$. The role of the normalizing constants c_λ is now taken over by γ_λ which, unlike c_λ, represent the values of the eigenfunctions not at $n = 1$, but at $n = N$, at the opposite edge of the interval $[0, a]$.

As for the bound states, $V(n)$ is reconstructed from the set $\{E_\lambda, \gamma_\lambda\}$. Now, however, $V(n)$ and $u_\lambda(n)$ must be constructed by going from one point to another in the opposite direction, from $n = N$ to $n = 0$, and in the recursion formulas (1.2.11) one must substitute $n + 1 \leftrightarrow n - 1$.

The parameters E_λ and γ_λ can be determined from the S-function since it is uniquely related to R. Indeed, by making use of the relations between R and $\mathscr{R}$ and between $\mathscr{R}$ and ψ we obtain

$$R(E) = \frac{\psi(E, N)}{\psi(E, N+1)} \frac{\Delta}{a}$$

$$= \frac{\exp(-i\theta N) - s(\theta)\exp(i\theta N)}{\exp[-i\theta(N+1)] - s(\theta)\exp[+i\theta(N+1)]} \frac{\Delta}{a} . \tag{1.2.14}$$

In accordance with (1.2.13) and (1.2.14) we find E_λ as the energies at which the denominator in (1.2.14), considered as a function of E, turns to zero, while γ_λ^2 represents the limit

$$\gamma_\lambda^2 = \lim_{E \to E_\lambda} [R(E)(E_\lambda - E)] . \tag{1.2.15}$$

We shall repeatedly turn to the R-matrix approach as it is extremely convenient for clarifying certain points of the theory. In practice, however, the R-matrix formalism is not applied often. The limitation to forces of a strictly finite range is not always desired, and the positions and widths of the R-resonances E_λ and γ_λ^2 may differ significantly from the parameters of the S-resonances. The values of E_λ and γ_λ depend on the choice of the boundary conditions for the auxiliary functions u_λ, although this dependence is not noticeably present in the expressions for the observed quantities (in numerical calculations this holds only up to the chosen approximation).

1.2.6 The Finite-difference Analogue of the R-matrix Scattering Theory

For forces of a finite range it suffices to know the wave function within the interval $[0, a]$ to describe the properties of the system since continuation of the ψ into the region of free motion ($r > a$) is trivial. Because of this, as pointed out above, it turns out to be possible by using a countable set of eigenfunctions on $[0, a]$, to parametrize in a *discrete* manner the *continuous* energy dependence of the logarithmic derivative of the wave function at point a and the scattering data at the same time. This is the essence of the R-matrix approach.

Now let us expand the wave function $\psi(E, n)$ over $u_\lambda(n)$:

$$\psi(E, n) = \sum_{\lambda=1}^{N} A_\lambda(E) u_\lambda(n) , \tag{1.2.16}$$

where, owing to the fact that the functions $u_\lambda(n)$ are orthogonal,

$$A_\lambda(E) = \sum_{n=1}^{N} \psi(E, n) u_\lambda(n) \Delta . \tag{1.2.17}$$

We now multiply (1.2.1a) for ψ by $u_\lambda(n)$ and the similar equation for $u_\lambda(n)$ by $\psi(E, n)$, then subtract the second one from the first and perform summation of the result over n. Then by using (1.2.17) and the boundary condition $u_\lambda(N+1) = 0$, we obtain:

$$A_\lambda(E) = \frac{\hbar^2}{2m\Delta} \frac{\psi(E, N+1) u_\lambda(N)}{E_\lambda - E} . \tag{1.2.18}$$

We now substitute (1.2.18) into (1.2.16):

$$\psi(E, n) = \frac{\hbar^2}{2m} \sum_{\lambda=1}^{N} \frac{u_\lambda(n) u_\lambda(N)}{E_\lambda - E} \psi(E, N+1) . \tag{1.2.19}$$

Introducing the notation

$$\gamma_\lambda = \frac{\hbar}{\sqrt{2ma}} u_\lambda(N) ,$$

we find, from (1.2.14,19) at $n = N$, the function $R(E)$ in the form of (1.2.13):

$$R(E) = \sum_{\lambda=1}^{N} \gamma_\lambda^2 / (E_\lambda - E) .$$

When bound states are present in the potential, the parameters $E_\lambda < 0$, and the respective γ_λ can be found from a nonlinear fit of the scattering data.

1.2.7 Conditions of Orthonormality and Completeness of the Eigenfunctions of the Finite-difference Schrödinger Operator on [0, a]

Let us multiply (1.2.1a) for u_λ by $u_{\lambda'}(n)\Delta$ and the similar equation for $u_{\lambda'}(n)$ by $u_\lambda(n)\Delta$, then we subtract one from the other and perform summation over n from 1 to N. Taking into account the boundary conditions, we obtain

$$(E_\lambda - E_{\lambda'}) \sum_{n=1}^{N} \Delta u_\lambda(n) u_{\lambda'}(n) = 0 . \tag{1.2.20}$$

Consequently, when $\lambda \neq \lambda'$, the eigenfunctions are orthogonal. Thus, after normalizing them we get

$$\sum_{n=1}^{N} \Delta u_\lambda(n) u_{\lambda'}(n) = \delta_{\lambda\lambda'} . \tag{1.2.21}$$

The conditions of completeness for $u_\lambda(n)$ are obtained by multiplying both sides of (1.2.21) by $u_{\lambda'}(m)$, then summing over λ' and changing the order of summation over n and λ'. This is quite legitimate because N is finite:

$$\sum_{n=1}^{N} \left\{ \Delta \sum_{\lambda'=1}^{N} u_{\lambda'}(m) u_{\lambda'}(n) \right\} u_\lambda(n) = u_\lambda(m) . \tag{1.2.22}$$

Indeed, the expression within braces in the left-hand side of (1.2.22) is similar to the Kronecker delta δ_{mn}:

$$\sum_{\lambda'=1}^{N} u_{\lambda'}(m) u_{\lambda'}(n) = \delta_{mn} / \Delta . \tag{1.2.23}$$

The relations of orthonormality and completeness for wave functions of bound states (1.2.7,8) are equivalent to (1.2.21,23).

1.2.8 Relations Between the Scattering Parameters $\{E_\lambda$ and $\gamma_\lambda\}$

Since values of the potential $V(n)$ at N points of the interval $[0, a]$ determine $2N$ spectral parameters $\{E_\lambda, \gamma_\lambda\}$, the latter must be interrelated. If, on the contrary, $\{E_\lambda, \gamma_\lambda\}$ are chosen arbitrarily, no local potential corresponds to them. In Sect. 1.3 it is shown that (within the difference approach) the set $\{E_\lambda, \gamma_\lambda\}$ corresponds to a quasi-local velocity-dependent potential.

No simple expressions have been obtained for the indicated relation. A straightforward method for the derivation of the quite cumbersome nonlinear relations between $\{E_\lambda\}$ and $\{\gamma_\lambda\}$ is presented in the review by *Zakhariev* et al (1977). In the case of a continuous coordinate the restrictions imposed on the denumerable set $\{E_\lambda, \gamma_\lambda\}$ are significantly weaker. Actually, they were not formulated directly for $\{E_\lambda, \gamma_\lambda\}$ but for the spectra $\{E_\lambda\}$ and $\{\gamma_\lambda\}$ for two Sturm–Liouville problems (Levitan and Gasymov 1964) on $[0, a]$. These involved the same potential but different boundary conditions, such as

$$u_\lambda(0) = u_\mu(0) = u_\lambda(a) = u'_\mu(a) = 0 \; .$$

The spectra should alternate,

$$-\infty < E_{\mu_1} < E_{\lambda_1} < E_{\mu_2} < E_{\lambda_2} < \dots$$

and exhibit the asymptotic behavior,

$$E_\lambda = \lambda^2 - 2A + \alpha_\lambda \; ; \quad E_\mu = (\mu - \tfrac{1}{2})^2 - 2A + \beta_\mu \; ,$$

where A is an arbitrary number and $\sum_i^\infty \alpha_i^2 < \infty \, ; \sum_i^\infty \beta_i^2 < \infty$ (Marchenko 1977).

1.2.9 Reconstruction of the Potential on the Semi-axis $0 \leqslant x < \infty$

Let us consider one more method of solving the inverse problem by making use of the completeness relation for the wave functions $\psi(n)$ on the unlimited interval $0 < n < \infty$:

$$\frac{1}{2\pi} \int_0^{2\hbar^2/\Delta^2 m} \psi(E, m)\psi^*(E, n)dE + \sum_\lambda^N \psi(E_\lambda, m)\psi(E_\lambda, n) = \frac{\delta_{mn}}{\Delta} \; . \qquad (1.2.24a)$$

Here summation is performed over bound states, if such exist. The peculiarity of the continuous spectrum of the finite-difference Schrödinger operator is that it has an upper limit

$$E_{\max} = 2\hbar^2/\Delta^2 m \; ;$$

therefore, integration over E in (1.2.24a) is performed over the finite interval $[0, 2\hbar^2/\Delta^2 m]$. This fact may be explained by the example of free motion. In problems with a continuous variable the complete set on the semi-axis $0 \leqslant x < \infty$ is composed of oscillating functions (such as $\sin kx$, for example), of all possible frequencies, stretching from zero to infinity. On the other hand, in the case of a discrete coordinate there exist no frequencies higher than the maximal one equal to $2/\Delta$, which corresponds precisely to $E_{\max} = 2\hbar^2/\Delta^2 m$.

The reconstruction procedure of $V(n)$ and $\psi(E, n)$ with the aid of (1.2.24a) differs from (1.2.11) only in that the integral over energy is added in the recursion relations and that the direction in which V and ψ are reconstructed goes from large n down to $n = 1$:

$$V(n) = \frac{\Delta}{2\pi} \int_0^{2\hbar^2/m\Delta^2} E|\psi(E, n)|^2 dE + \Delta \sum_{\lambda=1}^{N} E_\lambda \psi^2(E_\lambda, n) - \frac{\hbar^2}{m\Delta^2} ;$$

$$\psi(E, n-1) = \left\{ \frac{2m}{\hbar^2} \Delta^2 [V(n) - E] + 2 \right\} \psi(E, n) - \psi(E, n+1) .$$

In the inverse problem the S-function is known, so that the asymptotic behavior of the wave function (1.2.12a) is also known. The solution can be initiated from two adjacent points where ψ is of the form (1.2.12a). Instead of E, the variable θ can be used.

1.3 The Gelfand–Levitan–Marchenko Equations

The *Gelfand–Levitan* (1951) method arose as a continuous generalization of the corresponding results of the Jacobi matrix theory. At that time this theory seemed unsuitable for direct application to scattering problems. It concerns purely discrete quantities, while quantum collisions are described by wave functions dependent on the continuously changing coordinate and corresponding to states of the continuous spectrum of the Hamiltonian of the interacting particles.

However, as pointed out in Sect. 1.1, in the case of finite-range forces the finite-difference Schrödinger equation reduces to a finite set of algebraic equations with a three-diagonal coefficient matrix (i.e., the Jacobi matrix). There exists, therefore, a purely algebraic analogue of the Gelfand–Levitan–Marchenko theory in the physically justified quantum scattering model with discrete variables r_n, E_λ of space and energy. Thus, it becomes possible to return to the initial algebraic technique in the inverse scattering problem, but at a new level in the theory. In this section the integral equations of the inverse problem will be obtained in the limit $\Delta \to 0$ from finite-difference analogues, the derivation of which is more clear. The exact and approximate solutions of the inverse-problem equations will be dealt with in Chaps. 2, 3.

1.3.1 The Regular Solutions φ and $\mathring{\varphi}$

Consider the energy dependence of the solution introduced in Sect. 1.1 of the finite-difference Schrödinger equation (1.2.1a) with the boundary conditions (1.2.4a) and of the corresponding solution $\mathring{\varphi}$ for the equation without potential,

$\mathring{V}(n) = 0$:

$$-\frac{\hbar^2}{2m\Delta^2}[\mathring{\varphi}(E, n+1) - 2\mathring{\varphi}(E, n) + \mathring{\varphi}(E, n-1)] = E\mathring{\varphi}(E, n) \,. \qquad (1.3.1)$$

It is easy to verify that $\varphi(E, n)$ and $\mathring{\varphi}(E, n)$ are polynomials of the $(n-1)^{\text{th}}$ order in energy E. Indeed, in accordance with (1.2.4a) $\varphi(E, 1) = \mathring{\varphi}(E, 1) = \Delta$, i.e., they are independent of E (polynomials of the *zeroth order* in E). At the next point $n = 2$, both φ and $\mathring{\varphi}$ depend on E *linearly* since $\mathring{\varphi}(E, 2)$ and $\varphi(E, 2)$ are determined from (1.3.1) and (1.2.1a), or (1.2.3a), through the single energy factor E present in the right-hand side of (1.3.1, 2.3a), and the constants

$$\varphi(E, 0) = \mathring{\varphi}(E, 0) = 0 \quad \text{and} \quad \varphi(E, 1) = \mathring{\varphi}(E, 1) = \Delta \,.$$

At each next shift by one step the factor E will increase the order of polynomial energy dependence of φ and $\mathring{\varphi}$ by one. At the same n the polynomials φ and $\mathring{\varphi}$ differ only in their coefficients.

Construction of φ from $\mathring{\varphi}$ is easily achieved since the polynomial $\varphi(E, n)$ of order $(n-1)$ can be represented in the form of a linear combination of known polynomials $\mathring{\varphi}(E, m)$ of the orders $m - 1 = n - 1, n - 2, \ldots, 0$ with coefficients $K(n, m)$ containing all the information on the interaction:

$$\begin{aligned}
&\varphi(E, 1) = \Delta K(1, 1)\mathring{\varphi}(E, 1) \,; \\
&\varphi(E, 2) = \Delta K(2, 2)\mathring{\varphi}(E, 2) + \Delta K(2, 1)\mathring{\varphi}(E, 1) \,; \\
&\varphi(E, 3) = \Delta K(3, 3)\mathring{\varphi}(E, 3) + \Delta K(3, 2)\mathring{\varphi}(E, 2) + \Delta K(3, 1)\mathring{\varphi}(E, 1) \,; \\
&\ldots\ldots\ldots\ldots\ldots\ldots\ldots\ldots\ldots \qquad (1.3.2a) \\
&\varphi(E, n) = \Delta K(n, n)\mathring{\varphi}(E, n) + \Delta K(n, n-1)\mathring{\varphi}(E, n-1) + \ldots \\
&\qquad\quad + \Delta K(n, 1)\mathring{\varphi}(E, 1) \,; \\
&\ldots\ldots\ldots\ldots\ldots\ldots\ldots\ldots\ldots
\end{aligned}$$

The factors Δ in (1.3.2a) are conveniently singled out in all the sums over the coordinate variable, so that in the limit when $\Delta \to dx$ these sums transform into the corresponding integrals:

$$\varphi(E, x) = \mathring{\varphi}(E, x) + \int_0^x dx' K(x, x')\mathring{\varphi}(E, x') \,. \qquad (1.3.2)$$

For determining the functions φ it is now sufficient to find the coefficients $K(n, m)$ which will represent the solution of the inverse problem, since $V(n)$ are expressed through the K. Because the matrix of (1.2.1a) is three-diagonal and the potential terms $V(n)$ are present only on the diagonal, and because

$$\varphi(E, 1) = \mathring{\varphi}(E, 1) = \Delta \,,$$

we have for the diagonal values $K(n, n) = 1/\Delta$. Indeed, when $\varphi(E, n = 2, 3, \ldots)$ is constructed step by step, at higher orders of E the coefficient consists of terms that are independent of V. Consequently, the coefficients of E^{n-1} in $\varphi(E, n)$ and

$\mathring{\varphi}(E, n)$ are identical, i.e., $\Delta K(n, n) = 1$. Now we shall rewrite (1.3.2a) in a more compact form taking into account the *triangular form* (!) of the matrix of coefficients K:

$$\varphi(E; n) = \mathring{\varphi}(E, n) + \sum_{m=1}^{n-1} \Delta K(n, m)\mathring{\varphi}(E, m) .$$

1.3.2 The Algebraic Analogue of the Gelfand–Levitan Equations

As in Sect. 1.1, in order to determine the functions φ we shall use the relations of completeness (1.2.7) for the case of an infinite well $V(0 \leqslant n \leqslant N + 1)$, after first rewriting them in accordance with (1.2.5) in the form

$$\sum_{\lambda=1}^{N} c_\lambda^2 \varphi(E_\lambda, m)\varphi(E_\lambda, n) = \delta_{mn}/\Delta \tag{1.3.3a}$$

and similarly for $\mathring{\varphi}$:

$$\sum_{\lambda=1}^{N} \mathring{c}_\lambda^2 \mathring{\varphi}(\mathring{E}_\lambda, m)\mathring{\varphi}(\mathring{E}_\lambda, n) = \delta_{mn}/\Delta , \tag{1.3.3a$'$}$$

where $\mathring{E}_\lambda$ are the eigenvalues corresponding to (1.2.1a,2a) with $\mathring{V}(n) \equiv 0$, while

$$\mathring{c}_\lambda^2 = \left[\Delta \sum_{n=1}^{N} \mathring{\varphi}^2(\mathring{E}_\lambda, n)\right]^{-1} .$$

Relations (1.3.3a,3a$'$) represent the conditions of orthonormality with respect to the variable E_λ of the sets of vectors $\{\boldsymbol{\varphi}_n\}$ and $\{\mathring{\boldsymbol{\varphi}}_n\}$ with different weight multipliers c_λ^2 and $\mathring{c}_\lambda^2$, respectively. Thus, transition from the known basis $\{\mathring{\boldsymbol{\varphi}}\}$ in the N-dimensional vector space to the new basis of the desired $\boldsymbol{\varphi}$ may be considered as the conventional orthonormalization of the vectors (such as the Gram–Schmidt procedure) $\mathring{\boldsymbol{\varphi}}$ with new weight factors c_λ^2 being used instead of the $\mathring{c}_\lambda^2$ (Case et al 1973; Zakhariev et al 1977). From the conditions of orthonormalization we determine the desired coefficients $K(n, m)$. As the first vector of the new basis we shall simply take $\mathring{\boldsymbol{\varphi}}(E, 1)$ in accordance with (1.3.2). The coefficient $K(2, 1)$ for the second vector $\boldsymbol{\varphi}(E, 2)$ will be found from the requirement that $\boldsymbol{\varphi}(E, 2) \perp \boldsymbol{\varphi}(E, 1)$, which is equivalent to $\boldsymbol{\varphi}(E, 2) \perp \mathring{\boldsymbol{\varphi}}(E, 1)$, since $\boldsymbol{\varphi}(E, 1) = \mathring{\boldsymbol{\varphi}}(E, 1)$. The third vector $\boldsymbol{\varphi}(E, 3)$ must be orthogonal to $\boldsymbol{\varphi}(E, 1)$ and to $\boldsymbol{\varphi}(E, 2)$, which is equivalent to $\boldsymbol{\varphi}(E, 3) \perp \mathring{\boldsymbol{\varphi}}(E, 1)$ and $\boldsymbol{\varphi}(E, 3) \perp \mathring{\boldsymbol{\varphi}}(E, 2)$. Indeed, if $\boldsymbol{\varphi}(E, 3) \perp \boldsymbol{\varphi}(E, 1)$, then $\boldsymbol{\varphi}(E, 3) \perp \mathring{\boldsymbol{\varphi}}(E, 1)$ but since $\boldsymbol{\varphi}(E, 2)$ represents a combination of $\mathring{\boldsymbol{\varphi}}(E, 1)$ and $\mathring{\boldsymbol{\varphi}}(E, 2)$, from $\boldsymbol{\varphi}(E, 3) \perp \boldsymbol{\varphi}(E, 2)$ it follows that $\boldsymbol{\varphi}(E, 3) \perp \mathring{\boldsymbol{\varphi}}(E, 2)$.

The two conditions $\boldsymbol{\varphi}(E, 3) \perp \mathring{\boldsymbol{\varphi}}(E, 1)$ and $\boldsymbol{\varphi}(E, 3) \perp \mathring{\boldsymbol{\varphi}}(E, 2)$ determine $K(3, 1)$ and $K(3, 2)$. By continuing in this manner we find the coefficients $K(n, m)$ from the nth line of (1.3.2a), orthogonalizing $\boldsymbol{\varphi}(E, n)$ with respect to all $\boldsymbol{\varphi}(E, m)$ with $m < n$:

$$\sum_{\lambda=1}^{N} c_\lambda^2 \varphi(E_\lambda, n)\mathring{\varphi}(E_\lambda, m) = 0 \quad \text{when } m < n . \tag{1.3.4a}$$

Substituting $\varphi(E_\lambda, n)$ in the form of (1.3.2a) into (1.3.4a) we obtain the linear algebraic equations for $K(n, m)$ (algebraic analogue of the Gelfand–Levitan integral equation):

$$K(n, m) + Q(n, m) + \sum_{p=1}^{n-1} \Delta K(n, p) Q(p, m) = 0 \,, \quad m < n \,, \tag{1.3.5a}$$

where

$$Q(n, m) = \sum_{\lambda=1}^{N} c_\lambda^2 \mathring{\varphi}(E_\lambda, n) \mathring{\varphi}(E_\lambda, m) - \delta_{nm}/\Delta \,, \tag{1.3.6a}$$

or, making use of (1.3.3a′),

$$Q(n, m) = \sum_{\lambda=1}^{N} c_\lambda^2 \mathring{\varphi}(E_\lambda, n) \mathring{\varphi}(E_\lambda, m) - \sum_{\lambda=1}^{N} \mathring{c}_\lambda^2 \mathring{\varphi}(\mathring{E}_\lambda, n) \mathring{\varphi}(\mathring{E}_\lambda, m) \,. \tag{1.3.6a′}$$

Now we shall derive equations (1.3.5a) in a way which is convenient for subsequent transition to the limit $\Delta \to 0$. We multiply (1.3.2a) at $E = \mathring{E}_\lambda$ by $\mathring{\varphi}(E_\lambda, n')\mathring{c}^2$ and perform summation over λ, making use of (1.3.3a′). For the coefficients K we obtain

$$K(n, n') = \sum_{\lambda=1}^{N} \mathring{c}_\lambda^2 \varphi(\mathring{E}_\lambda, n) \mathring{\varphi}(\mathring{E}_\lambda, n') \,.$$

This equality is not violated if one subtracts the sum from (1.3.4a) that equals zero from its right-hand side. As a result, for K we obtain an expression similar in structure to (1.3.6a′):

$$K(n, m) = -\sum_{\lambda=1}^{N} c_\lambda^2 \varphi(E_\lambda, n) \mathring{\varphi}(E_\lambda, m) + \sum_{\lambda=1}^{N} \mathring{c}_\lambda^2 \varphi(\mathring{E}_\lambda, n) \mathring{\varphi}(\mathring{E}_\lambda, m) \,. \tag{1.3.6a″}$$

With the aid of the spectral function $\rho(E)$ this may be written in a more compact form:

$$K(n, m) = -\int_{-\infty}^{\infty} \varphi(E, n) \mathring{\varphi}(E, m) d[\rho(E) - \mathring{\rho}(E)] \,, \tag{1.3.6a‴}$$

$$\begin{cases} d\rho(E) = \sum_\lambda c_\lambda^2 \delta(E - E_\lambda) \,; \\ d\mathring{\rho}(E) = \sum_\lambda \mathring{c}_\lambda^2 \delta(E - \mathring{E}_\lambda) \,. \end{cases}$$

Substitution of φ from (1.3.2) into (1.3.6a″) or (1.3.6a‴) yields (1.3.5a). If the K are found from (1.3.5a), then the solution of the Schrödinger equation from (1.2.5) and (1.3.2a) is also known, and it is also not difficult to obtain the desired potential.

1.3.3 The Relationship Between $K(n, m)$ and the Potential $V(n)$

We substitute $\varphi(E_\lambda, n)$ in the form of (1.3.2a) into the finite-difference Schrödinger equation (1.2.1a), multiply both sides of the equation by $\mathring{c}_\lambda^2 \mathring{\varphi}(\mathring{E}_\lambda, n)$ and sum over λ. Taking into account (1.3.3a′) and that $\mathring{\varphi}$ satisfies (1.2.1a) with $V(n) = 0$ we have

$$V(n) = \frac{\hbar^2}{2m\Delta}[K(n+1, n) - K(n, n-1)] . \tag{1.3.7a}$$

Thus, solution of the inverse problem reduces to computation of $Q(n, m)$ in the form of (1.3.6a) by using the spectral parameters $c_{\lambda'}$ and $E_{\lambda'}$ and to subsequent solution of (1.3.5a). The obtained $K(n, m)$ give the potential by (1.3.7a) and the wave function $\psi(E, n) = c_\lambda \varphi(E, n)$ by (1.3.2a).

Equations (1.3.3a–3.6a) remain valid if one uses for $\mathring{\varphi}$ solutions with a known initial potential $\mathring{V}(n) \not\equiv 0$ and the known corresponding values of $\mathring{c}_\lambda$, $\mathring{E}_\lambda$. One must only add $+\mathring{V}(n)$ to the right-hand side of (1.3.7a).

1.3.4 The Gelfand–Levitan Formalism on a Finite Interval ($\Delta \to dx$)

Having already discussed the simple algorithm for reconstruction of $V(n)$, we can write out the formulas for the case of a continuous variable r by just allowing Δ to approach 0 [a traditional presentation may be found, for example, in *Chadan* and *Sabatier* (1977)].

The transformation (1.3.2a) of $\mathring{\varphi}$ into φ, known as the *generalized shift*, when the condition $\int_0^r r'|V(r')|dr' < \infty$ is satisfied assumes the form (Alfaro, Regge 1965; Newton 1982):

$$\varphi(E, r) = \mathring{\varphi}(E, r) + \int_0^r K(r, r')\mathring{\varphi}(E, r')dr' . \tag{1.3.2}$$

For a potential well of infinite depth and radius a we have, as $\Delta \to 0$, instead of (1.3.3a), the completeness relation:

$$\sum_\lambda^\infty c_\lambda^2 \varphi(E_\lambda, r)\varphi(E_\lambda, r') = \delta(r - r') , \tag{1.3.3}$$

and the same for $\mathring{\varphi}(E_\lambda, r)$.

The Gelfand–Levitan equation is obtained as $\Delta \to 0$ from (1.3.5a):

$$K(r, r') + Q(r, r') + \int_0^r K(r, r'')Q(r'', r')dr'' = 0 , \tag{1.3.5}$$

where, similarly to (1.3.6a′)

$$Q(r, r') = \sum_\lambda^\infty c^2 \mathring{\varphi}(E_\lambda, r)\mathring{\varphi}(E_\lambda, r') - \sum_\lambda^\infty \mathring{c}_\lambda^2 \mathring{\varphi}(\mathring{E}_\lambda, r)\mathring{\varphi}(\mathring{E}_\lambda, r') . \tag{1.3.6}$$

Instead of (1.3.7a), we obtain the relation between the potential $V(r)$ and the kernel $K(r, r')$ of the integral transformation for the generalized shift (1.3.2). In the right-hand side of (1.3.7a) upon the addition of $[-K(n, n) + K(n, n)]/2\Delta$, the terms rearrange into the sum of the difference derivatives of K with respect to the first and second arguments, respectively, which yields the total derivative of K as $\Delta \to 0$:

$$V(r) = \frac{\hbar^2}{m} \frac{d}{dr} K(r, r) . \tag{1.3.7}$$

Transition from (1.3.2–7) for a quantum system involving only bound states to scattering on potentials of finite radius within the framework of R-matrix theory is readily performed in the same manner as for the difference case.

1.3.5 The *R*-matrix Inverse Problem

This problem has not been previously addressed in the nonperiodical literature, therefore here we shall present the principal relations (Zakhariev et al 1974). They differ from (1.3.2–7) by the substitution of integration intervals $[r, a]$ for $[0, r]$, and of the reduced width γ_λ^2 for the normalizing factor c_λ^2, and also in that the boundary conditions (1.2.4) for φ and $\mathring{\varphi}$ are replaced by

$$\varphi(E, a) = \mathring{\varphi}(E, a) = 0 , \quad \varphi'(E, a) = \mathring{\varphi}'(E, a) = 1 .$$

As a result we have:

$$\varphi(E, r) = \mathring{\varphi}(E, r) + \int_r^a K(r, r')\mathring{\varphi}(E, r')dr' , \tag{1.3.2'}$$

$$\sum_\lambda^\infty \gamma_\lambda^2 \varphi(E_\lambda, r)\varphi(E_\lambda, r') = \delta(r - r') , \tag{1.3.3'}$$

and likewise for $\mathring{\varphi}(\mathring{E}_\lambda, r)$

$$K(r, r') + Q(r, r') + \int_r^a K(r, r'')Q(r'', r')dr'' = 0 ; \tag{1.3.5'}$$

$$Q(r, r') = \sum_\lambda^\infty \gamma_\lambda^2 \mathring{\varphi}(E_\lambda, r)\mathring{\varphi}(E_\lambda, r') - \sum_\lambda^\infty \mathring{\gamma}_\lambda^2 \mathring{\varphi}(\mathring{E}_\lambda, r)\mathring{\varphi}(\mathring{E}_\lambda, r') ; \tag{1.3.6'}$$

$$V(r) = -\frac{\hbar^2}{m} \frac{d}{dr} K(r, r) . \tag{1.3.7'}$$

To emphasize the similarities between the Gelfand–Levitan–Marchenko approach and other approaches to the inverse problem, we shall use the same letters K, Q for denoting analogous functions and only rarely assign other indices, such as GL, M, R . . .

1.3.6 The Inverse Problem on the Semi-axis Within the Gelfand–Levitan Approach

The title problem is formulated by applying the completeness condition to the set of wave functions corresponding to the discrete and continuous spectra as in (1.2.24a) (Case et al 1973):

$$\frac{(2m)^{1/2}}{\pi\hbar} - \int_0^\infty \frac{dE}{\sqrt{E}}\psi(E,r)\psi^*(E,r') + \sum_\nu \psi(E_\nu,r)\psi(E_\nu,r') = \delta(r-r') . \tag{1.2.24}$$

As in the case of bound states, $\psi(E>0,r)$ differs from $\varphi(E,r)$ only by a normalizing factor $c(E)$ which, like c_ν, plays an important role in the inverse problem. The scattering S-function is expressed directly through $c(E) = k/f^+(k)$, where $f^+(k)$ is known as the *Jost function*. Indeed, the solution φ can be represented in the form of a linear combination of the so-called Jost solutions, $f^\pm(k,r)$, $f^\pm(k_\nu,r)$ which assume the asymptotic form of $e^{\pm ikr}$, $e^{\mp\kappa_\nu r}$ where $-\kappa_\nu^2 = 2mE_\nu/\hbar^2$:

$$\lim_{r\to\infty}[f^\pm(k,r)e^{\mp ikr}] = 1 ; \tag{1.3.8}$$

$$\varphi(E,r) = \frac{i}{2k}\{f^+(k)f^-(k,r) - f^-(k)f^+(k,r)\} . \tag{1.3.9}$$

The coefficients $f^+(k)$ and $f^-(k)$ allow φ to satisfy the boundary conditions (1.2.4) at zero. From (1.2.5, 12, 3.9) we obtain

$$s(k) = f^-(k)/f^+(k) . \tag{1.3.10}$$

Now we shall rewrite the completeness condition (1.2.24) for the functions φ. To achieve a more compact form for the relations of completeness it is convenient to use the *spectral density* $\rho(E)\left(\int_0^\infty dE + \sum_{\text{bound states}} \to \int_{-\infty}^\infty d\rho(E)\right)$:

$$\int_{-\infty}^\infty \varphi(E,r)\varphi(E,r')d\rho(E) = \delta(r-r')$$

$$\frac{d\rho(E)}{dE} = \begin{cases} \dfrac{(2m)^{3/2}\sqrt{E}}{\hbar^3\pi|f(k)|^2}, & E \geqslant 0 ; \\ \sum_\nu c_\nu^2\delta(E-E_\nu), & E < 0 . \end{cases} \tag{1.3.11}$$

It is also possible to write (1.2.24) in a similar form for the difference problem:

$$\int_{-\infty}^\infty \varphi(E,n)\varphi(E,m)d\rho(E) = \delta_{nm}/\Delta ;$$

$$\frac{d\rho(E)}{dE} = \begin{cases} \dfrac{2\sqrt{\lambda^2-1}\,d\lambda}{\pi|f(E)|^2}, & \lambda = 1 - \dfrac{mE\Delta^2}{\hbar^2}; \quad 0 \leqslant E \leqslant 2\hbar^2/m\Delta^2 ; \\ \sum_\nu c_\nu^2\delta(E-E_\nu), & E < 0; \quad E > 2\hbar^2/m\Delta^2 . \end{cases} \tag{1.3.11a}$$

The expressions for kernels $K(Q)$ are reduced, with the aid of the spectral density, to the same unique form both in the case of infinitely deep potentials of finite radius and in the case of potentials where motion occurs along the semi-axis.

The continuous analogue of (1.3.6a″) for the operator of generalized shift $K(r, r')$ inducing reorthonormalization of the bases $\{\mathring{\varphi}\} \to \{\varphi\}$ through (1.3.2) is:

$$K(r, r') = -\int_{-\infty}^{\infty} \varphi(E, r)\mathring{\varphi}(E, r')d[\rho(E) - \mathring{\rho}(E)] . \qquad (1.3.12)$$

By substituting (1.3.2) for $\varphi(E, r)$ into (1.3.12) we obtain the Gelfand–Levitan equation (1.3.5) where

$$Q(r, r') = \int_{-\infty}^{\infty} \mathring{\varphi}(E, r)\mathring{\varphi}(E, r')d[\rho(E) - \mathring{\rho}(E)] . \qquad (1.3.6'')$$

Correspondingly, we obtain the expressions for $Q(n, m)$:

$$Q(n, m) = \int_{-\infty}^{\infty} \mathring{\varphi}(E, n)\mathring{\varphi}(E, m)d[\rho(E) - \mathring{\rho}(E)] .$$

The Gelfand–Levitan equations (1.3.5,3.5a) remain unchanged, only now the coordinates assume values on the whole semi-axis $0 \leqslant r < \infty; 0 < n < \infty$. The same holds for the relation between V and K (1.3.7,7a) and also for the relation between φ and $\mathring{\varphi}$ (1.3.2,3.2a). One must bear in mind that in the case of the semi-axis the condition

$$\int_{0}^{\infty} r|V(r)|dr < \infty$$

must be fulfilled. Equations (1.3.2–12) in which $\mathring{\varphi}$ occur, are also valid for $\mathring{V} \not\equiv 0$, only that $\mathring{V}$ must be added to the right-hand side in the equation for the relation between V and K. The formulas with $\mathring{V} \not\equiv 0$ are convenient when a potential $\mathring{V}$ close to V is known and it is economical to reconstruct only $\Delta V = V - \mathring{V}$.

1.3.7 The Algebraic Analogue of the Marchenko Method

In the Gelfand–Levitan approach reconstruction of the potential starts at the origin and proceeds towards the asymptotic values of r. By using the scattering function s as the initial function, it is more convenient to solve the inverse problem in the opposite direction, that is, starting from the asymptotics where the wave function is known.[1] Such a modification of the formalism was proposed by *Marchenko* (1955).

[1] Actually, cases exist in quark interactions, for example, when experimental data provide direct information on $\rho(E)$ rather than on the asymptotic behavior of ψ. This is because from lepton decays one can deduce the behavior of ψ at the origin.

In the Marchenko approach, instead of the auxiliary functions $\mathring{\varphi}(k,r)$, $\varphi(k,r)$, the Jost solutions, $\mathring{f}^{\pm}(k,r)$ and $f^{\pm}(k,r)$, are utilized. These exhibit the asymptotic behavior of (1.3.8) and are related to the physical wave function ψ directly through the scattering function s, instead of to the spectral density or the Jost functions

$$\psi(k,r) = \frac{\mathrm{i}}{2}\left[f^{-}(k,r) - s(k)f^{+}(k,r)\right] . \tag{1.3.13}$$

In the finite-difference approach [compare with (1.2.12a)]

$$\psi(\theta,n) = f^{-}(\theta,n) - s(\theta)f^{+}(\theta,n) ;$$
$$\lim_{n\to\infty}\left[f^{\pm}(\theta,n)\exp[\mp \mathrm{i}\theta n]\right] = 1 . \tag{1.3.13a}$$

In deriving the transformation (1.3.2a) of the generalized shift from $\mathring{\varphi}(E,n)$ to $\varphi(E,n)$ the polynomial dependence of these functions on the energy was utilized (Case 1973). Here the same part is attributed to the polynomial dependence $f^{\pm}(\theta,n)$ with respect to the new energy variable $z^{\pm 1} = \exp(\pm \mathrm{i}\theta)$. At the same time the solutions $\mathring{f}^{\pm}(\theta,n)$ are just equal to $z^{\pm 1}$ raised to the nth power.

Functions $\mathring{f}^{\pm}(\theta,n)$ represent discrete analogues of the free waves $\exp(\pm \mathrm{i}kr)$:

$$\mathring{f}^{\pm}(\theta,n) = z^{\pm n} = \exp(\pm \mathrm{i}\theta n) , \tag{1.3.14a}$$

where

$$z^{\pm 1} \equiv \lambda \pm \sqrt{\lambda^2 - 1} \; ; \; \lambda \equiv 1 - mE\Delta^2/\hbar^2 \equiv (z + z^{-1})/2 = \cos\theta . \tag{1.3.15a}$$

Indeed the functions $\mathring{f}^{\pm}$ satisfy the difference Schrödinger equation with zero potential, (1.3.1). This is readily verified by rewriting (1.3.1) in the form

$$\mathring{f}^{\pm}(z,n+1) + \mathring{f}^{\pm}(z,n-1) = (z + z^{-1})\mathring{f}^{\pm}(z,n) . \tag{1.3.16}$$

Since, when E varies within the limits of the continuous band of the spectrum of the difference problem

$$0 \leqslant mE/\hbar^2 < 2/\Delta^2$$

the parameter λ changes within the limits $-1 \leqslant \lambda \leqslant 1$, one can set

$$\lambda = \cos\theta ; \; z^{-1} = \lambda - \sqrt{\lambda^2 - 1} = \cos\theta - \mathrm{i}\sin\theta = \exp(-\mathrm{i}\theta) . \tag{1.3.17a}$$

From the equality $1 - \cos\theta = mE\Delta^2/\hbar^2$ it follows that as $\Delta \to 0$, $\theta \to 0$, too. But in this limit $E = \hbar^2 k^2/2m$, so that $\theta/\Delta \to k$ as $\Delta \to 0$.

As we have seen in the Gelfand–Levitan approach, the factor E in the right-hand side of the difference Schrödinger equation (1.2.1a,3.1) increases the power of E of the polynomial φ at each shift of the solution by a single step by one. In a similar way the energy factor $z + z^{-1}$ in the right-hand side of the same equation, but written in the form [compare with (1.3.16a)]

$$f^{\pm}(z,n+1) + v(n)f^{\pm}(z,n) + f^{\pm}(z,n-1) = (z + z^{-1})f^{\pm}(z,n) , \tag{1.3.18a}$$

will not spoil the polynomial dependence upon z of the solutions $f^{\pm}(z,n)$, assuming at large n the form $z^{\pm n}$. A shift of one step to the left may change the order of the polynomial in z by ± 1.

Let $V(n > N) = 0$. At the points N and $N+1$ $f^{\pm}(z, n) = \mathring{f}^{\pm}(z, n)$ and, in accordance with (1.3.14a)

$$f^{\pm}(z, N) = \mathring{f}^{\pm}(z, N) = z^{\pm N}; \quad f^{\pm}(z, N+1) = \mathring{f}^{\pm}(z, N+1) = z^{\pm(N+1)} . \tag{1.3.19a}$$

In accordance with (1.3.18a) with $n = N$ besides the term with $z^{\pm N}$, for $f^{\pm}(z, N-1)$ we also obtain another term with $z^{\pm(N-1)}$. At each consecutive step to the left the number of terms of different powers of z present in $f^{\pm}(z, n)$ will increase by *two* because of the appearance of terms with $z^{\pm n}$ and $z^{\pm(2N-n-1)}$.

Thus, similarly to (1.3.2a), it is possible to construct $f^{\pm}(z, n)$ as a linear combination of $f^{\pm}(z, m) = z^{\pm m}$:

$$f^{\pm}(z, n) = \mathring{f}^{\pm}(z, n) + \sum_{m=n+1}^{2N-n-1} \Delta K(n, m)\mathring{f}^{\pm}(z, m) , \tag{1.3.20a}$$

where the factors $K(n, m)$ ensure that the coefficients of terms of the same power of z on both sides of (1.3.20a) are equal.

Now, substituting (1.3.20a) into (1.3.13a) one can also represent ψ in the form of a linear combination of solutions $\mathring{\psi}$ of (1.3.16a):

$$\begin{aligned}\mathring{\psi}(z, n) &= \mathring{f}^{-}(z, n) - s(z)\mathring{f}^{+}(z, n) = z^{-n} - s(z)z^{n} \\ &= \exp(-\mathrm{i}\theta n) - s(\theta)\exp(\mathrm{i}\theta n) ;\end{aligned} \tag{1.3.21a}$$

$$\psi(z, n) = \mathring{\psi}(z, n) + \sum_{m=n+1}^{2N-n-1} \Delta K(n, m)\mathring{\psi}(z, m) . \tag{1.3.22a}$$

Function $\mathring{\psi}$ obeys the free wave equation, but it assumes the same asymptotic behavior as ψ, where the potential is nonzero.

As before, we find the coefficients $K(n, m)$ with the aid of the condition of completeness for the wave functions (that is, orthogonality with respect to the variable E):

$$\frac{1}{2\pi} \int_0^{2\hbar^2/m\Delta^2} \psi(E, n)\psi^*(E, m)dE + \sum_\nu \psi(E_\nu, n)\psi(E_\nu, m) = \frac{\delta_{nm}}{\Delta} , \tag{1.3.23a}$$

or

$$\frac{1}{2\pi \mathrm{i}} \oint \psi(z, n)\psi^*(z, m)\frac{dz}{z} + \sum_\nu \psi(z_\nu, n)\psi(z_\nu, m) = \frac{\delta_{nm}}{\Delta} ,$$

where the contour of integration is a circle of unit radius in the complex plane z.

Because $\psi(n)$ is orthogonal with respect to all $\psi(m > n)$, it follows that it is also orthogonal to all $\mathring{\psi}(m)$ with $m > n$ which compose the $\psi(m > n)$, in accordance with (1.3.22a) [compare with (1.3.4a) for φ]:

$$\frac{1}{2\pi} \int_0^{2\hbar^2/m\Delta^2} \psi(E, n)\mathring{\psi}^*(E, m)dE + \sum_\nu \psi(E_\nu, n)\mathring{\psi}(E_\nu, m) = \frac{\delta_{nm}}{\Delta} ,$$

or

$$\frac{1}{2\pi i}\oint \psi(z,n)\mathring{\psi}^*(z,m)\frac{dz}{z}+\sum_{\nu}\psi(z_\nu,n)\mathring{\psi}(z_\nu,m)=\frac{\delta_{nm}}{\Delta}\,. \tag{1.3.24a}$$

Substituting ψ in the form of (1.3.22a) into (1.3.24a) we obtain a set of linear equations for K:

$$K(n,m)+Q(n,m)+\sum_{p=n+1}^{2N-n-1}\Delta K(n,p)Q(p,m)=0\,, \tag{1.3.25a}$$

where

$$Q(n,m)\equiv Q(n+m)=\frac{1}{2\pi i}\oint \mathring{\psi}(z,n)\mathring{\psi}(z,m)\frac{dz}{z}+\sum_{\nu}\mathring{\psi}(z_\nu,n)\mathring{\psi}(z_\nu,m)\,;$$

$$\mathring{\psi}(z_\nu,n)=M_\nu \mathring{f}(z_\nu,n)=M_\nu z_\nu^n\,;\quad M_\nu^2=\left[\Delta\sum_{n=1}^{N}|f(z_\nu,n)|^2\right]^{-1}. \tag{1.3.26a}$$

In the general case in (1.3.25a) one assumes $N\to\infty$. This is exactly the equation which represents the algebraic analogue of the Marchenko integral equation.

The desired relation between the potential and $K(n,m)$ is similar to (1.3.7a):

$$V(n)=-\frac{\hbar^2}{2m\Delta}[K(n,n+1)-K(n-1,n)]\,. \tag{1.3.27a}$$

1.3.8 The Marchenko Equations for $\Delta\to dx$

As in the Gelfand–Levitan approach, we shall now derive main formulas of the Marchenko method in the limit $\Delta\to 0$ in (1.3.20a–27a). Assuming the condition $\int_0^\infty r|V(r)|dr<\infty$ (Newton 1982), we obtain

$$f^\pm(k,r)=\mathring{f}^\pm(k,r)+\int_r^\infty K(r,r')\mathring{f}^\pm(k,r')dr'\,; \tag{1.3.20}$$

$$\mathring{f}^\pm(k,r)=e^{\pm ikr}\,;$$

$$\psi(k,r)=\mathring{\psi}(k,r)+\int_r^\infty K(r,r')\mathring{\psi}(k,r')dr'\,; \tag{1.3.22}$$

$$K(r,r')+Q(r,r')+\int_r^\infty K(r,r'')Q(r'',r')dr''=0\,, \tag{1.3.25}$$

where

$$Q(r,r')=Q(r+r')=\frac{1}{2\pi}\int_{-\infty}^{\infty}[1-s(k)]e^{ik(r+r')}dk$$

$$+\sum_{\nu}^{N}M_\nu^2 e^{-\kappa_\nu(r+r')}\,. \tag{1.3.26}$$

and, finally,

$$V(r) = -\frac{\hbar^2}{m}\frac{d}{dr}K(r,r)\,. \tag{1.3.27}$$

The generalization of (1.3.20–27) to the case of a nonzero initial potential was considered by *Pivovarchik* and *Suzko* (1982) and *Moses* et al (1977, 1979). For this purpose $\mathring{f}^{\pm}$ must be understood to be the solution with $\mathring{V}(r) \not\equiv 0$, then one must add to the kernel Q the sum over the bound states in $\mathring{V}(r)$ and substitute $\mathring{s}(k)$ for unity in the integrand in (1.3.26):

$$\begin{aligned} Q(r,r') = \frac{1}{2\pi}\int_{-\infty}^{\infty} [\mathring{s}(k) - s(k)]\mathring{f}^{+}(k,r)\mathring{f}^{+}(k,r')dk \\ + \sum_{\nu}^{N} M_\nu^2 \mathring{f}^{+}(\mathrm{i}\kappa_\nu, r)\mathring{f}^{+}(\mathrm{i}\kappa_\nu, r') \\ - \sum_{\mathring{\nu}}^{\mathring{N}} \mathring{M}_\nu^2 \mathring{f}^{+}(\mathrm{i}\mathring{\kappa}_\nu, r)\mathring{f}^{+}(\mathrm{i}\mathring{\kappa}_\nu, r')\,, \end{aligned} \tag{1.3.26'}$$

where κ_ν, N, M_ν^2 and $\mathring{\kappa}_\nu$, N, M_ν^2 represent the energy of bound states, their number, and normalization constants for solutions with $V(r)$ and $\mathring{V}(r)$:

$$V(r) = \mathring{V}(r) - \frac{\hbar^2}{m}\frac{d}{dr}K(r,r')\,. \tag{1.3.27'}$$

Within the Marchenko and Gelfand–Levitan approaches with the constants M_ν and c_ν and auxiliary solutions φ and f one constructs the same normalized wave function:

$$\begin{aligned} &\psi(\mathrm{i}\kappa_j, r) = c_j\varphi(\mathrm{i}\kappa_j, r) = M_j f^{+}(\mathrm{i}\kappa_j, r)\,, \\ &\quad c_j^{-2} = \int_0^\infty \varphi^2(\mathrm{i}\kappa_j, r)dr\,; \quad M_j^{-2} = \int_0^\infty |f^{+}(\mathrm{i}\kappa_j, r)|^2 dr\,. \end{aligned} \tag{1.3.28}$$

1.3.9 Relation Between $V(n)$ and $K(m,n)$ in the Marchenko Approach

We shall adopt a reasoning similar to that used in the derivation of the conditions which must be satisfied by K for a continuous coordinate (Neimark 1969) in order for (1.3.20) to be solutions of the Schrödinger equation. We shall substitute into the difference equation (1.3.18a) the function $f^{\pm}(z,n)$ in the form of (1.3.20a) with $N \to \infty$. We shall use (1.3.16a) for $\mathring{f}^{\pm}(z,n)$ to cancel out the equal terms and to express $E\mathring{f}(z,n)$ through $\mathring{f}^{+}(z,n)$ and $\mathring{f}^{+}(z,n\pm 1)$. After simple transformations and noting that $\mathring{f}^{+}(z,n) = \exp(\mathrm{i}\theta n)$, we obtain

$$\sum_{m=n+1}^{N} \Delta A(m)\exp(\mathrm{i}\theta m) + B(n)\exp(\mathrm{i}\theta n) + C(n) = 0\,, \tag{1.3.29}$$

where

$$A(m) = -[K(n-1,m) - 2K(n,m) + K(n+1,m)]\hbar^2/2\Delta^2 m$$
$$-[K(n,m-1) - 2K(n,m) + K(n,m+1)]\hbar^2/2\Delta^2 m + V(n)K(n,m)\,;$$
$$B(n) = -[K(n-1,n) - K(n,n+1)]\hbar^2/2\Delta m + V(n)\,;$$
$$C(n) = K(n,N+1)\exp(\mathrm{i}\theta N)\hbar^2/2\Delta m - K(n,N)\exp[\mathrm{i}\theta(N+1)\hbar^2/2\Delta m\,. \tag{1.3.30}$$

Since (1.3.29) holds for all θ, all coefficients of $\exp(\mathrm{i}\theta n)$ must become zero for various m, i.e.,

$$A(m) = 0; \quad C(n) = 0; \quad B(n) = 0\,. \tag{1.3.31}$$

From the first and second equalities in (1.3.31) the following equations in partial differences and one boundary condition for $K(n,m)$ are obtained:

$$[K(n+1,m) - 2K(n,m) + K(n-1,m)]\hbar^2/2\Delta^2 m + V(n)K(n,m)$$
$$-[K(n,m+1) - 2K(n,m) + K(n,m-1)]\hbar^2/2\Delta^2 m = 0\,; \tag{1.3.32}$$

$$[K(n,N+1) - K(n,N)]/K(n,N) = [\mathring{f}^+(z,N+1) - \mathring{f}^+(z,N]/\mathring{f}^+(z,N)\,. \tag{1.3.33}$$

From the third equation (1.3.31) we obtain (1.3.27a).

1.4 Miscellaneous Direct and Inverse Problems

After having introduced in Sects. 1.1–3 the main concepts of the direct and inverse problems, we could have immediately proceeded with the solution of the Schrödinger and Gelfand–Levitan–Marchenko equations. Cases when the solutions are obtained in exact analytical form will be considered in Chap. 2. Chapter 3 deals with approximate methods of computation of wave functions from given potentials and of reconstruction of potentials from spectral parameters and scattering data.

Here, in order to develop a deeper understanding of the concepts of Sects. 1.1, 2 and to extend their scope of application, we shall discuss various modifications of the formulation of one-particle one-channel problems. For instance, in the discrete coordinate variable approach it is better to write the Schrödinger differential equation by a difference equation of *higher* than second order, see *Zakhariev* et al (1979). This requires rejection of the conventional solution procedures. The specific character of the problem of motion along the entire axis $-\infty < x < \infty$ makes it quite attractive. It exhibits features of multi-channel and multi-dimensional processes. Thus, threshold phenomena are possible. This problem was used by Faddeev in his development of the formalism of complicated multi-dimensional problems.

We shall demonstrate how the algorithms for the ordinary case of $V(r)$ must be altered to reconstruct potentials dependent on the energy $V(k, r)$, the angular momentum $V(\lambda, r)$, where $\lambda = l + \frac{1}{2}$, and on the momentum operator $V(p^2, r)$ (or velocity).

If we allow the potentials to be symmetric then we simplify the problem by reducing the input data. With the theorem of two spectra one can determine the form of an infinitely deep well $V(x) = V(-x)$ by using only the energy levels E_ν, and in the scattering problem only the positions of R-resonances (without further data on the spectral parameters c_ν, M_ν, or γ_ν). In the inverse problem one may also make use of the position of the poles of the S-function in the complex plane k (resonances, bound states) as initial information.

The material presented in this section has not been presented in previous monographs. Those issues that were discussed by other authors are presented in a novel manner. The formalism for reconstructing potentials from phase shifts at a given energy, which is presented in detail by *Chadan* and *Sabatier* (1989), will only be briefly touched on, only to make the discussion in following chapters easier.

Before we continue to the main topic of Sect. 1.4 we shall make a few comments on the propagation of waves in quantum mechanics.

1.4.1 Stationary Solutions and the Propagation of Waves

In the stationary formulation of scattering problems when neither the potential nor the boundary conditions depend on time, in the Schrödinger equation $\mathrm{i}\hbar \frac{\partial \psi}{\partial t} = H\psi$ the variable t is separated from the coordinate x, and the wave function can be separated into

$$\psi(x, t) = \exp(-\mathrm{i}Et)\psi(x) ,$$

the factor $\exp(-\mathrm{i}Et)$ is dropped, i.e., only $\psi(x)$ is considered. However, the language used often implies the development of processes in time. For example, the solution $\exp(ikx)$ of the equation with zero potential is called a wave "propagating" to the right, even though the function $\exp(ikx)$ itself remains unaltered with time. To avoid such inaccuracies from giving rise to further misunderstandings, we would like to review a few principles from wave mechanics.

We shall make up a packet of free solutions $\exp(ikx)$, corresponding to an energy close to a certain value E_0, and show that such a packet actually does move to the right with a speed $v = \hbar k/2m$. We integrate $\exp(ikx - \mathrm{i}Et)$ over E with a weight $A(E - E_0)$ that decreases rapidly as $|E - E_0|$ increases:

$$\int A(E - E_0)\exp(ikx - \mathrm{i}Et/\hbar)dE . \tag{1.4.1}$$

The maximum of the packet at a fixed t is situated at the point where all the waves composing the packet have the *same phase* and therefore do not dampen each other by interference. The position of this point (with a phase constant in E) is obtained by setting the derivative with respect to E of the phase $(ikx - \mathrm{i}Et)$ of the integrand in (1.4.1) equal to zero, where A is chosen to be a real function not contributing to the phase:

$$d(ikx - \mathrm{i}Et/\hbar)/dE = \mathrm{i}x\sqrt{m/2E\hbar^2} - \mathrm{i}t/\hbar = 0\ , \quad \text{i.e., } x \sim vt\ . \tag{1.4.2}$$

Thus, the center of the packet moves to the right with time with a speed v. A packet composed of waves $\exp(-ikx)$ will move to the left. As a result of scattering by a potential, the waves acquire a phase shift $\delta(k)$. This leads to displacement of the center of the packet by $\Delta x = -v d\delta(k)/dE$, which may be interpreted as a delay of the waves in the interaction region by a time $\Delta t = d\delta(k)/dE$ when $d\delta(k)/dE < 0$, or as an acceleration (advance) relative to the free motion of the packet when $d\delta(k)/dE > 0$.

We note that although the waves of the packet represent quantum mechanical objects (they satisfy the Schrödinger equation), the packet center moves according to the laws of classical mechanics. Because the Schrödinger equation is linear, individual components of the packet move independently of each other at different energies and velocities, letting the packet spread out in time. However, because the energies of the waves composing the packet are close to each other, this spreading out occurs so slowly that it can usually be neglected. (If a spread packet is taken to as the initial condition in the time inversion problem, it will first become more compact and, only upon having passed through the stage of maximum localization, will it start spreading out again.)

Some solutions of the nonlinear Schrödinger equation are represented by waves that do not spread out but retain their form in free motion, for example, solitons or certain Hartree–Fock solutions of the multi-particle time-dependent problem which will be considered in Chaps. 2, 7.

As an example of the motion of a wave packet, we shall consider its scattering from a rectangular well (Good 1972; Goldberg 1967). Figure 1.1 shows the momentum distribution of the packet $\psi(p = \hbar k, t)$ and its space localization $\psi(x, t)$ at various points in time.

Until the packet reaches the well ($t \leqslant t_0$), $\psi(p, t)$, and $\psi(x, t)$ represent single peaks of regular form. When the packet partly enters the well, $\psi(p, t)$ breaks up into several peaks. Peak 4 in Fig. 1.1c ($t = t_2$) situated at the position of the initial peak characterizes the momentum distribution of that part of the packet which has either not yet entered the well or has already left it. Peak 2 corresponds to the same momentum as peak 4 but is moving in the opposite direction. It is produced by waves reflected from the well to the left. Peaks 5 and 1 correspond to motion inside the well to the right and to the left. The absolute values of the momenta of waves inside the well are, naturally, higher than outside. Peak 3 is due to the interference of waves moving to the right from both sides of the well in regions A and B. The appearance of peak 3 may be explained

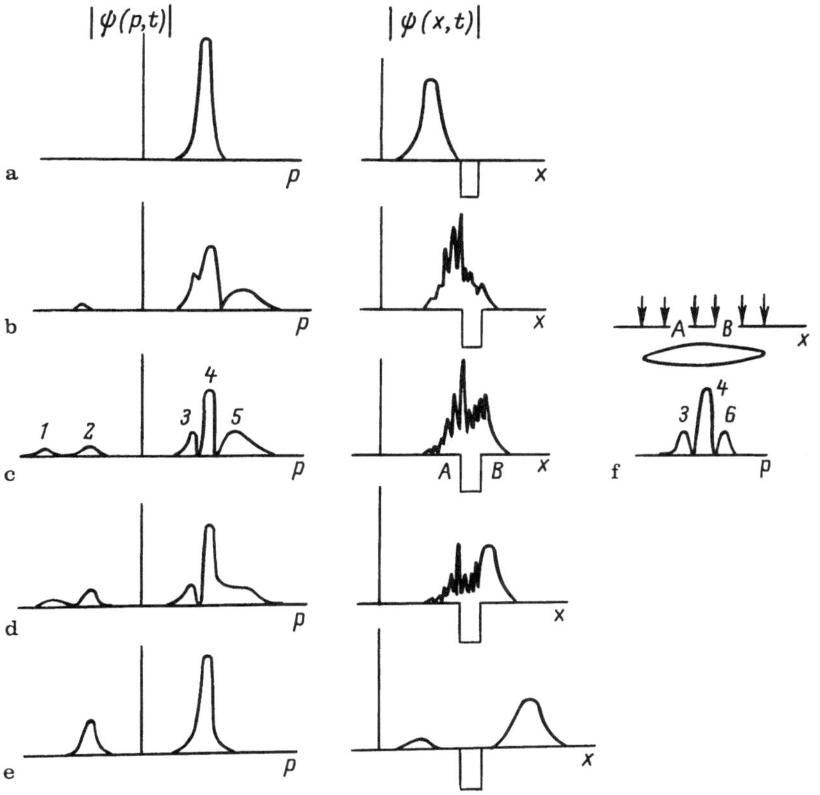

Fig. 1.1a–f. Motion of a wave packet in the field of a rectangular potential well; the mean initial energy of the packet equals half the depth of the well. **a–e** Momentum and space distributions at different points in time $t_0, t_1, \ldots, t_4$. *Right*—passage of the packet through the well and reflection from it. **f** Optical analogue of the momentum distribution for $t = t_2$ (compare with **c**)

with the aid of the diffraction pattern (Fig. 1.1f) of light passing through two slits situated at a distance a from each other, equal to the width of the potential well and through a lens. The two light beams A and B (Fig. 1.1f) correspond to waves moving in the same direction and separated by the well (regions A and B in Fig. 1.1c). The Fraunhofer maximum 3 in Fig. 1.1f represents the analogue of peak 3 in Fig. 1.1c, while 6 merges with 5 in Fig. 1.1c.

Now consider the form of the packet $\psi(x, t)$. The sharp oscillations arising when the packet is found within the range of action of the potential is the result of interference of waves moving in opposite directions. To the right of the well, where only waves that have already passed are present, no oscillations are observed. Inside the well oscillations are more frequent. Because of the high frequency of the interfering waves such subtle details are indistinguishable in Fig. 1.1.

Another example of wave interference is given in Fig. 1.2 where we see the difference in the time evolution of wave packets of different shapes (Ulyanov 1987). In the case of an initially rectangular packet (Fig. 1.2a) the cross-flows from the opposite ends of the packet give a jagged probability distribution. The spreading of a Gaussian packet (Fig. 1.2c) goes through smooth forms. Two Gaussian packets (Fig. 1.2d) have some additional shape oscillations due to the interference of cross-flows.

Numerically the time evolution of wave packets is somewhat at variance with the solution of the stationary Schrödinger equation; for details see *Goldberg* (1967). The technique of the long-time evolution calculations was investigated by *Girard* et al (1988).

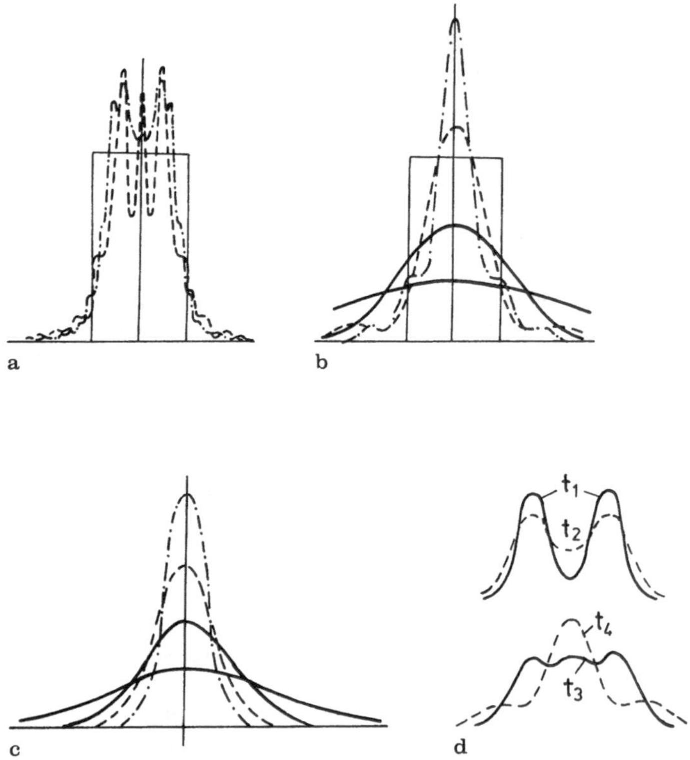

Fig. 1.2. Spreading of wave packets of different forms: rectangular **a** ($t_{i=1,2,3}$), **b** ($t_{i=1,4,5,6,7}$), **c** Gaussian, **d** two separated Gaussian packets

1.4.2 Penetrability of Potential Barriers

The notion of the tunnel effect represents an essential component in our understanding of the microcosm. In the simplest one-channel nonresonance case it was understood very soon after the discovery of quantum mechanics. Many related problems still remain uninvestigated, however, and a number of well established facts are not sufficiently widely known. This particularly concerns the passage through barriers of systems consisting of several particles. In such systems the phenomena of *penetrability enhancement* and *symmetry violation* of tunnelling in opposite directions are revealed.

Multi-particle problems will be considered in detail in Chap. 7. Here we shall only touch on a few aspects of wave motion in a field, drawing special attention to limiting cases in which the features of the tunnelling process are exceptionally starkly revealed.

Consider a wave incident on a barrier from one side. The probability $|\psi(x)|^2$ will oscillate as a result of interference of waves in the region where the incident and reflected fluxes encounter each other. The value of $|\psi(x)|^2$ changes periodically, with x, between the maximum $|\psi|^2_{\max} = (|A| + |B|)^2$ and the minimum $|\psi|^2_{\min} = (|A| - |B|)^2$, where $\psi(x) = A\exp(ikx) + B\exp(-ikx)$. On the other side of the barrier where only the wave that has passed is present $\psi(x) = C\exp(ikx)$, we have $|\psi(x)|^2 = |A|^2 - |B|^2 = |C|^2$. The nontrivial point is that the probability of finding the particle at certain points in front of the barrier may turn out to be smaller than that of finding it behind the barrier, even in the case of an arbitrarily weak penetrability susceptibility $|\psi(x)|^2_{\min} < C^2$ (Draper 1980).

Examples of limiting cases are totally transparent and totally opaque fields. Potentials which are transparent (not reflecting) at any arbitrary energy include soliton potentials obtained by the inverse-problem method (Zakharov et al 1980). Such attractive potentials pertaining to exactly solvable models will be considered in Chap. 2.

Another remarkable case of transparent potential barriers is represented by *resonance tunnelling*. Let there be two barriers at a certain distance from each other. Each one taken separately is very weakly penetrable at the energy indicated by the dot-dashed line in Fig. 1.3. The first impression may be that it will be more difficult to overcome two such barriers than one. Actually, if the energy of the incident wave is close to the energy of the quasi-stable state in which waves in the region between the potential walls x_1, x_2 (Fig. 1.3a) are standing waves, the *transmission coefficient may turn out to be close to unity.*

The validity of this may be verified by solving the corresponding Schrödinger equation. The exact solution of a sample problem will be considered a bit later, but first we shall present some arguments to facilitate understanding of this astonishing phenomenon.

First we shall explain what happens on the left barrier in Fig. 1.3a. A corresponding picture may be obtained by the operation of complex conjugation of the solution ψ for the scattering of a wave incident on one of the barriers

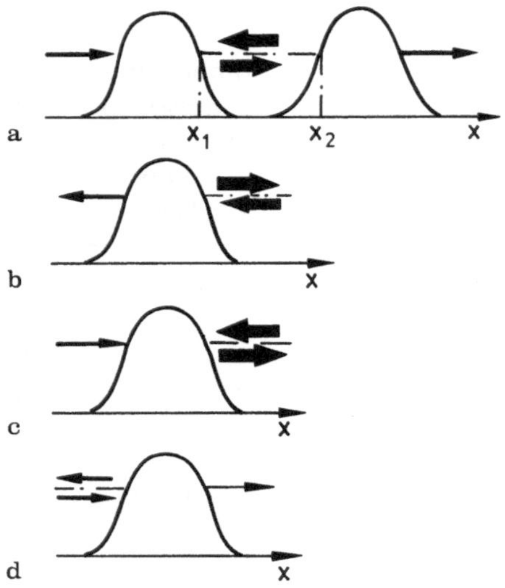

Fig. 1.3a–d. Resonance tunnelling at the energy of a quasi-stable state (oscillation between barriers). The thickness of the arrows indicates the particle flux value

from the right, as shown in Fig. 1.3b. Indeed, in the conjugate solution ψ^* the particles move in the opposite direction with respect to ψ (Fig. 1.3c). Figure 1.3c also coincides with the left-hand side of Fig. 1.3a. The absence of the reflected wave to the left in Figs. 1.3a,c is due to mutual cancelling of the two waves opposite in phase and equal in amplitude (the wave having passed from right to left and the wave reflected to the left), as happens when adding wave functions of processes shown in Figs. 1.3b,d.

The right-hand part of Fig. 1.3a represents conventional tunnelling. It only remains to get the functions of the two halves to join smoothly. This is achieved by suitable choice of the wave amplitudes and the separation between the barriers which at the given energy must allow the appearance of a standing wave (quasi-stationary state)[2]. This stationary state could be established as a result of prolonged irradiation of the system, until in the region between the barriers the probability density in the quasi-stable state $|\psi(x)|^2$ reaches such a value that the flux tunnelling towards the incident waves totally suppresses the reflected waves as a result of interference.

At energies in the interval between the quasi-stable states where waves between the barriers suppress each other, that is, the condition for the formation

[2] Let us present another line of reasoning. The general solution of the Schrödinger equation contains two arbitrary constants. Let us require two homogeneous boundary conditions to be satisfied on the different sides of the system of barriers: for instance, let waves travelling to the left be absent. For the constants this will be a set of two homogeneous algebraic equations having a solution at an energy when the determinant D of the coefficient matrix becomes zero. This yields the energies of quasi-stable states. To increase the transparency it is sufficient that D assume a value close to zero.

of standing waves is violated, the penetrability is sharply decreased compared to the penetrability where interference is not taken into account.

There exists an exactly solvable model of the resonance passage of two δ-barriers,

$$V(x) = V_0[\delta(x + a) + \delta(x - a)] \,, \tag{1.4.3}$$

situated at a distance $L = 2a$ from each other (Lapidus 1982). In this case an explicit expression for the penetrability coefficient T is obtained:

$$T = \{1 + [(1/\varepsilon^2)(\sin kL + \varepsilon \cos kL]\}^{-1} \,, \tag{1.4.4}$$

where $\varepsilon = E/E_0$; $E_0 = \text{const}$. From (1.4.4) it can be seen that total transparency of the barrier $T = 1$ is achieved when $\tan kL = -\varepsilon$ (Fig. 1.4). It is remarkable that the same results are also obtained for δ-wells. The effect is independent of the sign of V_0; $\delta(x \pm a)$-wells yield intense over-barrier reflection. In the limit $L \to 0$ where the two barriers merge into one the resonances vanish and

$$T = E/(E + E_0) \,. \tag{1.4.5}$$

The penetrability of a system of N identical potential barriers is shown in Fig. 1.5. We see how the conduction bands, just as in a crystal lattice, are gradually formed with increasing N.

We now consider another limiting case. Potentials having bound states in the *continuous spectrum* represent totally reflecting potentials at a certain energy. A wide class of exactly solvable problems with such potentials exists. They will be discussed in greater detail in Chap. 2. Here we shall restrict ourselves to their qualitative characteristics.

In Fig. 1.6 one can see how the form of a potential changes when the energy level of a bound state is shifted from the region of negative to positive values (Stillinger, Herrik 1975) while all other spectral characteristics are left unchanged. The higher the energy of the bound state *embedded in the continuum*, the sharper are the oscillations of the form of the potential. The wave is confined and does not escape towards infinity [$\psi(x)$ is square integrable] owing to coherent reflection from an infinite series of barriers with heights decreasing

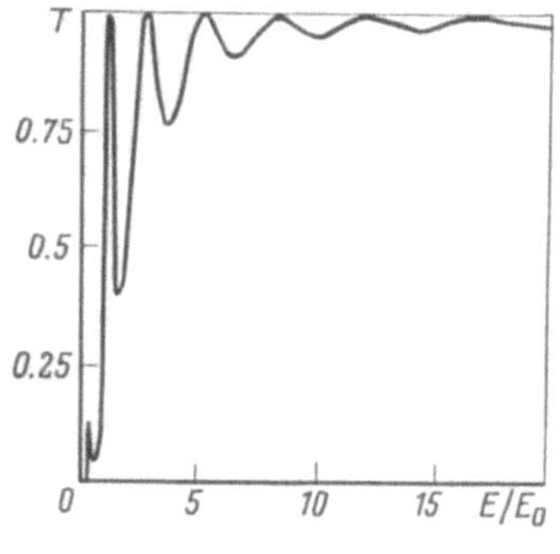

Fig. 1.4. Pentrability coefficient of two δ-barriers separated by a distance of $L = 5h^2/mv$. Resonance penetrability ($T = 1$) is observed at energies corresponding to quasi-stationary states

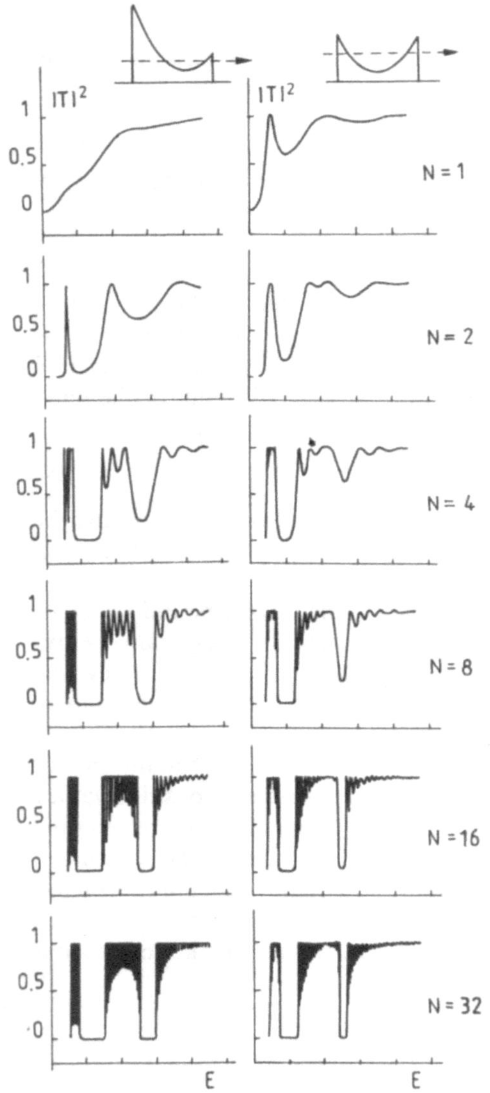

Fig. 1.5. Penetrability $|T|^2$ of N barriers as a function of energy for different values of N (Kovalsky et al 1987)

inversely proportional to the distance. As $x \to \infty$, the wave function also decreases with the same speed.

It is interesting to compare the form of potentials having a single level in the continuous spectrum, $E = 1$ or $E = 4$ (Fig. 1.7a,b), with a potential possessing two bound states $E_1 = 1$, $E_2 = 4$, (Meyer–Vernet 1982) (Fig. 1.7c). In the latter case the oscillations of the potential are more complicated. Particularly unexpected is the existence of discrete levels situated higher than the maximum energy of the oscillations of $V(x)$.

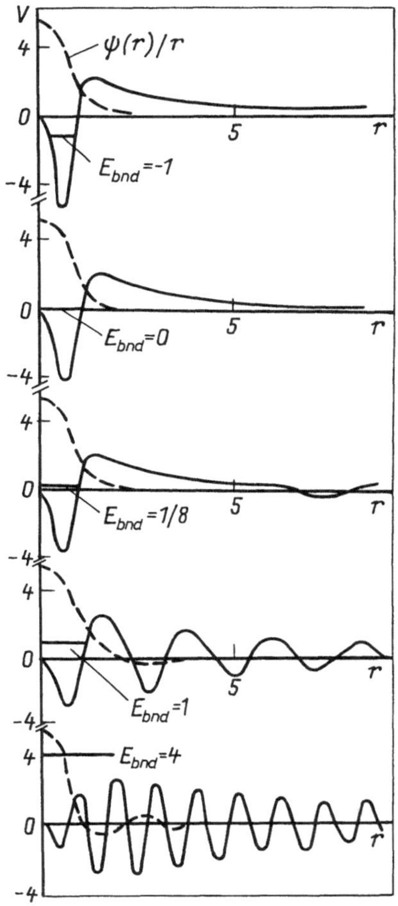

Fig. 1.6. Change of a Bargmann-type potential as the bound state level is raised from the region of negative energies to the continuous spectrum (embedding into continuum)

The property of wave confinement at a given energy is mainly due to the asymptotic behavior of the potentials at large x. One can arbitrarily alter $V(x)$ within any finite interval $[0, a]$ only if the new solution on $[0, a]$, smoothly joined with the previous one at point a, becomes zero at the origin of the coordinate system.

The examples of bound states embedded in the continuum may serve as models for real physical systems, for example, bound states of electrons on the surface of a crystal which are embedded in the conductive band. These examples also permit a physical check of our quantum intuition.

Multi-channel models with $E_{\text{bound}} > 0$ are discussed in Chap. 7. The confinement of waves inside the continuum in multi-channel systems can be achieved without long-range potential oscillations.

a

b

c

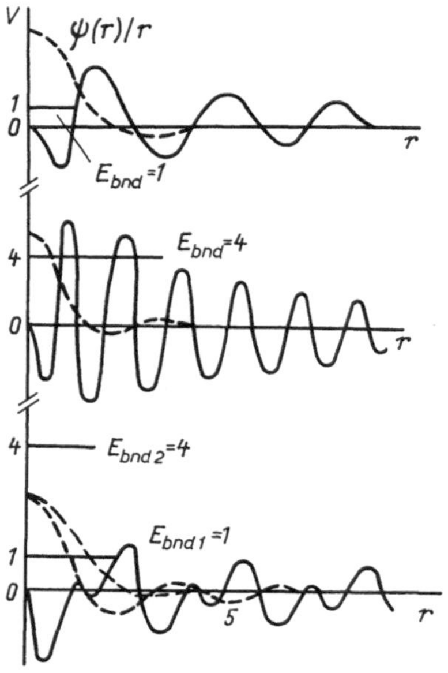

Fig. 1.7a–c. Bound states embedded in the continuum: **a,b** potentials with a single level in the continuous spectrum; **c** potentials with two levels (Meyer–Vernet, 1982)

1.4.3 The Inverse Problem on the Whole Axis

We have considered above a number of examples of the direct problem on the whole axis. We shall now proceed to discuss aspects of the corresponding inverse problem starting with the purely discrete spectrum.

Reconstruction of symmetric potentials from eigenvalues

One of the most important issues in the inverse problem theory is that of uniquely determining the form of a potential with a minimum of spectral information. The following example illustrates with what economy this can be achieved.

In the traditional formulation of the inverse problem, for a purely discrete spectrum of bound states or positions of R-matrix resonances, it is necessary to have a double set of parameters: Besides the energy eigenvalues E_λ, one must also know the normalizing constants c_λ or the amplitude of the reduced width γ_λ. For symmetric potentials $V(x) = V(-x)$, a knowledge of $\{E_\lambda\}$ is sufficient and no additional initial information is required (Zakhariev et al 1979; Barcilon 1974).

This can be explained in a nonrigorous way. When $V(x)$ consists of two symmetric halves, if we find one of them, we immediately know the other. Thus only half of the potential must be reconstructed, therefore, only half of the spectral data is utilized.

This is demonstrated more rigorously by the *theorem on two spectra* (Levitan 1976; Levitan, Gasymov 1964). This theorem establishes the connection between the parameters $\{E_\mu^{\mathrm{I}}, c_\mu^{\mathrm{I}} \text{ or } \gamma_\mu^{\mathrm{I}}\}$ and $\{E_\mu^{\mathrm{II}}, c_\mu^{\mathrm{II}} \text{ or } \gamma_\mu^{\mathrm{II}}\}$ of two problems relative to the eigenvalues of one and the same equation with the same potential but with two different sets of boundary conditions. The normalizing constants c_μ or γ_μ of one of the sets may be expressed through the two sets $\{E_\mu^{\mathrm{I}}\}$ and $\{E_\mu^{\mathrm{II}}\}$, i.e., the set of quantities $\{E_\mu^{\mathrm{I}}\}$ and $\{E_\mu^{\mathrm{II}}\}$ is equivalent to the set $\{E_\mu^{\mathrm{I}}, c_\mu^{\mathrm{I}} \text{ or } \gamma_\mu^{\mathrm{I}}\}$. So the knowledge of two spectra is sufficient for reconstruction of the potential even when it is not symmetric.

The spectrum $\{E_\lambda\}$ of bound state levels of a symmetric well determined *on the entire axis* (Fig. 1.8) may be considered as the combination of the two spectra $\{E_\mu^{\mathrm{I}}\}$ and $\{E_\mu^{\mathrm{II}}\}$ for the Schrödinger equation on the semi-axis $x \geqslant 0$. Indeed, the odd and even levels of the problem defined on the entire axis (with boundary conditions $\psi(x) \to 0$ as $x \to \pm\infty$) coincide with the eigenvalues of the same equation of motion that correspond to two different sets of boundary conditions *on the semi-axis* [either the function $\psi(x)$ or its derivative are equal to zero at $x = 0$]:

$$\text{I.}\quad \psi_\mu^{\mathrm{I}\prime}(0) = 0; \quad \psi_\mu^{\mathrm{I}}(x) \underset{x\to\infty}{\longrightarrow} 0; \quad E_{\lambda=2\mu-1} = E_\mu^{\mathrm{I}}, \quad \mu = 1, 2 \ldots;$$

$$\text{II.}\quad \psi_\mu^{\mathrm{II}}(0) = 0; \quad \psi_\mu^{\mathrm{II}}(x) \underset{x\to\infty}{\longrightarrow} 0; \quad E_{\lambda=2\mu} = E_\mu^{\mathrm{II}}, \quad \mu = 1, 2, \ldots \tag{1.4.6}$$

Thus, knowledge of a single spectrum of the symmetric problem defined on the entire axis gives us two spectra on the semi-axis. In accordance with the theorem on two spectra we obtain, for example, c_μ^{II} from $\{E_\mu^{\mathrm{I}}, E_\mu^{\mathrm{II}}\}$

$$c_\mu^{\mathrm{II}^2} = (E_\mu^{\mathrm{II}} - E_\mu^{\mathrm{I}}) \prod_k^{\infty}{}' \frac{E_k^{\mathrm{I}} - E_\mu^{\mathrm{II}}}{E_k^{\mathrm{II}} - E_\mu^{\mathrm{II}}}, \tag{1.4.7}$$

where the prime on the product sign indicates that $k \neq \mu$. As a result, in solving the inverse problem II we find the potential $V(x)$ on $-\infty < x < \infty$ following the chain:

$$\{E_\lambda\} \underset{\text{entire axis}}{\longrightarrow} \{E_\mu^{\mathrm{I}}, E_\mu^{\mathrm{II}}\} \underset{\text{semi-axis}}{\longrightarrow} \{E_\mu^{\mathrm{II}}, c_\mu^{\mathrm{II}}\} \underset{\text{semi-axis}}{\longrightarrow} V(x \geqslant 0) \to$$

$$\to V(-\infty < x < \infty)\,.$$

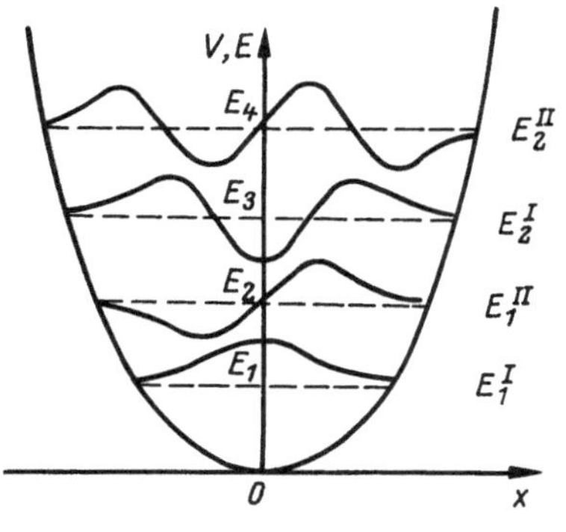

Fig. 1.8. Spectrum of a symmetric well, $E_1, E_2 \ldots$ This spectrum can be composed of two spectra of problems defined on the semi-axis $[0, \infty)$, $E_1^{\mathrm{I}}, E_2^{\mathrm{I}} \ldots$ and $E_1^{\mathrm{II}}, E_2^{\mathrm{II}} \ldots$ corresponding to different boundary conditions at zero. For reconstruction of $V(x)$ it suffices to know $E_1, E_2 \ldots$ without the normalizing constants

The derivation of (1.4.7) is quite difficult (Levitan 1976; Levitan, Gasymov 1964), but one can take advantage (following *Mel'nikov*) of a finite-dimensional quantum model to render the proof of (1.4.7) more clear.

We shall derive the *theorem on two spectra for the difference operator* in the R-matrix scattering theory. We shall take for the two spectra the eigenvalues of the difference operator $\{E_\mu^{\mathrm{I}}, E_\mu^{\mathrm{II}}\}$ corresponding to the boundary conditions

$$\text{I. } u_\mu^{\mathrm{I}}(N+1)=0\,, \quad u_\mu^{\mathrm{I}}(0)=0\,;$$

$$\text{II. } u_\mu^{\mathrm{II}}(N+1)=u_\mu^{\mathrm{II}}(N+2)\,, \quad u_\mu^{\mathrm{II}}(0)=0\,,$$

where at the edge of the interaction interval $r=a=(N+1)\Delta$, either the eigenfunction u or its difference derivative become zero. In the second case we obtain for the R-matrix an expression similar to (1.2.13):

$$R(E)=\frac{\hbar^2}{2ma}\sum_\mu^{N+1}\frac{u_\mu^{\mathrm{II}^2}(N+1)}{E_\mu^{\mathrm{II}}-E}\equiv\sum_\mu^{N+1}\frac{\gamma_\mu^{\mathrm{II}^2}}{E_\mu^{\mathrm{II}}-E}\,.$$

Note the relation between $R(E)$ and the Green function of the Schrödinger difference equation

$$(H-E)G(n,m)=\delta_{nm}/\Delta\,.$$

The right-hand side of this equality can be substituted by $\sum_\mu^{N+1}u_\mu^{\mathrm{II}}(n)\,u_\mu^{\mathrm{II}}(m)$ owing to the completeness relation of (1.2.23). Since the action of $(H-E)^{-1}$ on the eigenfunctions u_μ^{II} simply reduces to multiplication by $(E_\mu^{\mathrm{II}}-E)^{-1}$, we immediately have

$$G(n,m)=\sum_\mu^{N+1}u_\mu^{\mathrm{II}}(n)\,u_\mu^{\mathrm{II}}(m)/(E_\mu^{\mathrm{II}}-E)\,,$$

and, comparing this expression with $R(E)$, we obtain

$$G(N+1,N+1)=R(E)\,.$$

At the same time, solving directly the set of algebraic equations

$$(H-E)G(n,N+1)=\delta_{n,N+1}/\Delta\,; \quad n=1,2,\ldots,N+1\,,$$

we obtain

$$G(N+1,N+1)=A(E)/D(E)\,,$$

where D is the determinant of the $(N+1)\times(N+1)$ matrix of the (1.2.3a) kind, equal to $\prod_\lambda^{N+1}(E-E_\lambda^{\mathrm{II}})$, and $A(E)$ is the corresponding cofactor, coinciding with the $N\times N$ matrix corresponding to problem I: $A(E)=\prod_\lambda^N(E-E_\lambda^{\mathrm{I}})$. Writing (Korn, Korn 1968)

$$G(N+1,N+1)=\sum_\lambda A(E_\lambda^{\mathrm{II}})\,[D'(E)|_{E=E_\lambda^{\mathrm{II}}}(E-E_\lambda^{\mathrm{II}})]^{-1}$$

and comparing it with the expression for $R(E)$, we find the desired relation

$$\gamma_\lambda^{\mathrm{II}2} = A(E_\lambda^{\mathrm{II}})/D'(E)_{E=E_\lambda^{\mathrm{II}}} = \prod_\mu^N (E_\mu^{\mathrm{II}} - E_\mu^{\mathrm{I}}) \Big/ \prod_\nu^{N+1} (E_\lambda^{\mathrm{II}} - E_\nu^{\mathrm{II}})$$

$$\xrightarrow[\Delta\to 0]{N\to\infty} (E_\lambda^{\mathrm{II}} - E_\lambda^{\mathrm{I}}) \prod_\mu^{\infty}{}' (E_\mu^{\mathrm{II}} - E_\mu^{\mathrm{I}})/(E_\lambda^{\mathrm{II}} - E_\mu^{\mathrm{II}}) .$$

Integral equations of the inverse problem with $V(x) \neq 0$ on the entire axis. Several versions of the algorithm for reconstruction of the potential on the entire axis exist (Faddeev 1976; Moses et al 1977, Newton 1980, Tsai 1976). The mathematical machinations are in many aspects similar to those applied to the case of the semi-axis. Therefore, we shall present the formulas without derivation within the most general approach involving $\mathring{V}(x) \neq 0$. This has not yet been addressed in the review literature.

A complete set of functions on the entire axis, unlike the problem of radial motion, is composed of the wave functions of the continuous spectrum of *two* scattering processes involving waves incident on the interaction region on the right (ψ_1) and on the left (ψ_2) and the functions of bound states:

$$\frac{1}{\pi}\int_0^\infty [\psi_1(k,x)\psi_1^*(k,y) + \psi_2(k,x)\psi_2^*(k,y)]dk$$

$$+ \sum_\nu^N \psi(\mathrm{i}\kappa_\nu, x)\psi(\mathrm{i}\kappa_\nu, y) = \delta(x-y) , \tag{1.4.8}$$

where, as $x \to -\infty$,

$$\begin{cases} \psi_1(k,x) \to s_{11}(k)\exp(-ikx) ; \\ \psi_2(k,x) \to \exp(ikx) + s_{21}(k)\exp(-ikx) . \end{cases}$$

While as $x \to +\infty$,

$$\begin{cases} \psi_1(k,x) \to \exp(-ikx) + s_{12}\exp(ikx) ; \\ \psi_2(k,y) \to s_{22}(k)\exp(ikx) . \end{cases}$$

The completeness relation (1.4.8) can be rewritten with the aid of the Jost solutions of one kind

$$f_1^\pm(k,x)\left[\exp(\mp ikx) \lim_{x\to+\infty} f_1^\pm(kx) = 1\right],$$

since $f_1^\pm$ form a fundamental set and both ψ_1 and ψ_2 can be expressed through them:

$$\frac{1}{2\pi}\int_{-\infty}^{\infty} dk[f_1(k,x)f_1^*(k,y) + f_1(k,x)f_1(k,y)s_{12}(k)]$$

$$+ \sum_\nu^N M_{1\nu}^2 f_1(\mathrm{i}\kappa_\nu, x) f_1(\mathrm{i}\kappa_\nu, y) = \delta(x-y) . \tag{1.4.8'}$$

Here s_{12} is the reflection coefficient for waves incident from the right. A similar relation can be written for the other Jost solutions $f_2^{\pm}(k, x)$,

$$\lim_{x \to -\infty} \exp(\pm ikx) f_2^{\pm}(k, x) = 1 ,$$

except that s_{12} and $M_{1\nu}$ must be replaced by s_{21} and $M_{2\nu}$. The norms $M_{1\nu}$ and $M_{2\nu}$ represent coefficients of the decreasing exponentials in the asymptotic behavior of the wave functions of bound states as $x \to \pm\infty$.

The inverse-problem equation for the generalized shift operator $K_1(x, y)$, corresponding to (1.4.8), has the form

$$K_1(x, y) + Q_1(x, y) + \int_x^{\infty} K_1(x, z) Q_1(z, y) dz = 0 , \tag{1.4.9}$$

where the kernel $Q_1(x, y)$ is calculated from known reflection coefficients $\mathring{s}_{12}(k)$, $s_{12}(k)$, the levels of bound states $\mathring{E}_\nu$, E_ν, and the normalizing constants $\mathring{M}_{1\nu}$, $M_{1\nu}$ for the potentials $\mathring{V}(x)$ and $V(x)$, respectively:

$$Q_1(x, y) = \frac{1}{2\pi} \int_{-\infty}^{\infty} \mathring{f}_1(k, x) \mathring{f}_1(k, y) [\mathring{s}_{12}(k) - s_{12}(k)] dk$$
$$+ \sum_{\nu}^{N} M_{1\nu}^2 \mathring{f}_1(i\kappa_\nu, x) \mathring{f}_1(i\kappa_\nu, y) - \sum_{\nu}^{\mathring{N}} \mathring{M}_{1\nu}^2 \mathring{f}_1(i\mathring{\kappa}_\nu, x) \mathring{f}_1(i\mathring{\kappa}_\nu, y) . \tag{1.4.10}$$

Equations similar to (1.4.9,10) are obtained for K_2 and Q_2. Because the inverse problem on the entire axis is over-determined, it is sufficient to use equations with only one of the indices (1 or 2).

The potential and solutions are determined from any one of the $K_i(x, y)$:

$$V(x) = \mathring{V}(x) + \Delta V(x) ; \quad \Delta V(x) = -\frac{\hbar^2}{m} \frac{d}{dx} K_1(x, x) = \frac{\hbar^2}{m} \frac{d}{dx} K_2(x, x) ; \tag{1.4.11}$$

$$f_1(k, x) = \mathring{f}_1(k, x) + \int_x^{\infty} K_1(x, y) \mathring{f}_1(k, y) dy . \tag{1.4.12}$$

For $i = 2$ integration in (1.4.9,12) is performed over $(-\infty, x)$. The conditions to be satisfied by the scattering matrix corresponding to one-dimensional motion along the entire axis have been stated by *Faddeev* (1974).

Potentials Exhibiting Different Asymptotic Behavior at Infinity. One-dimensional motion on the entire axis exhibits certain features characteristic of a *two-*

channel system. The coefficients of reflection and penetrability in both directions form a 2 × 2 scattering matrix:

$$\mathbf{S} = \begin{pmatrix} s_{11} & s_{12} \\ s_{21} & s_{22} \end{pmatrix}.$$

Actually, unlike the genuine multi-channel case, for reconstruction of the one-dimensional potential $V(-\infty < x < \infty)$ we only need to know a single element of the S-matrix (one of the reflection coefficients, as may be seen, for example, from (1.4.9,10). This is possible because of the relations between the elements of the S-matrix in the one-dimensional problem. Naturally, spectral parameters of states belonging to the discrete spectrum are also needed. One can do without the data on bound states if one uses the entire S-matrix (Newton 1980). In addition, the levels of a one-dimensional problem with a nonsingular potential cannot be degenerate, while in a truly multi-channel problem the degree of degeneracy may be equal to the number of channels (Chap. 5).

A particle in the field of a potential tending towards different finite limits V_+ and V_- as $x \to \pm\infty$ represents the simplest physical model for studying threshold phenomena: It is as if a "second channel" opens up at an energy equal to the largest of the two quantities V_+ and V_-. Interestingly, when an energy threshold $E_{\mathrm{thresh}} = \sup\{V_+, V_-\}$ exists, below which the propagation of waves from one side is "closed", it is possible to find the whole scattering S-matrix from the sole reflection coefficient only in the case of waves incident from the lower value of $\{V_+, V_-\}$.

Potentials Depending on the Velocity, $V(\hat{p}^2, r)$. Potentials depending on the momentum operator $\hat{p} = -i\hbar\nabla$ are very important for example, in the meson theory of nuclear forces. They are also of interest from a methodological point of view for the inverse problem in the difference approximation because they permit establishing correspondence between the number of interaction parameters and the number of spectral data (Zakhariev et al 1977). Usually $V(\hat{p}^2, r)$ is written in the form

$$V(\hat{p}^2, r) = V(r) + \left[\frac{\hat{p}^2}{2m} V_1(r) + V_1(r)\frac{\hat{p}^2}{2m}\right] \Big/ 2\,, \tag{1.4.13}$$

where $V(r)$ and $V_1(r)$ are different functions of r. Such potentials lead to a Schrödinger equation in which the second derivative has a variable coefficient. This coefficient is said to determine the effective mass which depends on the coordinate. This mass, for instance, is introduced to account for dissipation of energy in quantum systems.

In the difference approximation the equation of motion with such interaction has the form[3]:

$$-\frac{\hbar^2}{2m\Delta^2}[1+V_1(n)/2]\{\psi(E,n+1)-2\psi(E,n)+\psi(E,n-1)\}$$

$$+V(n)\psi(E,n)-\frac{\hbar^2}{4m\Delta^2}\{V_1(n+1)\psi(E,n+1)$$

$$-2V_1(n)\psi(E,n)+V_1(n-1)\psi(E,n-1)\}=E\psi(E,n)\,. \qquad (1.4.14)$$

Consider the case of a purely discrete spectrum when homogeneous boundary conditions are defined at $r=0$ and $r=a$. The proof that the set of eigenfunctions $u(E_\mu, n)\equiv u_\mu(n)$ is complete and orthonormalized is exactly as in the case of an ordinary potential. Equations (1.3.2a, 3a, 4a, 6a) also remain valid. However, the coefficients $K(n,m)$ for orthogonalization of the functions φ and $\mathring{\varphi}$ are found from the set of equations

$$\Delta K(n,n)Q(n,m)+K(n,m)+\sum_{p=n+1}^{N}\Delta K(n,p)Q(p,m)=0,\, n>m\,;$$

$$Q(n,n)+\sum_{p=n+1}^{N}K(n,p)Q(p,n)K^{-1}(n,n)=K^2(n,n)/\Delta^3,\, n=m\,, \qquad (1.4.15)$$

which differ from (1.3.5a) by the additional equation for $K(n,n)$ obtained by substitution of (1.3.2a) into (1.3.3a) at $n=m$. This minor complication of the equations for K is due to the fact that in the case of the potential $V(\hat{p}^2,n)$ the coefficients of higher powers of E in the polynomials $\varphi(E,n)$ and $\mathring{\varphi}(E,n)$ are no longer equal, i.e., $K(n,n)\neq 1/\Delta$.

The relation between V and K is determined, instead of by (1.3.7a), by the recursion relations

[3] In deriving (1.4.14) one must bear in mind the finite-difference derivative of a product: "the spreading" of the arguments. We shall denote the operators of the right and left difference differentiation by $\delta_\pm$:

$$\delta_+f(n)=[f(n+1)-f(n)]/\Delta\,;\quad \delta_-f(n)=[f(n)-f(n-1)]/\Delta\,,$$

then

$$\delta_-[f(n)g(n)]=f(n-1)\delta_-g(n)+g(n)\delta_-f(n)\,;$$

$$\delta_+[f(n)g(n)]=f(n+1)\delta_+g(n)+g(n)\delta_+f(n)\,.$$

The second derivative assumes the form

$$\delta_+\delta_-[f(n)g(n)]=\{f(n+1)g(n+1)-2f(n)g(n)+f(n-1)g(n-1)\}/\Delta^2=\delta_-\delta_+[f(n)g(n)]\,.$$

$$V_1(n-1) = 2K(n,n)/K(n-1,n-1) - 2 - V_1(n)\,;$$

$$V(n) = -\frac{\hbar^2}{2m\Delta^2}\left[\frac{K(n,n+1)}{K(n,n)} - \frac{K(n-1,n)}{K(n-1,n-1)}\right] - \frac{\hbar^2 V_1(n)}{m\Delta^2}\,. \tag{1.4.16}$$

The first equation in (1.4.16) is obtained by substitution of $\varphi(E,n)$ as given in (1.3.2a) into (1.4.14), then by multiplication of the latter by $\mathring{\varphi}(E,n-1)$ and integration over the spectral measure $d\mathring{\rho}$, after accounting for (1.3.2a). The second equation in (1.4.16) is derived in a similar way, only that (1.4.14) is multiplied by $\mathring{\varphi}(E,n)$. Other equations, similar to (1.4.16) have been obtained by *Zakhariev* and *Suzko* (1975) and can also be utilized.

Because the difference potential $V(\hat{p}^2,n)$ is defined at the points $n = 1, 2, \ldots, N$ by $2N$ values of the functions $V(n)$ and $V_1(n)$ and because the number of spectral parameters $\{E_\mu, \gamma_\mu\}$ is also equal to $2N$, the latter are *not bound by any additional conditions*, as was the case for the ordinary potential. This fact is essential for any practical solution of inverse problems. *Berezansky* (1965) has proved the more general statement: for spectral function satisfying the condition $\int d\rho(E) = 1$ there correspond some V, V_1.

In the continuous limit $\Delta \to 0$ the equation of motion with $V(\hat{p}^2, r)$ can be reduced, with the aid of the Liouville transformation, by transforming the variables to the conventional Schrödinger equation (1.2.1). Therefore, from the inverse problem one determines instead of the individual V and V_1, a single new potential related in such a way to V and V_1 that they can be chosen with some degree of arbitrariness.

1.4.4 Reconstruction of Potentials from Resonance Parameters

The inverse problem within the framework of the R-matrix theory has already been mentioned above. In the case of the modified R-matrix scattering theory within the Kapur–Peierls approach the eigenfunctions $u(k,r)$ are used as the basis states. Besides $u(k,0) = 0$, these functions also satisfy the condition imposed on the logarithmic derivative at point $r = a$:

$$u'(k,r)/u(k,r)|_{r=a} = ik_0\,, \tag{1.4.17}$$

where $k_0 > 0$ is a certain fixed value of the wave number. Unlike the usual R-matrix theory the eigenvalues E_λ which correspond to an imaginary value of the logarithmic derivative of the function u at point a, are *complex*. The proof of the completeness of the set $\{u_\lambda(r)\}$ is presented by *Kravitsky* (1968). These functions are biorthogonal to the conjugate set $\{u_\lambda^*(r)\}$:

$$\int_0^a (u_\lambda^*(r))^* u_{\lambda'}(r)dr = \int_0^a u_\lambda(r)u_{\lambda'}(r)dr = \delta_{\lambda\lambda'}\,. \tag{1.4.18}$$

In other respects, the direct and inverse problems in the Kapur–Peierls approach are solved in the same way as in the usual R-matrix theory.

If k, the argument of $u(k, r)$ is substituted for k_0 in the boundary condition (1.4.17):

$$u'(k, a)/u(k, a) = \mathrm{i}k \,, \tag{1.4.19}$$

then either true resonances, or bound and antibound states are produced. Relation (1.4.19) is equivalent to the requirement that in $u(k, r)$ there exist for $r > 0$ only a single outgoing wave or only an exponentially decreasing (or increasing) solution with $k = \pm\mathrm{i}\kappa$.

The completeness of the set of these states was demonstrated by *Regge* (1958), while the convergence of the expansion over these states of functions defined on the finite interval $[0, a]$ as well as their uniqueness was proved by *Kravitsky* (1968). *Bang* et al (1978–1981) discussed their application to the direct problem. The expansion of functions of complex quantum systems over such states has both merits, such as the discreteness of the expansion and the possibility of physical interpretation of its individual terms and debits, such as the fact that determining complex eigenvalues is significantly more difficult than finding real E_λ.

The peculiarity of the set $\{u_\lambda(r)\}$ corresponding to (1.4.19) is that eigenfunctions with different λ are not orthogonal to each other:

$$\int_0^a u_\lambda(r)u_{\lambda'}(r)dr + \frac{\mathrm{i}u_\lambda(a)u_{\lambda'}(a)}{k_\lambda + k_{\lambda'}} = \delta_{\lambda\lambda'} \,. \tag{1.4.20}$$

Nevertheless, the inverse problem can be formulated in the usual form

$$Q(r, r') + K(r, r') + \int_r^a K(r, r'')Q(r'', r')dr'' = 0 \,, \tag{1.4.21}$$

where

$$Q(r, r') = \sum_\lambda \mathring{\varphi}(E_\lambda, r)\mathring{\varphi}(E_\lambda, r')u_\lambda^2(a) - \delta(r - r') \,;$$

$$\varphi(E, r) = \mathring{\varphi}(E, r) + \int_r^a K(r, r')\mathring{\varphi}(E, r')dr' \,;$$

$$\varphi(E, r = a) = \mathring{\varphi}(E, r = a) = 1 \,;$$

$$\varphi'/\varphi|_{r=a} = \mathring{\varphi}'/\mathring{\varphi}|_{r=a} = \mathrm{i}k_\lambda \,; \quad u_\lambda(r) = \varphi(E_\lambda, r)u_\lambda(a) \,. \tag{1.4.22}$$

The Reaction Matrix $\mathscr{K}$ in the Direct and Inverse Problems. In scattering theory various methods are utilized for defining boundary (asymptotic) conditions for wave functions in terms of the matrices S, $T = S - 1$, R, $\mathscr{K}$. The matrices S and R have already been discussed as scattering characteristics in Sect. 1.2. The matrix $\mathscr{K}$ corresponds to the following behavior of ψ at large $r (l = 0)$:

$$\psi(k, r) \underset{r\to\infty}{\sim} \sin kr + \mathscr{K} \cos kr \,. \tag{1.4.23}$$

This is the asymptotic condition that is usually chosen, for two reasons: When the potential is real, together with (1.2.1, 4.23) define a real ψ, consequently, in numerical computations there is no need to apply complex arithmetic. In addition, no matter with what errors $\mathscr{K}$ has been determined, the scattering S-function that is expressed through it,

$$S = (1 - \mathrm{i}\mathscr{K}/2)/(1 + \mathrm{i}\mathscr{K}/2) \tag{1.4.24}$$

always turns out to be unitary. The physical information contained in ψ is, of course, independent of the form in which it is written. It is obvious that the potential, together with the parameters of bound states, can also be reconstructed from $\mathscr{K}$. To this end it is sufficient to use S in the form of (1.4.24) in the kernels of the Marchenko equation. The inverse scattering problem for standing wave solutions was considered by *Somersalo* (1987).

1.4.5 Construction of Potentials from Spectral Data at a Single Energy and Differing Angular Momenta

The inverse problem is overdetermined if more scattering data are available than is necessary for reconstruction of the interaction potential. In the case of a local spherically symmetrical potential, such a surplus is due to the potential depending on *one* scalar space variable $|r|$, while the scattering data are functions of two variables, for instance, energy and the scattering angle θ, or energy and angular momentum l. Therefore, two equally legitimate formulations of the inverse problem that utilize only part of the scattering information exist. Either $V(r)$ is sought from the phase shifts for all $E \geqslant 0$ and parameters of bound states corresponding to a single l, or for one value of E the entire set of partial phase shifts δ_l with all possible l is used, which corresponds to some angular distribution of the scattering amplitude for a fixed E. In the discussion above we have only dealt with the first approach, we shall now proceed to consider the second.

Although the two formalisms exhibit essential differences, several aspects are common to both. In the $E = \text{const}$ approach the Gelfand–Levitan–Marchenko type equations are used just as for $l = \text{const}$, but they are obtained from other completeness relations: from the orthogonality of state vectors under summation (integration) over the variable l or $\lambda = l + \frac{1}{2}$ instead of the momentum variable k.

The similarity of the procedures to find the eigenvalues λ^2 and k^2 becomes more evident if one transforms the radial Schrödinger equation by the change of variables $r = \exp(\rho)$ and the function $\psi_l(k, r) = \Phi_E(\lambda, \rho)\exp(\rho/2)$:

$$-\psi_l''(k, r) + \left[\frac{l(l+1)}{r^2} + \frac{2m}{\hbar^2} V(r)\right]\psi_l(k, r) = k^2\psi_l(k, r)\,; \tag{1.4.25}$$

$$-\Phi_E''(\lambda, \rho) + \left[\frac{2m}{\hbar^2} V(r) - k^2\right]\exp(2\rho)\Phi_E(\lambda, \rho) = -\lambda^2\Phi_E(\lambda, \rho)\,. \tag{1.4.26}$$

Actually, the similarity between (1.4.25,26) is not complete. The semi-axis $0 \leqslant r < \infty$ becomes the entire axis $-\infty < \rho < \infty$ or, to be more precise, the interval $0 < r \leqslant 1$ stretches out along the left semi-axis $-\infty < \rho \leqslant 0$, while $1 \leqslant r < \infty$ becomes the right semi-axis $0 \leqslant \rho < \infty$ undergoing logarithmic stretching ($\rho = \ln r$). The coefficients of $\psi_l(k, r)$ and $\Phi_E(\lambda, \rho)$ are also very different. In addition, because the signs in the right-hand sides of (1.4.25,26) are different from one another, the λ situated on the *imaginary* axis in the complex λ plane correspond to the continuous spectrum of real k.

From a simple example (Levitan 1976) one can see the analogy between the completeness of the set of states with imaginary values of the angular momentum variable $\lambda = \mathrm{i}\tau(-\infty < \tau < \infty)$ and the completeness of the basis of the ordinary Fourier transformation. Let $k^2 = 2mE/\hbar^2 = 0$; $V(r) = 0$. The solutions $\mathring{\varphi}|_{k_2=0}$ of equation (1.4.25),

$$\mathring{\varphi}/\sqrt{r} = r^{\mathrm{i}\tau} = \exp(\mathrm{i}\tau \ln r)$$

oscillate with infinitely increasing frequency as $r \to 0$. These oscillations correspond to motion of a particle in a singular attractive field ("overturned centrifugal barrier"), in which the so-called falling onto the center is possible. When, on the other hand, we transform from r to ρ, we obtain an ordinary free wave: As the vicinity of the $r = 0$ is stretched out into the semi-axis with respect to the variable ρ, the oscillations of the function $\exp(\mathrm{i}\tau \ln r)$ (with the frequency increasing infinitely as $r \to 0$) become uniform when it is considered as a function $\exp(\mathrm{i}\tau\rho)$ dependent on ρ. The direct and inverse Fourier transformations

$$\mathscr{F}(\tau) = \int_{-\infty}^{\infty} f(\rho)\exp(-\mathrm{i}\tau\rho)d\rho\ ;$$

$$f(\rho) = \frac{1}{2\pi}\int_{-\infty}^{\infty} \mathscr{F}(\tau)\exp(\mathrm{i}\tau\rho)d\tau \tag{1.4.27}$$

corresponding to the complete set of waves $\exp(\mathrm{i}\tau\rho)$ become the Mellin transformation when $\ln r$ is substituted for ρ:

$$\mathscr{F}(\tau) = \int_0^{\infty} g(r)r^{-\mathrm{i}\tau}\frac{dr}{r}; \quad g(r) = \frac{1}{2\pi}\int_{-\infty}^{\infty} d\tau\mathscr{F}(\tau)r^{-\mathrm{i}\tau}\,. \tag{1.4.28}$$

Equation (1.4.28) corresponds to the orthonormalization relation

$$\frac{1}{2\pi}\int_{-\infty}^{\infty}\frac{r^{-\mathrm{i}\tau}r'^{\mathrm{i}\tau}}{\sqrt{rr'}}d\tau = \delta(r - r')\,, \tag{1.4.29}$$

which can be written as the condition of completeness for the solutions of (1.4.25) with $V \equiv 0$, $k = 0$:

$$\mathring{\varphi}(\mathrm{i}\tau, r) = r^{\mathrm{i}\tau}/(2^{\mathrm{i}\tau}r\Gamma(1 + \mathrm{i}\tau))\,,$$

making use of the formula $\Gamma(1 + \mathrm{i}\tau)\Gamma(1 - \mathrm{i}\tau) = \tau\pi/\sinh \pi\tau$,

$$\frac{1}{2}\int_{-\infty}^{\infty}\mathring{\varphi}(r, \mathrm{i}\tau)\mathring{\varphi}^*(r', \mathrm{i}\tau)\frac{\tau d\tau}{rr'\sinh \pi\tau} = \delta(r - r')\,. \tag{1.4.30}$$

In the more general case of $E \neq 0$, $V \neq 0$ in addition to the continuous spectrum along the imaginary axis, discrete values of λ also contribute to the completeness relation. How this occurs will be clarified below.

Two principal modifications of the inverse problem with $E = \text{const}$ exist. Like the Gelfand–Levitan and Marchenko approaches in the case of $l = \text{const}$, they are based on the use of various auxiliary solutions: solutions regular at zero, φ, and Jost solutions f.[4]

Equations of the Inverse Problem Based on the Solutions $f(\lambda, r)$ for k = const. The completeness relation for the Jost solutions at $E = \text{const}$ was obtained by *Burdet* et al (1965, 1966) in the same way as described by *Newton* (1982) for $l = \text{const}$. In deriving the relation one must only perform a change of variables $k \leftrightarrow \lambda$ and of functions $f \leftrightarrow \varphi$.

The Parceval equality for the solutions $f(\lambda, r)$ for $E = \text{const}$ assumes a form similar to the completeness relation for (1.3.11):

$$\int_0^{i\infty} \frac{f^+(\lambda, r) f^+(\lambda, r') \lambda^2 d\lambda}{rr' f^+(\lambda) f^+(-\lambda)} + \sum_j^N M_j^2 \frac{f^+(\alpha_j, r) f^+(\alpha_j, r')}{rr'} = \delta(r - r') , \tag{1.4.31}$$

where integration is performed along the imaginary semi-axis λ (to be more precise, in integrating one must go round the Regge poles lying on the imaginary axis) and M_j^2 represents the complex normalizing constant for the Jost solutions at the Regge poles $\lambda = \alpha_j$ (the zero values of the Jost functions $f^+(\lambda)$ for $\mathrm{Re}\{\lambda\} > 0$)

$$M_j^{-2} = \int_0^\infty \frac{f^{+2}(\alpha_j, r)}{r^2} dr .$$

Besides the usual condition at large r, at the points $\lambda = \alpha_j$ the Jost functions behave as $r^{\lambda + \frac{1}{2}} = r^l$ when $r \to 0$.

Starting from the orthonormalization (1.4.31) we obtain an equation similar to (1.3.25):

$$K(r, r') + Q(r, r') + \int_r^\infty K(r, t) Q(t, r') dt = 0 , \tag{1.4.32}$$

[4] Just as $\varphi(k, r)$ is an even function of k (the equation for φ contains only k^2, while the boundary conditions $\varphi(k, 0) = 0$, $\varphi'(k, 0) = 1$ are totally independent of k), so is $f(\lambda, r)$ an even function of λ (the coefficients of equation (1.4.26) contain λ^2, while the asymptotic behavior of $f(\lambda, r) \underset{r \to 0}{\sim} \exp(ikr)$ is independent of λ). Functions φ and f are analytic functions in the K- and λ-planes, respectively.

where the kernel Q is defined similarly to (1.3.26′):

$$Q(r,r') = \int_0^{i\infty} \frac{\mathring{f}^+(\lambda,r)\mathring{f}^+(\lambda,r')}{rr'}\left[\frac{1}{f^+(\lambda)f^+(-\lambda)} - \frac{1}{\mathring{f}^+(\lambda)\mathring{f}^+(-\lambda)}\right]\lambda^2 d\lambda$$
$$+ \sum_j^N M_j^2 \mathring{f}^+(\alpha_j,r)\mathring{f}^+(\alpha_j,r')/rr'$$
$$- \sum_j^{\mathring{N}} \mathring{M}_j^2 \mathring{f}^+(\mathring{\alpha}_j,r)\mathring{f}^+(\mathring{\alpha}_j,r')/rr' \,, \qquad (1.4.33)$$

while the potential is expressed through $K(r,r')$ almost as in (1.3.7):

$$V(r) = \mathring{V}(r) - \frac{\hbar^2}{m}\frac{1}{r}\frac{d}{dr}[rK(r,r)] \,. \qquad (1.4.34)$$

Equations of the Inverse Problem Based on the Solutions $\varphi(\lambda,r)$ *for* $k = \text{const.}$ For the derivation of the completeness relations for the solutions $\varphi(\lambda,r)$ we refer the reader to *Levitan* (1976). The completeness of the states pertaining to the continuous spectrum in λ was already dealt with for the special case of $E=0$ and $V(r)=0$. Usually, however, besides the continuous spectrum there also exists an *infinite discrete* spectrum of values of λ_n situated on the real axis. These values do not coincide with the physical values $\lambda = l + \frac{1}{2}$, where l are positive integers and are not Regge poles.

Orthonormalization of φ with respect to the variable λ (the condition of completeness) assumes the form

$$\sum_{n=1}^{\infty} \frac{1}{rr'} c_n^2 \varphi(\lambda_n,r)\varphi(\lambda_n,r') + \frac{1}{2}\int_{-\infty}^{\infty} [\varphi(i\tau,r)$$
$$- \frac{\mu(-i\tau)}{\mu(i\tau)}\varphi(-i\tau,r)]\,\varphi(-i\tau,r')\frac{\tau d\tau}{rr' \sinh \tau\pi} = \delta(r-r') \,, \qquad (1.4.35)$$

where $c_n^{-2} = \int_0^\infty \frac{\varphi^2(\lambda_n,k,r)}{r^2} dr$. (The appearance in completeness relations of weight factors quadratic in r^{-1} is explained in Chap. 6, in connection with Sturm functions.)

This equality is similar in structure to (1.4.8′). Here the Jost functions are replaced by $\mu(i\tau)$, except that the usual Jost functions represent the values of the Jost solutions at $r=0$, while $\mu(\pm i\tau)$ are the values of the solutions $\varphi(\pm\lambda,k,r)$ for *large* $r(r = x_0$; Chadan 1986). The ratio $\mu(-i\tau)/\mu(i\tau)$ still remains the S-function ($S = f^-(k)/f^+(k)$). The relation between $\mu(i\tau)$ and the scattering phases is

$$\mu(i\tau) = A(-i\tau)\sin(x_0 + \delta(-i\tau) + \pi i\tau/2) \,.$$

Like $f(\pm k,r)$ with $l = \text{const}$, the solutions $\varphi(\pm\lambda,k,r)$ form a fundamental set for a given k (the dependence on λ of the boundary conditions for $\varphi(\lambda,k,r) \xrightarrow[r\to 0]{} r^{\lambda+\frac{1}{2}}$ corresponds to the dependence of the asymptotic behavior of the functions $f(\pm k,r)$ on k).

When $V(r) \equiv 0$, (1.4.35) assumes the form

$$\sum_{n=1}^{\infty} 2nJ_{2n}(r)J_{2n}(r') + \frac{1}{2}\int_{-\infty}^{\infty} [J_{i\tau}(r) + J_{-i\tau}(r)]J_{-i\tau}(r')\frac{\tau d\tau}{\sinh \tau\pi} = \delta(r - r') , \tag{1.4.36}$$

where J is a Bessel function of the first kind.

The inverse-problem equation corresponding to (1.4.35) coincides with the usual Gelfand–Levitan equation (1.3.5) when the change $dr' \to dr'/r'$ is performed, only that the kernel Q is calculated by

$$Q(r, r') = \sum_{n=0}^{\infty} C_n^2 J_{\lambda n}(r)J_{\lambda n}(r') - \sum_{n=1}^{\infty} 2nJ_{2n}(r)J_{2n}(r')$$
$$- \frac{1}{2}\int_{-\infty}^{\infty} \left[\frac{\mu(-i\tau)}{\mu(i\tau)} + 1\right]J_{-i\tau}(r)J_{-i\tau}(r')\frac{\tau d\tau}{\sinh \pi\tau} . \tag{1.4.37}$$

When $\mathring{V}(r) \equiv 0$, the kernel Q depends on the product rr' (compare with the Marchenko approach for $l = \text{const}$ (1.3.26), where Q depends on $r + r'$ which is related to the logarithmic transformation of the coordinates in going from (1.4.25) to (1.4.26).

Sabatier (1968) and *Levitan* (1976, 1984) have shown that for potentials permitting analytic continuation of the function $rV(r)$ to the whole complex r-plane Q can be sought in the form of a purely discrete expansion:

$$Q(r, r') = \sum_{l=1}^{\infty} c_l^2 J_l(r)J_l(r') + \sum_{l=0}^{\infty} c_{l+\frac{1}{2}}^2 J_{l+\frac{1}{2}}(r)J_{l+\frac{1}{2}}(r') , \tag{1.4.38}$$

from which in the special case of $c_l = 0$ for integer l the kernel is obtained by the Newton–Sabatier method (Newton 1982; Chadan, Sabatier 1977, 1989):

$$Q(r, r') = \sum_{l=0}^{\infty} c_{l+\frac{1}{2}}^2 J_{l+\frac{1}{2}}(r)J_{l+\frac{1}{2}}(r') . \tag{1.4.39}$$

The corresponding function $K(rr')$ is

$$K(r, r') = \sum_{l=0}^{\infty} c_{l+\frac{1}{2}}^2 \varphi(l + \tfrac{1}{2}, r)J_{l+\frac{1}{2}}(r') \tag{1.4.40}$$

and $\varphi(l + \frac{1}{2}, r)$ are determined by an infinite set of algebraic equations

$$\varphi(l + \tfrac{1}{2}, r) = J_{l+\frac{1}{2}}(r) + \sum_{l}^{\infty} L_{ll'}(r)c_{l'+\frac{1}{2}}^2 \varphi(l' + \tfrac{1}{2}, r) , \tag{1.4.41}$$

where

$$L_{ll'}^{(r)} = \int_0^r J_{l+\frac{1}{2}}(r')J_{l'+\frac{1}{2}}(r')\frac{dr'}{r'}$$
$$= \frac{J_{l+\frac{1}{2}}(r)J'_{l'+\frac{1}{2}}(r) - J'_{l+\frac{1}{2}}(r)J_{l'+\frac{1}{2}}(r)}{(l'-l)(l'+l+1)} \tag{1.4.42}$$

The exact solution of these equations is considered in Chap. 2, and the results of approximate reconstruction of potentials by the Newton–Sabatier method are given in Chap. 3. This method provides a unique solution within the class of potentials decreasing more rapidly than $r^{-3/2}$ towards infinity.

Potentials Depending on k and l: $V(k,r)$, $V(\lambda,r)$

The reconstruction of $V(k,r)=V(r)\pm 2kV_1(r)$ has already been discussed by *Chadan* and *Sabatier* (1977). Since then, the work of *Jaulent* and *Jean* (1972–1982) has extended the class of $V(k,r)$ for which the formalism of the inverse problem was developed. A similar formalism was proposed for reconstruction from the phase shift at a given energy of the potentials

$$V(\lambda,r)=V(r)\pm\lambda V_1(r)\,;\quad \lambda=l+\tfrac{1}{2}\,,$$

arising when the spin-angular momentum coupling is taken into account (Pivovarchik 1980). Pivovarchik has suggested that these methods for the reconstruction of $V(k,r)$ and $V(\lambda,r)$ can also be derived by following the derivation of the Marchenko equations. This should make the presentation easier to follow.

First, the generalized shift operation is performed from the free waves

$$\mathring{f}^{+}(k,r)=\exp(ikr)$$

to the solutions of a simple equation involving only part of the interaction $[V_1(r)]$:

$$\begin{aligned} &f_1^{\pm\prime\prime}(k,r)+k^2f_1^{\pm}(k,r)\mp 2kV_1(r)f_1^{\pm}(k,r)\\ &\quad -V_1^2(r)f_1^{\pm}(k,r)\pm \mathrm{i}V_1'(r)f_1^{\pm}(k,r)=0\,. \end{aligned}\tag{1.4.43}$$

(Here the signs "$\pm$" in $f^{\pm}$ correspond to the signs in $V(r)\pm 2kV_1(r)$.)

Equation (1.4.43) can be derived as follows. In the quasi-classical expression for the Jost solutions with the potential $V(k,r)$

$$\begin{aligned} f^{\pm}(k,r)&=\sqrt[4]{k^2/(k^2\mp 2kV_1(r)-V(r))}\\ &\quad\times\exp\left(\mathrm{i}\int_r^\infty dt\sqrt{k^2\mp 2kV_1(t)-V(t)}\right), \end{aligned}$$

the dependence on $V(r)$ vanishes in the limit of large k:

$$f^{\pm}(k,r)\xrightarrow[k\to\infty]{} f_1^{\pm}(k,r)=\exp\left[(ikr)\mp \mathrm{i}\int_r^\infty V_1(t)dt\right].$$

Equation (1.4.43) for $f_1^{+}(k,r)$ is obtained if $f_1^{+}(k,r)$ is differentiated twice.

Although V_1 in (1.4.43) is not known a priori, one of the solutions of this equation can nevertheless be written immediately in its explicit form:

$$f_1^{\pm}(kr)=\exp\left[ikr\pm \mathrm{i}\int_r^\infty V_1(t)dt\right]=\exp(ikr)F^{\pm}(r)\,,\tag{1.4.44}$$

i.e., the action of the generalized shift operator, $\mathring{f}^{\pm}(k,r) \to f_1^{\pm}(k,r)$ simply reduces to multiplication by $\exp\left(\pm \mathrm{i} \int_r^\infty V_1(t)dt\right) = F^{\pm}(r)$.

The functions $f_1^{\pm}(k,r)$ will now play the part of the reference solutions $\mathring{f}^{\pm}(k,r)$ in the usual Marchenko procedure for constructing the solution $f^{\pm}(k,r)$ of the Schrödinger equation with the total potential $V(k,r)$:

$$f^{\pm\prime\prime}(k,r) + k^2 f^{\pm}(k,r) \pm 2kV_1(r) f^{\pm}(k,r) - V(r) f^{\pm}(k,r) = 0 \,. \tag{1.4.45}$$

Indeed, (1.4.43) differs from (1.4.45) by the conventional potential perturbation

$$\Delta V(r) = V(r) + V_1^2(r) \pm \mathrm{i} V_1'(r)$$

which depends only on r but not on k. This is readily verified by adding and subtracting in (1.4.43) the terms $[V(r) + 2V_1(r) + V_1'(r)]f^{+}(k,r)$.

If, as before, we denote the kernel of the operator of the generalized shift $f_1^{+}(k,r) \to f^{+}(k,r)$ by K^{M},

$$f^{\pm}(k,r) = f_1^{\pm}(k,r) + \int_r^\infty K^{\mathrm{M}}(r,t) f_1^{+}(k,t)dt \,,$$

then by substituting in (1.4.44) we obtain the transformation of the free wave $\exp(\mathrm{i}kr)$ into $f^{+}(k,r)$:

$$f^{\pm}(k,r) = F^{\pm}(r)\exp(\mathrm{i}kr) + \int_r^\infty K^{\mathrm{JJ}\pm}(r,t)\exp(\mathrm{i}kt)dt \,, \tag{1.4.46}$$

where $K^{\mathrm{JJ}\pm}(r,t) = K^{\mathrm{M}}(r,t) F^{\pm}(t)$ is the Jaulent–Jean kernel. A similar relation is obtained for $f^{\pm}(-k,r)$. The completeness of the solutions (1.4.46), was demonstrated by *Jaulent* and *Jean* (1976) in Appendix C. For $K^{\mathrm{JJ}\pm}$ the following set of equations is obtained

$$K^{\mathrm{JJ}\pm}(r,r') + F^{\pm}(r) Q^{\mathrm{JJ}\pm}(r+r') + \int_r^\infty K^{\mathrm{JJ}\mp}(r,t) \times Q^{\mathrm{JJ}\pm}(t+r')dt = 0 \,, \quad t \geqslant r \geqslant 0 \,; \tag{1.4.47}$$

$$F^{+}(r) - F^{-}(r) = \int_r^\infty [K^{\mathrm{JJ}+}(r,t) - K^{\mathrm{JJ}-}(r,t)dt \,;$$

$$F^{+}(r) F^{-}(r) = 1 \,; \quad \lim_{r\to\infty} F^{\pm}(r) = 1 \,, \tag{1.4.48}$$

where

$$Q^{\mathrm{JJ}\pm} = \lim_{R\to\infty} \frac{1}{2\pi} \int_{-R}^{R} \{[F^{\pm}(0)]^2 - S^{\pm}(k)\} \mathrm{e}^{-\mathrm{i}kr} dk + \sum_n^{N\pm} M_n^{\pm 2} \exp(-k_n r) \,; \quad \lim_{|k|\to\infty} S^{\pm}(k) = [F^{\pm}(0)]^2 \,. \tag{1.4.49}$$

Summation in (1.4.49) is performed over bound states. Equations (1.4.47) differ from the usual Marchenko equations by additional unknown factors $F^{\pm}(r)$. But $K^{\mathrm{JJ}\pm}(r,r')$ can be expressed explicitly through $F^{\pm}$ applying auxiliary solutions

$a^{\pm}_{1,2}(r, r')$ of equations of the Marchenko type obtained from (1.4.47) by substitution for $F^{\pm}(r)$ of 1 and $\pm$ i:

$$K^{JJ\pm}(r, r') = F^{\mp}(r)\alpha^{\pm}(r, r') + F^{\pm}(r)\beta^{\mp}(r, r'), \quad r' \geqslant r \geqslant 0, \tag{1.4.50}$$

where

$$\alpha^{\pm}(r, r') = [a_1^{\pm}(r, r') \mp i a_2^{\pm}(r, r')]/2;$$

$$\beta^{\mp}(r, r') = [a_1^{\pm}(r, r') \pm i a_2^{\pm}(r, r')]/2.$$

Substituting (1.4.50) into (1.4.48) we have

$$F^{+}(r)\left\{1 - \int_r^{\infty} [\alpha^{-}(r, r') - \beta^{-}(r, r')]dr'\right\}$$

$$= F^{-}(r)\left\{1 - \int_r^{\infty} [\alpha^{+}(r, r') - \beta^{+}(r, r')]dr'\right\},$$

and making use of $F^{+} F^{-} = 1$ we find from (1.4.48),

$$[F^{\pm}(r)]^2 = \left\{1 - \int_r^{\infty} [\alpha^{\pm}(r, r') - \beta^{\pm}(r, r')]dr'\right\} \Big/$$

$$\times \left\{1 - \int_r^{\infty} [\alpha^{\mp}(r, r') - \beta^{\mp}(r, r')]dr'\right\},$$

from which we uniquely determine $F^{\pm}(r)$, taking into account that the sign in front of $F^{\pm}$ is determined by the condition $\lim_{r\to\infty} F^{\pm}(r) = 1$ from (1.4.48). The potentials $V(r)$ and $V_1(r)$ are found from

$$V(r)F^{\pm}(r) = F^{\pm\prime\prime}(r) - 2[K^{JJ\mp}(r, r)]' \pm 2V_1(r)K^{JJ\mp}(r, r);$$

$$V_1(r) = i[F^{\mp}(r)]'/F^{\mp}(r). \tag{1.4.51}$$

Matrix generalization of this formalism allows reconstruction of potentials exhibiting a more complicated dependence on energy (Jaulent and Jean 1976):

$$V(E, r) = \sum_{p=0}^{n} (\sqrt[2n]{E})^p V_p(r).$$

1.5 Notes on the Literature

Of all the numerous manuals on scattering theory we should like to draw the reader's attention to the monograph by *Newton* (1982), the books by *Taylor* (1972) and *Wu* (1962) and the recent monographs by *Mentkovsky* (1982), *Lendel* and *Salack* (1983), and by *Berezin* and *Shubin* (1983). The history of the development of the inverse-problem formalism has been presented by *Chadan* and *Sabatier* (1989); see also reviews by *Faddeev* (1963, 1976).

Although phase shift analysis (the determination of the S-matrix from scattering cross sections) is not discussed here, a presentation has been given in *Nichitiu* (1983) and also *Chadan* and *Sabatier* (1989), Chap. 10. The stability of phase shift analyses was considered by *Stefanescu* (1988).

It is often more convenient to investigate inverse problems on the basis of the more general Riemann problem (Takhtajan, Faddeev 1986; Newton 1989).

Various types of inverse problems, not only involving the Schrödinger equation, have been dealt with by *Buchgeim* (1983). Detailed reports are also available in the Proceedings of Interdisciplinary Workshops on Inverse Problems held annually in Montpellier (France). It should be noted, however, that by using the Liouville transformation one can reduce equations of the form

$$-\frac{d}{dx}\left\{p(x)\frac{dy}{dx}\right\} + q(x)y = \mu r(x)y$$

to the Schrödinger equation (Levitan, Sargsyan 1970). Such equations are encountered in various areas of physics for example, in the determination of the ionospheric electron density (Pechenick 1983), of the permeability and permittivity, from scattering data (Suzko et al 1986; Coen 1981), as well as in problems on the propagation of waves in an absorbing inhomogeneous medium (Jaulent, Jean 1972–82). In these cases the potentials $V(k, r)$ that were considered in Sect. 1.4 are obtained. *Denisov* (1982) proposed the R-matrix analogue of the inverse problem with a discrete set of initial spectral data for the coefficient of heat conduction. The technique of path-integration for the derivation of the inverse problem equations was used by *Kac* (1974). The reconstruction of potentials from scattering data using nonlinear equations (phase function-type approach) was proposed by *Abramov* (1988). The off-energy-shell scattering matrix can be continued from on-shell data, as shown by *Babenko* and *Petrov* (1988). They take into account the existence of a hard core in the potential.

The completeness of eigenfunctions used in scattering theory was first proved for the Sturm–Liouville problem by *Steklov* (1907–9). Various other proofs can be found in the works of *Levitan* and *Sargsyan* (1970) and *Newton* (1982). For the case of *non-self-conjugate* operators, when the set of eigenfunctions is supplemented by adjoint functions satisfying $(H - E)^n u = 0$ the reader is referred to *Neimark* (1969). The concept of multiple completeness is encountered in investigations of the direct and inverse problems for resonance states (Bang et al 1978–81; Kravitsky 1968; Regge 1958) corresponding to the boundary conditions $u_\nu(0) = u'_\nu(a) + iku_\nu(a) = 0$; the set $\{u_\nu\}$ also contains bound and antibound states. It is complete on $[0, 2a]$ and doubly overcomplete on $[0, a]$. Products of eigenfunctions may also form a complete set. The proof of the completeness of products of eigenfunctions and a thorough bibliography on this issue may be found in *Christov* (1981). The completeness of the Jost solutions for $V(\lambda, r)$ was verified by *Pivovarchik* et al (1984). The conditions for the existence of bound states in the continuous spectrum has been discussed by *Skriganov* (1973) and *Albeverio* (1972).

In addition to the publications mentioned above, *Guseinov* (1976–82) and *Manakov* (1974–80) address the inverse problem within the finite-difference approach. The corresponding technique is useful in the theory of orthogonal polynomials (Case 1974). The inverse problem with two spectra as initial data for identical boundary conditions at the origin but different behavior as $r \to \infty$ was solved by *Gasymov* and *Guseinov* (1989).

A transformation between the Schrödinger, the Klein–Gordon, and the Dirac equations was given by *Leon* (1981). About the reduction of the non-relativistic (supersymmetric) Schrödinger equation to relativistic Dirac and Klein–Gordon equations see *Sokatchev* (1987), *Aneva* (1989).

The infinite period limit of the inverse problem formalism for periodic potentials is considered by *Venakides* (1988).

Extensive bibliographies on quantum tunnelling are given in *Rabotnov* (1983) and *Kojushner* (1983), the optical analogue of sub-barrier penetrability is described by *Tsai* (1976), while a similar effect in wave guides is discussed by *Campi* and *Harrison* (1966–67). Vacuum tunnelling of electrons permits investigation of surface structures with great accuracy and with an uncertainty of 0.01–0.02 nm [(Binnig, Rohrer 1987)–1986 Nobel prize]. Resonance tunnelling through two different delta-like barriers is considered by *Lessie* and *Spadaro* (1986). Rather than to tunnelling, *Cohen* (1965) attributed the penetrability of potential barriers to passage above the barrier of that part of the waves which acquire sufficient energy to do so owing to the uncertainty principle (fluctuations). The tunnelling penetrability of a quasi-classically impenetrable barrier was discussed by *Exner* et al (1985). The time dependence of wave packet tunnelling through a system of barriers was considered by *Jauho* (1986–7).

Reconstruction of periodical potentials was considered by *Marchenko* (1977), while the inverse problem for local perturbation in a periodic field was dealt with by *Firsova* (1975, 1986) and *Newton* (1985). Equivalence classes of periodic potentials yielding the same dependence $E(k)$ were discussed by *Trubovitz* and *Deift* (1984).

Newton (1980) proposed a solution of the inverse problem on the whole axis utilizing for the initial information the entire 2×2 scattering matrix rather than the individual matrix element S_{21} (the reflection coefficient). Later, *Newton* (1982) applied this formalism to reconstruct multi-dimensional potentials.

Among the Newton–Sabatier potentials there exist "transparent" ones for which all the scattering phase shifts δ_l at a given fixed energy are equal to zero (Sabatier 1966).

The mathematical aspects of the problems of bound states in the continuous spectrum were studied by *Eastham* and *Kalf* (1982), *Reed* and *Simon* (1978–9). The solutions of the nonlinear *KdV* equation corresponding to the potentials with bound states in the continuum (as in Fig. 1.5) were obtained by *Novikov* (1984). A generalization of the inverse-problem formalism to nonlocal potentials was proposed by *Muzafarov* (1985–8). The limiting transition from the quantum to the classical inverse problem is considered by *Bogdanov* (1985).

The inverse problem involving a variable coupling constant was developed by *Chadan* et al (1984–9) and with a variable charge, for fixed E and l, to reconstruct the potential from scattering data in the complex plane of the Coulomb coupling constant, by *Poplavsky* (1986) and *Popushoy* (1985–6). The inverse scattering problem on noncompact graphs was considered by *Gerasimenko* (1988).

The quantum motion on manifolds such as a semi-axis and plane joining at one point and three semi-axes with joined ends (branching one-dimensional wave guide) were considered by *Exner* et al (1987). For the description of such systems the extension theory of symmetric to self-conjugate operators was used. The same technique was used by *Kuperin* et al (1986) for developing models describing the relation between the nonrelativistic motion of the centers of mass of colliding complexes and their intrinsic motion, which may be governed by different parameters such as relativistic effects or energy-dependent potentials.

The principle of maximum entropy was applied to the solution of the inverse problem by *Soroko* (1981). Explicit formulas for the potential and the wave function expressed through scattering data were obtained by *Yegikyan* and *Zhidkov* (1985–7). For the first steps in the development of p-adic quantum mechanics see the paper by *Zelenov* (1989) and references therein. The relativistic inverse problem was dealt with by *Amirkhanov* et al (1971–81), *Solovtsev* (1984), *Malyarov* et al (1979) and within the theory of general relativity (gravity) by *Bogdanov* and *Demkov* (1987). The entire issue of the Proc. IEEE, *74*, N3 (1986) was devoted to reviews on the inverse problems applied in seismology.

The inverse problem for fixed $aE + bl(l + 1)$ was considered by *Rudyak* et al (1984). They and *Suzko* (1985) presented exact solutions. The classification of nonlinear equations soluble within the inverse-problem approach (the complete list of integrable systems) was given by *Mikhailov* et al (1987).

1.6 Exercises

1. Describe the qualitative behavior of the energy levels of bound states as a function of the width and depth of the potential well and upon the mass of the particle in the bound state.
2. How does the potential change if one alters the norm (C or M) of the ith bound state while the other spectral parameters remain fixed? (Hint: for the ground state see Fig. 2.1 which corresponds to the exactly solvable case.)
3. Describe the behavior of the limiting transition of the wave function of a resonance state over a rectangular barrier (Fig. 1.9) to a bound state as the height of the barrier increases to infinity but the height of the resonance level above the upper plateau of the barrier remains fixed.
4. Describe the spectrum of bound states above a potential barrier with the form of a well turned upside down and having exponentially rising walls [see formula 8.492.1 in *Gradstein* and *Ryzhik* (1971)]. Hint: see *Reed* and *Simon* (1979) vol. 2, ch. 2.

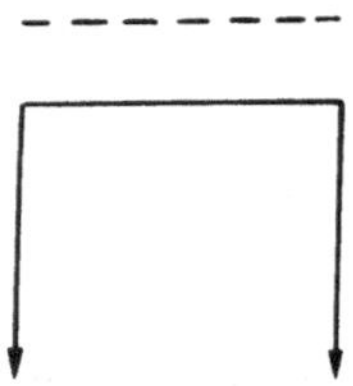

Fig. 1.9. Quasi-bound state above the potential barrier

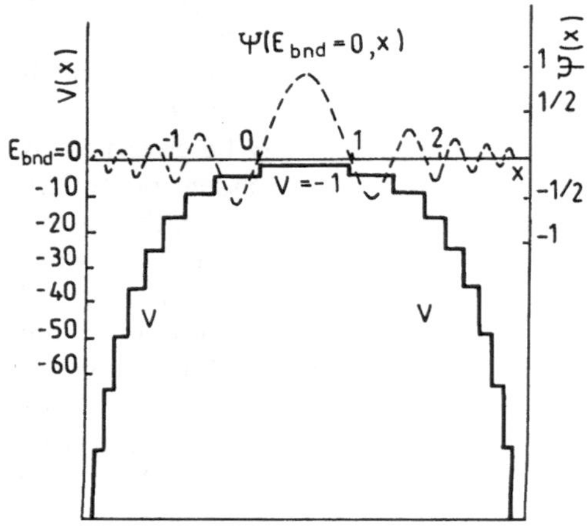

Fig. 1.10. The "cloud" of the bound state embedded into the continuum above the top of a stepped potential (Pivovarchik et al, 1986)

5. Extend the potentials and wave functions of bound states embedded in the continuous spectrum and depicted in Figs. 1.4, 5 to the entire axis $-\infty < x < \infty$.
6. Explain the existence of a bound state above the peak of a potential symmetrical "pyramid" $V(x)$:

$$V(x > 0) = -n^2 \text{ for } x_n < x < x_{n+1}\ ; \quad x_{n+1} - x_n = \frac{1}{n}\ ;$$

$$\psi(E_{\text{bound}} = 0;\, x > 0) = (-1)^{n-1}\left(\frac{1}{n}\right)\sin\left[n(x - x_n)\right],$$

which are shown in Fig. 1.10.

7. Consider the finite-difference Schrödinger equation with the linear potential $V_n = Cn$ (homogeneous "electric field"). Obtain the bound states in the inclined allowed zone. Hint: consider the recursive relations for Bessel functions J_n, $J_{n\pm1}$.

Chapter 2

Exactly Solvable Models: Bargmann Potentials V^{B}

2.1 General Comments

The search for exact solutions to the Schrödinger equation was initiated immediately after its publication. Naturally, the first to be found were the solutions for the most simple potentials which reduce the equation of motion to differential equations for known elementary and special functions. Examples are provided by the potentials $V(r) = c_n r^n$ where $n = 2$ represents the oscillatory, $n = 1$ the linear, $n = 0$ the constant, $n = -1$ the Coulomb, and $n = -2$ the centrifugal potentials (Flugge 1971).

The simplest potentials for the integral analogue of the Schrödinger equation, that is, the Lippmann–Schwinger equation, are separable nonlocal potentials

$$V(r, r') = \sum_{i}^{N} V_i(r) V_i(r') \, .$$

These potentials correspond to *degenerate kernels* of the integral equation, because of which the determination of the exact wave functions reduces to purely algebraic operations (Zubarev 1981).

The increasing number of exactly solvable models has been stimulated by the inverse problem theory. Choosing the kernels of the integral equations in a separable form yields new families of such models with potentials of the Bargmann type. (It should be noted that the separability of the potentials, the kernels of the Lippmann–Schwinger equation, and that of the kernels Q of the inverse-problem equations should not be confused. Nonseparable, local V correspond to separable Q.) These models are also obtained by factorization of the Hamiltonian as proposed by Schrödinger (Infeld, Hull 1951).

The same exact models are equally suitable for the direct and inverse problems. They explicitly establish the mutual relationship $V \leftrightarrow S$ between the interaction and the spectral characteristics. Some of these models were easier to find by studying the Schrödinger equation, while others are simpler when the Gelfand–Levitan–Marchenko equations are applied. This chapter is devoted mainly to the latter.

Now a general approach based on the Riemann problem, instead of traditional Gelfand–Levitan–Marchenko equations, is often used to get exact solutions not only in nonrelativistic quantum physics, but also in quantum field theory and statistical physics (e.g. Takhtajan 1986; Baxter 1982; Dubrovin 1988).

Bargmann models may be classified as follows:

I. by the region in which the potentials are defined: on the entire axis $(-\infty < x < \infty)$, on the semi-axis $(0 \leqslant r < \infty)$, on a finite interval $(0 \leqslant r \leqslant a < \infty)$;

II. by restriction of the manifold of parameters $\{E, l\}$, fixed energy or fixed angular momentum l.

There are also special groups:

III. models with a purely discrete energy spectrum with bound states in infinitely deep potential wells (for example, quark confining fields) or resonances in the R-matrix scattering theory;

IV. systems with discrete space variables: multi-channel, multi-dimensional, multi-particle systems;

V. models obtained from the Crum–Krein–Marchenko (Darboux) transformations.

Some of the listed models are considered in Part Two.

When studying novel phenomena the explicit analytic form of potentials and wave functions is a priceless advantage. They also lend themselves readily to the demonstration of known properties of quantum systems.

Among the indicated models are several with remarkable potentials:

spectrum- or *phase-equivalent*, possessing the same spectra or scattering characteristics;

transparent at all energies; these play an important role for the solution of nonlinear equations within the framework of the inverse problems formulation. Certain of these potentials coincide in form with the so-called soliton (solitary waves) potentials;

potentials possessing *bound states embedded in the continuous spectrum* and potentials which are totally reflecting above a given energy barrier.

Exact models are not only important in their own right, but can also serve as approximate solutions of general inverse problems when the kernels Q are not degenerate. Approximation of an arbitrary potential by Bargmann potentials corresponds to approximation of the scattering function S by the fractional-rational expressions

$$S(k) = \prod_j \frac{(k - \mathrm{i}a_j)(k + \mathrm{i}b_j)}{(k - \mathrm{i}b_j)(k + \mathrm{i}a_j)} .$$

Thus, regularization of the solution of an ill-posed problem occurs due to the narrowing of the interaction space to a set of potentials depending on a finite number of parameters. This point will be dealt with in greater detail in Chap. 3.

2.2 Simplest Examples of $V^{\mathrm{B}}(0 \leqslant r < \infty)$ for $l=0$

In this section potentials that correspond to factorized kernels

$$Q(r, r') = Q_1(r) Q_2(r') ,$$

of the inverse-problem equations are considered. The generalization to the case of multi-term separable kernels

$$Q(r,r') = \sum_i Q_{1i}(r)Q_{2i}(r')$$

and $l \neq 0$ is presented in Sect. 2.3. It makes sense to examine the solutions with single-term Q^{GL} and Q^{M} in more detail so as to later avoid making the general expressions too cumbersome. It is also useful to note the differences in the construction of a given model in the Gelfand–Levitan and Marchenko approaches. For example, the single-term Q^{GL} given below in (2.2.1) corresponds to the multi-term Q^{M} (the opposite situations are also possible).

2.2.1 Potentials with a Single Bound State

Factorized Kernel Q^{GL}. To obtain the simplest factorized kernel G^{GL}, one has to retain in [compare with (1.3.6) where $\mathring{V}$ has no bound states]

$$Q(r,r') = \int_0^\infty \mathring{\varphi}(k,r)\mathring{\varphi}(k,r')d(\rho(k) - \mathring{\rho}(k)) + \sum_\lambda c_\lambda^2 \mathring{\varphi}(\mathrm{i}\kappa_\lambda, r)\mathring{\varphi}(\mathrm{i}\kappa_{\lambda'}, r')$$

only a single term in the sum over bound states, by assuming that all the c_λ, except, for example, c_1, are equal to zero, and that the contribution from the continuous spectrum is zero, by setting $\rho(k) \equiv \mathring{\rho}(k)$:

$$Q^{\mathrm{GL}}(r,r') = c^2 \mathring{\varphi}(\mathrm{i}\kappa, r)\mathring{\varphi}(\mathrm{i}\kappa, r') \,, \tag{2.2.1}$$

where $\kappa^2 = -2mE_{\mathrm{b}}/\hbar^2$ (E_{b} is the energy of the bound state). With such a degenerate kernel the Gelfand–Levitan equation (1.3.5) reduces to a single algebraic equation. We note that together with Q^{GL} the dependence on the coordinates r and r' in $K(r,r')$ also factorizes. This can be seen when (1.3.5) is rewritten by making use of (2.2.1) for

$$K^{\mathrm{GL}}(r,r') = -c^2\{\mathring{\varphi}(\mathrm{i}\kappa, r) + \int_0^r K^{\mathrm{GL}}(r,r'')\mathring{\varphi}(\mathrm{i}\kappa, r'')dr''\}\mathring{\varphi}(\mathrm{i}\kappa, r'), \tag{2.2.2}$$

and the expression within the braces is $\varphi(\mathrm{i}\kappa, r)$ in accordance with (1.3.2):

$$K^{\mathrm{GL}}(r,r') = -c^2 \varphi(\mathrm{i}\kappa, r)\mathring{\varphi}(\mathrm{i}\kappa, r') \,. \tag{2.2.2'}$$

The yet unknown function $\varphi(\mathrm{i}\kappa, r)$ is determined from the simple algebraic equation obtained from (2.2.2) by substitution of (2.2.2'):

$$\varphi(\mathrm{i}\kappa, r) = \mathring{\varphi}(\mathrm{i}\kappa, r) - c^2\varphi(\mathrm{i}\kappa, r)\int_0^r \mathring{\varphi}^2(\mathrm{i}\kappa, r'')dr'' \,, \tag{2.2.3}$$

from which we immediately find

$$\varphi(\mathrm{i}\kappa, r) = \mathring{\varphi}(\mathrm{i}\kappa, r) \Bigg/ \left[1 + c^2 \int_0^r \mathring{\varphi}^2(\mathrm{i}\kappa, r'')dr''\right] \tag{2.2.4}$$

and, in agreement with (2.2.2′)

$$K^{\mathrm{GL}}(r,r') = -c^2\mathring{\varphi}(\mathrm{i}\kappa,r)\mathring{\varphi}(\mathrm{i}\kappa,r')\Big/\left[1+c^2\int_0^r \mathring{\varphi}^2(\mathrm{i}\kappa,r'')dr''\right]. \qquad (2.2.5)$$

The potential V^{B} which is an exact solution of the inverse problem has in this case, in accordance with (1.3.7), the form

$$V^{\mathrm{B}}(r) = \frac{\hbar^2}{m}\frac{d}{dr}K^{\mathrm{GL}}(r,r) + \mathring{V}(r) = -\frac{\hbar^2}{m}\frac{d^2}{dr^2}\ln p(r) + \mathring{V}(r)\,;$$

$$p(r) = 1 + c^2\int_0^r \mathring{\varphi}^2(\mathrm{i}\kappa,r')dr'\,.$$

For $\mathring{V}(r)\equiv 0$

$$V^{\mathrm{B}}(r) = \frac{\hbar^2}{m}\frac{[2\kappa(r/2 - c^{-2}\kappa^2) - 2\sinh^2\kappa r]}{\left[c^{-2}\kappa^2 + \left(\dfrac{\sinh 2\kappa r}{2\kappa} - r\right)\Big/2\right]^2}\,. \qquad (2.2.6)$$

The regular solution φ of the Schrödinger equation with this potential is obtained from (1.3.2) and (2.2.5):

$$\varphi(k,r) = \mathring{\varphi}(k,r) - c^2\mathring{\varphi}(\mathrm{i}\kappa,r)\int_0^r \mathring{\varphi}(\mathrm{i}\kappa,r')\mathring{\varphi}(k,r')dr'\,p^{-1}(r)\,, \qquad (2.2.7)$$

or

$$\varphi(k,r) = \sin kr/k - c^2\frac{\sinh\kappa r(k\cosh\kappa r\sin kr - k\sinh\kappa r\cos kr)}{k(\kappa^2+k^2)\left(\kappa^2 + \dfrac{c^2}{2}(\sinh(2\kappa r)/2\kappa - r)\right)} \qquad (2.2.7')$$

when $\mathring{V}(r)\equiv 0$.

In (2.2.7) one must distinguish between the roles of the energy parameter κ and the wave number k occurring in the argument of function φ. The choice of κ (and of c) fixes the potential (2.2.6) for which (2.2.7) gives solutions for any energy $E = k^2\hbar^2/2m$ including also the case of $k = \mathrm{i}\kappa$ for the bound state (2.2.4).

Regular solutions $\varphi(k,r)$ differ from the physical wave functions $\psi(k,r)$ only by a normalization constant, $\psi = c\varphi$, as already noted in Sect. 1.2. Indeed, one of the boundary conditions for φ and ψ is common to both: $\psi(k,0) = \varphi(k,0) = 0$, therefore, as solutions of a linear ordinary differential equation of the second order, they are equal to each other up to a constant factor. For a bound state this factor is determined from the condition of normalization to unity:

$$\int_0^\infty \psi^2(\mathrm{i}\kappa,r)dr = \int_0^\infty c_\kappa^2\varphi^2(\mathrm{i}\kappa,r)dr = 1\,,$$

$$c_\kappa^{-2} = \int_0^\infty \varphi^2(\mathrm{i}\kappa,r)dr\,, \qquad (2.2.8)$$

while for $E>0$, from comparison of the asymptotic behavior of φ and ψ [see (1.3.8,9,13) where $f^{\pm}(k,r)$ and $f^{\pm}(k)$ are the Jost solutions and the Jost functions]:

$$\varphi(k,r) \xrightarrow[r\to\infty]{} \frac{\mathrm{i}}{2k}\{f^+(k)\exp(-\mathrm{i}kr) - f^-(k)\exp(\mathrm{i}kr)\}, \ \mathrm{Im}\{k\} = 0\ ; \quad (2.2.9)$$

$$\psi(k,r) \xrightarrow[r\to\infty]{} \frac{\mathrm{i}}{2}[\exp(-\mathrm{i}kr) - S(k)\exp(\mathrm{i}kr)], \ \mathrm{Im}\{k\} = 0\ . \quad (2.2.10)$$

From (2.2.9,10) we obtain (taking account of $s(k) = f^-(k)/f^+(k)$

$$\psi(k,r) = \frac{k}{f^+(k)}\varphi(k,r)\ ; \quad E > 0\ . \quad (2.2.11)$$

We find the Jost functions $f^{\pm}(k)$ in accordance with (2.2.9) as the coefficient of $\exp(\mp \mathrm{i}kr)$, going to the asymptotic limit $r \to \infty$ in (2.2.7′) and cancelling the increasing factors in the numerator and denominator:

$$f^{\pm}(k) = (k \mp \mathrm{i}\kappa)/(k \pm \mathrm{i}\kappa)\ , \quad (2.2.12)$$

i.e.,

$$S(k) = (k + \mathrm{i}\kappa)^2/(k - \mathrm{i}\kappa)^2\ . \quad (2.2.13)$$

As a result, we have obtained the two-parameter family of potentials (2.2.6) with exact wave functions and the scattering S-function (2.2.13). All these potentials possess a single bound state each, the energy levels of which, $E = -\hbar^2\kappa^2/2m$, as well as the normalizing constants c^2 can be chosen arbitrarily. Since c^2 is changed, a whole set of phase-equivalent potentials arises (Fig. 2.1), when c^2 increases, the well narrows and becomes deeper.[1] For c^2 one can choose any value from 0 to ∞.

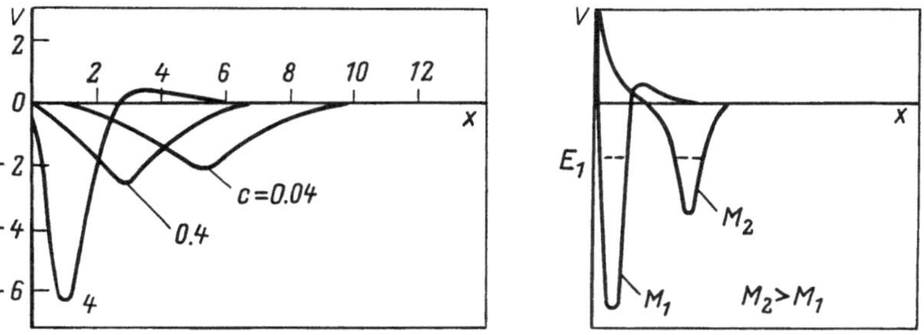

Fig. 2.1. Phase-equivalent potentials (Bargmann 1949; Klein 1976). A change of the normalization of the bound state (a decrease of c or an increase of M) leads to a shift of the well to the right; the dotted curve indicates the level E_{bnd}

[1] To increase the derivative ψ' at point $r = 0$ [$c = \psi'(\kappa, 0)$] at a constant total energy E, one must increase the "local frequency" of oscillations of the wave function, i.e., the kinetic energy near $r = 0$. Therefore, V is made deeper. But at the origin of the coordinate system the width of the well at $E = -\kappa^2 h^2/2m$ must decrease so as not to permit the appearance of nodes in the wave function (as it should be for the sole eigenstate).

In the Marchenko approach, of course the same family of potentials is obtained, if starting from $S(k)$ in the form (2.2.13) and from an arbitrary normalization M^2 of the bound state one calculates $Q^{\mathrm{M}}(r, r')$ from (1.3.26). The difference, however, is revealed in the intermediate computations. Thus, Q^{GL} was especially chosen with the integral term in (1.3.6″) equal to zero, while in Q^{M} the continuous spectrum gives a contribution to the integral (1.3.26). Therefore, derivation of V and of ψ from Q^{M} turns out to be more complicated here than from Q^{GL}. It appears that even the expressions for the potentials and wave functions derived from Q^{GL} and Q^{M}, are different, since in one case they are constructed from $\mathring{\varphi}(k, r)$ and in the other case from $\mathring{f}^{\pm}(k, r)$. In essence, however, the expressions for V and ψ in the two formalisms are equivalent, as can be verified by simple transformations.

2.2.2 Potentials Without Bound States: $S(k)$ with a Single Pole in the Upper k Half-plane

Factorization of the kernel Q can also be achieved by methods other than reducing to a single bound state. The factorized kernel Q^{M} is also obtained with the initial Jost function

$$f(k) = (k + \mathrm{i}a)/(k + \mathrm{i}b)\,; \quad \mathrm{Im}\{a\} = \mathrm{Im}\{b\} = 0\,, \tag{2.2.14}$$

where $a > 0$ (to eliminate bound states); $b > 0$ (for the Jost function $f^{+}(k)$ to be analytic in the upper half-plane, to which corresponds the potential satisfying the condition $\int_0^\infty r|V(r)|dr < \infty$ (Chadan, Sabatier 1977)). When $a = -b = -\kappa$, (2.2.14) coincides with (2.2.12).[2] Closing the integration contour in the upper half-plane in the formula for Q and making use of the residue theorem, we have

$$\begin{aligned} Q^{\mathrm{M}}(r, r') &= \frac{1}{2\pi}\int_{-\infty}^{\infty}\left[1 - \frac{f(-k)}{f(k)}\right]\mathrm{e}^{\mathrm{i}k(r+r')}dk \\ &= \frac{1}{2\pi}\int_{-\infty}^{\infty}\left[1 - \frac{(k - \mathrm{i}a)(k + \mathrm{i}b)}{(k - \mathrm{i}b)(k + \mathrm{i}a)}\right]\mathrm{e}^{\mathrm{i}k(r+r')}dk \\ &= 2b\,\frac{b - a}{b + a}\exp[-b(r + r')]\,. \end{aligned} \tag{2.2.15}$$

With such Q^{M} after denoting $(b - a)/(b + a) = \mathrm{e}^{2br_0}$; $p(r) = 1 + \exp[-2b(r - r_0)]$ we obtain from the Marchenko equation,

---

[2] If one sets $a > 0$ and $b < 0$, then $S(k)$ will have no residues in the upper half-plane k, meaning that

$$Q^{\mathrm{M}}(r, r') = 0 \rightarrow K^{\mathrm{M}}(r, r') = 0 \rightarrow V(r) = 0\,,$$

which contradicts $S(k) \neq 1$, corresponding to (2.2.14).

$$K^M(r,r') = -2b\exp[-b(r+r'-2r_0)]p^{-1}(r);$$

$$V(r) = -\frac{4b^2\hbar^2}{m}p^{-2}(r)\exp[-2b(r-r_0)]$$

$$= -\frac{\hbar^2}{m}b^2/c^{0s}h^2[b(r-r_0)];$$

$$f^{\pm}(k,r) = \exp(\pm ikr)\left\{1 + 2b\frac{\exp[-2b(r-r_0)]}{(\pm ik - b)}p^{-1}(r)\right\}. \quad (2.2.16)$$

For $a=0$, $b>0$ we have $V(r) = -\hbar^2 b^2/mc^{0s}h^2(br)$.

Comparison with the Gelfand–Levitan Approach. Yet another comparison between the Marchenko and the Gelfand–Levitan approaches can be made. With the function $f^+(k)$ from (2.2.12) corresponding to a single bound state, it is easier to derive the formulas for V and the wave functions by making use of Q^{GL} (2.2.1–7). In contrast, with $f^+(k)$ from (2.2.14) when no discrete spectrum is present, it is simpler to do so by using Q^M (2.2.15,16).

Although the formulas for potentials and functions within the two approaches appear to be different, for example, $V(r)$ in (2.2.16) looks not the same as

$$V(r) = -\frac{\hbar^2}{m}\frac{d^2}{dr^2}\ln\frac{a\sinh bre^{iar} + ib\cosh bre^{iar}}{ib(a^2-b^2)}. \quad (2.2.17)$$

found in the Gelfand–Levitan approach, they actually exactly coincide (Eckart potential). It is that they are expressed through different elementary functions.

2.2.3 Potentials with $S(k)$ with Two Poles in the Upper k Half-plane

Assume that in $f(k) = (k+ia)/(k+ib)$, $b>0$, as in (2.2.14) but $a = -\kappa < 0$ to provide for the existence of a single bound state with $E = -\hbar^2\kappa^2/2m$. Now, factorization of Q^M is achieved by special choice of the normalization constant M_κ^2. We substitute the corresponding scattering function

$$S(k) = (k+i\kappa)(k+ib)/(k-i\kappa)(k-ib) \quad (2.2.18)$$

into expression (1.3.26) for Q^M:

$$Q^M(r+r') = \frac{1}{2\pi}\int_{-\infty}^{\infty}dk\left[1 - \frac{(k+i\kappa)(k+ib)}{(k-i\kappa)(k-ib)}\right]$$
$$\times\exp[ik(r+r')] + M_\kappa^2\exp[-\kappa(r+r')]. \quad (2.2.19)$$

Closing the contour of integration by a semicircle of infinite radius in the upper half-plane of complex k, we find that the integral equals the sum of the residues of the expression

$$-S(k)\exp[ik(r+r')] \quad (2.2.20)$$

at points $k = \mathrm{i}b$ and $k = \mathrm{i}\kappa$:

$$Q^{\mathrm{M}}(t = r + r') = \frac{b+\kappa}{b-\kappa} 2b\mathrm{e}^{-bt} - \mathrm{i}\,\mathrm{Res}\,S(\mathrm{i}\kappa)\mathrm{e}^{-\kappa t} + M_\kappa^2 \mathrm{e}^{-\kappa t} . \tag{2.2.21}$$

One can make the second and third terms in the right-hand side of (2.2.21) cancel out if the norm M_κ^2 is chosen to be equal to i Res $S(k)$ at point $k = \mathrm{i}\kappa$:

$$M_\kappa^2 = \mathrm{i}\,\mathrm{Res}\,S(k)|_{k=\mathrm{i}\kappa} = -2\kappa \frac{\kappa+b}{\kappa-b} . \tag{2.2.22}$$

As a result, instead of three, only a single term remains in Q^{M}:

$$Q^{\mathrm{M}}(t) = 2b \frac{b+\kappa}{b-\kappa} \mathrm{e}^{-bt} . \tag{2.2.23}$$

(A similar situation is obtained in the Gelfand–Levitan approach, only there the normalizing constant c_κ^2 is expressed through Res $[f^+(k)f^-(k)]^{-1}$:

$$c_\kappa^2 = 4\mathrm{i}\kappa^2 [\dot{f}^+(\mathrm{i}\kappa) f^-(\mathrm{i}\kappa)]^{-1} , \tag{2.2.24}$$

where $\dot{f}^+$ is the derivative of f^+ with respect to k.)

With this Q^{M} we obtain formulas in the same form as (2.2.16), just like as in the absence of a bound state when there exists only one pole in $S(k)$ at $k = \mathrm{i}b$:

$$V(r) = -\frac{4\hbar^2}{m} b^2 \exp[-2b(r-r_0)]\{1 + \exp[-2b(r-r_0)]\}^{-2}$$

$$= -\frac{\hbar^2}{m} b^2 \frac{1}{\cosh^2[b(r-r_0)]} ; \quad \frac{b+\kappa}{b-\kappa} = \exp(2br_0) ; \tag{2.2.25}$$

$$f^\pm(k,r) = \exp(\pm \mathrm{i}k, r)\left\{1 + \frac{\exp[-2b(r-r_0)]2b/(\pm \mathrm{i}k - b)}{1 + \exp[-2b(r-r_0)]}\right\} ; \tag{2.2.26}$$

$$f^+(\mathrm{i}\kappa, r) = \exp(-\kappa r)\left\{\frac{1 - \exp(-2br)}{1 + \exp[-2b(r-r_0)]}\right\} . \tag{2.2.27}$$

But the potential well V is always deeper in (2.2.25) than in (2.2.16). This is natural, since in the well of (2.2.25) there are found both a pole of the function $S(k = \mathrm{i}b)$ corresponding to the pole of the function $f^-(k = \mathrm{i}b)$ as in (2.2.16) and a bound state [a zero of the function $f^+(k = \mathrm{i}\kappa)$ which is absent in the case of V from (2.2.16)]. The potentials become equal only when $a = \kappa = 0$ which is the critical value of the parameter when a level appears. Expressions (2.2.25–27) turn into (2.2.16,17) when κ is replaced by $-a$. We shall use the described method for choosing normalization constants in calculations involving multi-term kernels Q^{M}.

Bound States in a Continuous Spectrum

The factorized kernels Q^{GL} and Q^{M} in (2.2.1,15) were constructed from the solutions $\mathring{\varphi}$ and $\mathring{f}$ corresponding to negative energies,

$$E = -\kappa^2\hbar^2/2m < 0 .$$

It turns out that interesting models are also obtained with $E^b > 0$ (Gazdy 1977):

$$Q^{\mathrm{GL}}(r, r') = c^2 \mathring{\varphi}(k_1, r)\mathring{\varphi}(k_1, r') = \frac{c^2}{k_1^2} \sin k_1 r \sin k_1 r' . \tag{2.2.28}$$

From the Gelfand–Levitan equation (1.3.5) with the degenerate kernel (2.2.28) we find the function $K^{\mathrm{GL}}(r, r')$ determining the potential $V(r)$ and the corresponding solution of the Schrödinger equation

$$V(r) = -\frac{\hbar^2}{m} \frac{d^2}{dr^2} \ln p(r) ; \quad p(r) = 1 + \frac{c^2}{k_1^2} \int_0^r \sin^2 k_1 t dt ; \tag{2.2.29}$$

$$\varphi(k, r) = \frac{\sin kr}{k} - c^2 \frac{\sin k_1 r}{k_1^2 k} \int_0^r \sin k_1 t \sin kt dt p^{-1}(r) . \tag{2.2.30}$$

It is remarkable that at $k = k_1$ the terms in $\varphi(k, r)$ that do not damp out at infinity, cancel out, resulting in a square-integrable wave function

$$\psi(k, r) = c\varphi(k, r)$$

describing a bound state embedded in the continuum. The forms of $V(r)$ and $\psi(k, r)/r$ are shown in Figs. 1.6,7, respectively. Such solutions were first found by *Neuman* and *Wigner* (1929). Cases in which several such bound states exist were considered by *Meyer–Vernet* (1982). In these cases the general technique for constructing potentials of the Bargmann type obtained with the aid of multi-term kernels Q may be utilized. This is the subject of the next section.

In the multi-channel case (Chap. 5) bound states which are embedded in the continuous spectrum are often obtained with short-range forces that fall off exponentially at large r, together with the corresponding wave functions.

2.3 More General Models

We shall extend the family of exactly solvable problems by taking into consideration degenerate kernels Q in the form of a sum of several terms with a factorized coordinate dependence,

$$Q(r, r') = \sum_j^N Q_j(r) Q_j(r') .$$

Such kernels are obtained within the Gelfand–Levitan and the Marchenko approaches when the rational Jost functions

$$f^+(k) = \prod_j^N \frac{(k + ia_j)}{(k + ib_j)} \tag{2.3.1}$$

and the respective scattering S-functions or spectral densities $\rho(k)$:

$$s(k) = \prod_j^N \frac{(k - ia_j)(k + ib_j)}{(k - ib_j)(k + ia_j)} ;$$

$$d\rho(k) = \frac{2k^2 dk}{\pi f^+(k) f^-(k)} = \frac{2k^2 dk}{\pi} \prod_j^N \frac{k^2 + b_j^2}{k^2 + a_j^2} \qquad (2.3.2)$$

are chosen. The same number of factors in the numerator and denominator of $f^+(k)$ yields the asymptotic behavior: $f^+_{|k| \to \infty}(k) \to 1$ when $\operatorname{Im}\{k\} > 0$, which must be satisfied by the Jost function, if the potential belongs to the class

$$\int_0^\infty |V(r)| r\, dr < \infty .$$

For such potentials $\operatorname{Re}\{b_j\} > 0$, while the zeros $k = -ia_j$ of the function $f^+(k)$ in the upper complex k half-plane lie on the imaginary axis (bound states: $\kappa_j = -a_j > 0$). For real potentials the zeros $k = -ia_j$ and the poles $k = -ib_j$ of function $f^+(k)$ must be situated symmetrically with respect to the imaginary axis.

The possibility of infinitely increasing the number of factors N in (2.3.1,2) is utilized for the approximation of arbitrary scattering data (non-Bargmann type as well) in the approximate solution of inverse problems with the aid of Bargmann models (Chap. 3).

We noted above that in the inverse problem one need not proceed from

$$\mathring{V}(r) \equiv 0, \quad \mathring{\varphi}(k,r) = \sin kr/k, \quad \mathring{f}^\pm = \exp(\pm ikr) .$$

If $\mathring{V}(r) \not\equiv 0$ belongs to the set of potentials for which explicit solutions are allowed, then with the degenerate kernels Q expressed through $\mathring{\varphi}(k,r)$ and $\mathring{f}^\pm(k,r)$ it is possible to construct $\Delta V(r)$ such that $V(r) = \mathring{V}(r) + \Delta V(r)$ will correspond to new, exactly solvable, models. In this case the Bargmann potentials themselves are suitable for $\mathring{V}(r)$ and therefore, rectangular well, Coulomb, centrifugal, and other potentials as well.

Both methods for extending the classes of V (either multi-term Q or $\mathring{V}(r) \not\equiv 0$) are closely related to each other. Thus, in the case of $f^+(k)$, it is possible to obtain solutions applying the technique $\mathring{V}(r) \not\equiv 0$ from (2.3.1). Starting from $\mathring{f}(k) = 1$, $\mathring{V}(r) \equiv 0$ we find

$$V_1(r), \quad \varphi_1(k,r), \quad f_1^\pm(k,r) ,$$

corresponding to the first factor

$$f_1^+(k) = (k + ia_1)/(k + ib_1)$$

in (2.3.1). Then, assuming $V_1(r)$ as the initial potential, we take into account the second factor in $f(k)$ and so on, until we find

$$V(r) = V_1(r) + \Delta V_2(r) + \ldots + \Delta V_N(r)$$

and the respective $\varphi(k,r), f^\pm(k,r)$. But there are still other methods.

2.3.1 Multi-term Degenerate Kernels of the Inverse-Problem Equations

We already obtained the double-term kernel Q^{M} (2.2.21) where two last terms can be unified, when we chose the Jost function to be of the form

$$f^{+}(k)=\frac{k-\mathrm{i}\kappa}{k+\mathrm{i}b} \quad \text{with} \quad \operatorname{Im}\{\kappa\}=\operatorname{Im}\{b\}=0\,, \quad \kappa, b>0\,.$$

However, then the special value of (2.2.25) was fixed for the normalization M_κ^2, so as to make Q^{M} be single-term. We shall now let M_κ^2 be arbitrary. We shall reduce solution of the problem to the successive application of the procedures described above for single-term Q^{M}. This allows us to obtain a family of phase-equivalent potentials in a simpler manner than *Jost* and *Kohn* (1952) (Zakhariev, Suzko 1988; Chadan, Sabatier 1989, p. 58). The Marchenko approach is more suitable for this scattering problem.

For the initial potential we shall use the expression (2.2.27) obtained with

$$\mathring{M}_\kappa^2=\mathrm{i}\operatorname{Res} S(k)|_{k=\mathrm{i}\kappa}\,.$$

Since the scattering S-function is independent of the choice of the normalization constant we have $S(k)=\mathring{S}(k)$. As a result, when $\mathring{V}(r)\neq 0$ the integral term in the generalized expression for Q^{M} (1.3.26′) vanishes. Since, on the other hand, both in $V(r)$ and in $\mathring{V}(r)$ the sole bound states have the same energy $-\kappa^2\hbar^2/2m$ and different normalization factors M_κ^2 and $\mathring{M}_\kappa^2$, respectively, we find

$$Q^{\mathrm{M}}(r,r')=(M_\kappa^2-\mathring{M}_\kappa^2)\mathring{f}(\mathrm{i}\kappa,r)\mathring{f}(\mathrm{i}\kappa,r')\,. \tag{2.3.3}$$

Discarding the integral over the continuous spectrum in the formula for Q^{M} with $\mathring{V}(r)\equiv 0$ is illegitimate since it violates the Levinson theorem. Now, however, in deriving (2.3.3) with $\mathring{V}(r)\not\equiv 0$, this operation was performed correctly since the numbers of bound states in $V(r)$ and $\mathring{V}(r)$ are equal. In complete analogy with the procedure (2.2.1–8) in the Gelfand–Levitan approach, from (2.3.3) we obtain

$$f^{+}(k,r)=\mathring{f}^{+}(k,r)-(M_\kappa^2-\mathring{M}_\kappa^2)\mathring{f}(\mathrm{i}\kappa,r)\int_r^\infty \mathring{f}(\mathrm{i}\kappa,r')\mathring{f}^{+}(k,r')dr'p^{-1}(r)\,;$$

$$f^{+}(\mathrm{i}\kappa,r)=\mathring{f}^{+}(\mathrm{i}\kappa,r)p^{-1}(r);$$

$$V(r)=\mathring{V}(r)-\frac{\hbar^2}{m}\frac{d^2}{dr^2}\ln p(r)\,;$$

$$p(r)=1+(M_\kappa^2-\mathring{M}_\kappa^2)\int_r^\infty \mathring{f}^{+2}(\mathrm{i}\kappa,r')dr'\,. \tag{2.3.4}$$

This represents a family of phase-equivalent potentials since the condition $S(k)=\mathring{S}(k)$ is fulfilled and $V(r)$ in (2.3.4) contains the arbitrary normalization factor M_κ^2. This follows directly from (2.3.4): assuming $r=0$ in $f^{+}(k,r)$ and taking into account $\mathring{f}(\mathrm{i}\kappa,0)=0$, we obtain $f^{\pm}(k)=\mathring{f}^{\pm}(k)$ and $S(k)=\mathring{S}(k)$.

The potentials (2.2.6) and the scattering data found earlier can be obtained from (2.3.4) as the limiting case of $b\to \mathrm{i}\kappa$. However, in doing so, $\mathring{M}_\kappa^2\to\infty$ and it becomes necessary to compute uncertainties such as $0 \cdot \infty$, for example, in the expression for $p(r)$. Therefore, for

$$f(k) = (k - \mathrm{i}\kappa)/(k + \mathrm{i}\kappa)$$

it is more convenient to adopt the Gelfand–Levitan approach.

Zakhariev et al (1982) and *Chadan* and *Sabatier* (1977) describe a less convenient method for solving the problem with

$$f^{+}(k) = (k - \mathrm{i}\kappa)/(k + ib)$$

within the Gelfand–Levitan approach by applying the following procedure. Having multiplied and divided the Jost function by $k + \mathrm{i}\kappa$, we represent $f^{+}(k)$ as

$$f^{+}(k) = \frac{k + \mathrm{i}\kappa}{k + ib}\,\frac{k - \mathrm{i}\kappa}{k + \mathrm{i}\kappa} = \mathring{f}(k)\,\frac{k - \mathrm{i}\kappa}{k + \mathrm{i}\kappa}\,. \tag{2.3.5}$$

The solution is achieved in two stages. First, the functions and potentials corresponding to

$$\mathring{f}(k) = (k + \mathrm{i}\kappa)/(k + ib)$$

without bound states are found following (2.2.17–22). Then the second factor in the Jost functions is taken into account and, together with it a bound state is introduced:

$$E = -\hbar^2\kappa^2/2m = \hbar^2 a^2/2m\,,$$

as in (2.2.1–7). Because (2.2.17–22) are cumbersome, a solution following the Marchenko procedure is preferable. The approaches with Q^{M} and Q^{GL} may be combined. At the first stage one can find Q^{M} from $f^{+}(k)$ by using (2.3.5) and the respective solutions, and at the second stage the bound state is deduced by following the Gelfand–Levitan approach. The normalization constants M_n and c_n for one state $f^{+}(\mathrm{i}\kappa, r)$ are expressed through each other:

$$M_n^2 = 4\kappa_n^2/[c_n^2 \cdot f(\mathrm{i}\kappa_n, 0)]\,.$$

Solution of the inverse problem with a more complicated Jost function (2.3.1) can be performed in several ways in different combinations of the schemes described above. We shall describe the most simple and direct way.

Let there be N_{bnd} bound states

$$\mathrm{i}a_j = -\mathrm{i}\kappa_j\,; \quad \kappa_j > 0\,; \quad j = 1, 2, \ldots, N_{\mathrm{bnd}}$$

corresponding to the Jost function (2.3.1) and the scattering functions (2.3.2) and let the following conditions be fulfilled for the remaining singularities S:

$$\mathrm{Re}\{a_j\} > 0\,, \quad j = N + 1, N + 2, \ldots, N\,;$$

$$\mathrm{Re}\{b_j\} > 0\,, \quad j = 1, 2, \ldots, N\,.$$

In addition, for real potentials the points $\mathrm{i}a_j$ and ib_j must be situated symmetrically relative to the imaginary axis in the complex k plane.

We shall find Q^{M} from (1.3.26) with S from (2.3.2) by closing the integration contour in the upper half-plane. From the theory of residues we have

$$Q^{\mathrm{M}}(r,r') = \sum_{n}^{N+N_{\mathrm{bnd}}} A_n \mathrm{e}^{-d_n(r+r')} . \tag{2.3.6}$$

In this expression the contributions of bound states and poles of the S-function at the points $k = \mathrm{i}b_j$ are combined. By d_n we denote all the poles of S inside the contour of integration; $d_n = \kappa_n$ for $n = N+1, \ldots, N+N_{\mathrm{bnd}}$, $d_n = b_n$ for $n = 1, \ldots, N$, and

$$A_{n+N} = \frac{2\kappa(\kappa_n + b_n)}{\kappa_n - b_n} \prod_{n' \neq n}^{N_{\mathrm{bnd}}} \frac{(\kappa_n + b_n)(\kappa_n - a_{n'})}{(\kappa_n + a_{n'})(\kappa_n - b_{n'})} + M_n^2 . \tag{2.3.7}$$

at $n = 1, 2, \ldots, N_{\mathrm{bnd}}$ and

$$A_n = \frac{2b_n(b_n - a_n)}{(b_n + a_n)} \prod_{n' \neq n}^{N} \frac{(b_n + b_{n'})(b_n - a_{n'})}{(b_n + a_{n'})(b_n - b_{n'})} \tag{2.3.8}$$

at $n = 1, \ldots, N$.

Substituting (2.3.6) into the Marchenko equation, we obtain

$$K(r,r') + \sum_{n}^{N+N_{\mathrm{bnd}}} A_n \left\{ \mathrm{e}^{-d_n r} + \int_r^{\infty} K(r,r'')\mathrm{e}^{-d_n r''} dr'' \right\} \mathrm{e}^{-d_n r'} = 0 , \tag{2.3.9}$$

where the expression in braces is the Jost solution for the desired $V(r)$, so that $K(r,r')$ has a form similar to that of $Q(r,r')$ from (2.3.6) with a separabilized dependence on r and r', but with the Jost solution $f^+(\mathrm{i}d_n r)$ instead of $\mathrm{e}^{-d_n r}$:

$$K(r,r') = -\sum_{n}^{N+N_{\mathrm{bnd}}} A_n f^+(\mathrm{i}d_n, r)\mathrm{e}^{-d_n r'} . \tag{2.3.10}$$

Substituting this expression into the formula for the generalized shift $\mathring{f} \to f$ (1.3.20) we obtain a set of equations for $f^+(\mathrm{i}d_n, r)$:

$$\sum_{n'}^{N+N} f(\mathrm{i}d_{n'}, r) \left[\delta_{nn'} + A_{n'} \int_r^{\infty} \mathrm{e}^{-(d_{n'}+d_n)r'} dr' \right] = \mathrm{e}^{-d_n r} , \tag{2.3.11}$$

solution of which yields

$$f^+(\mathrm{i}d_{n'} r) = \sum_{n'}^{N+N_{\mathrm{bnd}}} \mathrm{e}^{-d_{n'} r} (P^{-1}(r))_{n'n'} , \tag{2.3.12}$$

where P^{-1} is the inverse of the matrix P:

$$P_{n'n}(r) = \delta_{n'n} + A_{n'} \exp[-(d_{n'} + d_n)r]/(d_{n'} + d_n) . \tag{2.3.13}$$

Knowing the Jost solutions for fixed values of k, we find $K(r,r')$ from (2.3.10) and determine the potential and Jost solutions for any k:

$$V = -\frac{\hbar^2}{m} \frac{d^2}{dr^2} \ln \det P(r) ; \tag{2.3.14}$$

$$f^{\pm}(k, r) = \mathrm{e}^{\pm ikr} + \sum_{nn'}^{N+N_{CB}} A_n(P^{-1}(r))_{n'n}\, \mathrm{e}^{-(d_n + d_{n'} \mp ik)r}/(d_n \mp ik) \,. \tag{2.3.15}$$

By assuming $r=0$ in (2.3.15) we obtain (2.3.1) for the initial Jost function.

Since S from (2.3.2) corresponding to the potential (2.3.14) is independent of the normalizing constants M_n^2, (2.3.14) represents a family of phase-equivalent potentials depending on N_{bnd} parameters. Among them is a potential with

$$M_n^2 = \mathring{M}_n^2 = \mathrm{i}\,\mathrm{Res}\, S(\mathrm{i}\kappa_n) \,,$$

when the contributions to Q from bound states [from the integral and sum in (1.3.26)] cancel out: $A_{n+N} = 0$; $n = 1, \ldots, N_{\mathrm{bnd}}$. Only N terms remain in Q, which simplifies the derivation of the formulas for $V(r)$ and $f^{\pm}(k, r)$.

Instead of (2.3.6–15), one can do the calculations in two stages. For example, at the first stage one can reconstruct $\mathring{V}$ corresponding to the constants $\mathring{M}_n^2$ and then by using the corresponding $f(k, r)$ as the initial solutions, obtain (2.3.14,15) for arbitrary constants M_n^2. This is as in (2.3.3,4) with

$$Q^{\mathrm{M}}(r, r') = \sum_{n}^{N_{\mathrm{bnd}}} (M_n^2 - \mathring{M}_n^2)\mathring{f}(\mathrm{i}\kappa_n, r)\mathring{f}(\mathrm{i}\kappa_n, r') \,,$$

when the integral term in (1.3.26′) no longer contributes to Q^{M} because $S(k) = \mathring{S}(k)$.

Chadan and *Sabatier* (1980) discuss a similar, but more cumbersome, solution scheme: In the Jost function (2.3.1) one singles out factors of the type given by (2.2.15) corresponding to bound states

$$f^{+}(k) = f^{+(1)}(k) \prod_{n}^{N_{\mathrm{bnd}}} \frac{(k - \mathrm{i}\kappa_n)}{(k + \mathrm{i}\kappa_n)} \,.$$

First, the Gelfand–Levitan equation is solved with Q^{GL} obtained from $f^{(1)}(k)$ in accordance with (2.2.17–22) (complicated enough as that is) whereas bound states are introduced at the second stage by using $\varphi^{(1)}$ as the reference functions. In Q^{GL} there remains then only the contribution of the sum over the bound states, since $d[\rho(k) - \rho^{(1)}(k)] = 0$ for $E > 0$:

$$Q^{\mathrm{GL}}(r, r') = \sum_{n}^{N_{\mathrm{bnd}}} c_n^2 \varphi^{(1)}(\mathrm{i}\kappa_n, r)\varphi^{(1)}(\mathrm{i}\kappa_n, r') \,.$$

These methods of solving the inverse problem in two stages are examples of the application of the technique of nonzero base potentials $\mathring{V}(r) \not\equiv 0$ for construction of Bargmann models.

In the base potentials one can also shift the bound states $\mathring{E}_n \to E_n$ or change their number $\mathring{N}_{\mathrm{bnd}} \to N_{\mathrm{bnd}}$ and normalization constants $\mathring{c}_n^2 \to c_n^2$, leaving $d[\rho(E) - \mathring{\rho}(E)] = 0$ for $E > 0$, to which corresponds

$$Q^{\mathrm{GL}}(r, r') = \sum_{n}^{N_{\mathrm{bnd}}} c_n^2 \mathring{\varphi}(E_n, r)\mathring{\varphi}(E_n, r') - \sum_{n}^{\mathring{N}_{\mathrm{bnd}}} \mathring{c}_n^2 \mathring{\varphi}(\mathring{E}_n, r)\mathring{\varphi}(\mathring{E}_n, r') \,. \tag{2.3.16}$$

The inverse problem for a nonself-adjoint Schrödinger operator was formulated by *Lyance* (1967), see also his addendum in the book by *Neimark* (1969) and *Abraham* et al (1980), *DeFacio* et al (1980). In the general case, instead of normalizing constants c_n, e.g., as in (3.16), there must be normalizing polynomials. This is one of significant peculiarities of the formalism and is due to the fact that the complete set of functions includes, besides the eigenfunctions, also adjoint functions.

2.3.2 Models of One-dimensional Motion on the Whole Axis

Now we shall consider two special cases, when only bound states contribute to the kernel Q and when no bound states exist and the reflection coefficient that determines Q is a rational function of k (Faddeev 1976). Other exact solutions are obtained through combinations of these two ways of constructing $V(x)$ and the wave functions.

2.3.3 Reflectionless Potentials

If the reflection coefficient is chosen to be equal to zero at all energies $S_{12} = 0$ (simultaneously the reflection coefficient for waves incident from the other side is also zero), then the integral in (1.4.10) for $Q_1(x, y)$ vanishes and only the sum over bound states ($\mathring{V}(x) = 0$, $\mathring{S}_{12} = 0$) remains:

$$\begin{aligned} Q_1(x+y) &= \sum_j^{N_{\text{bnd}}} M_{1j}^2 \mathring{f}_1^+(\mathrm{i}\kappa_j, x) \mathring{f}_1^+(\mathrm{i}\kappa_j, y) \\ &= \sum_j^{N_{\text{bnd}}} M_{1j}^2 \exp[-\kappa_j(x+y)] . \end{aligned} \tag{2.3.17}$$

With this kernel, we reduce in a standard manner the integral equation of the inverse problem to a set of algebraic equations with the matrix of coefficients:

$$\begin{aligned} P_{1ij} &= \delta_{ij} + M_{1i}^2 \int_x^\infty \mathring{f}_1^+(\mathrm{i}\kappa_i, t) \mathring{f}_1^+(\mathrm{i}\kappa_j, t)dt \\ &= \delta_{ij} + M_{1i}^2 \exp[-(\kappa_i + \kappa_j)x]/(\kappa_i + \kappa_j) . \end{aligned} \tag{2.3.18}$$

Similarly to (2.3.12–15) we obtain

$$\begin{aligned} V(x) &= -\frac{\hbar^2}{m}\frac{d^2}{dx^2} \ln \det P_1(x) ; \\ f_1^+(k, x) &= \mathring{f}_1^+(k, x) - \sum_{ij}^{N_{\text{bnd}}} M_{1i}^2 \mathring{f}_1^+(\mathrm{i}\kappa_i, x) \\ &\quad \times (P_1^{-1}(x))_{ij} \int_x^\infty \mathring{f}_1^+(\mathrm{i}\kappa_j, t) \mathring{f}_1^+(k, t)dt . \end{aligned} \tag{2.3.19}$$

Thus,

$$f_1^+(\mathrm{i}\kappa_j, x) = \sum_i^{N_{\mathrm{bnd}}} \mathring{f}_1(\mathrm{i}\kappa_i, x)\,(P_1^{-1}(x))_{ij} \,. \tag{2.3.19'}$$

For a completely transparent potential transmission coefficient $S_{11}(K)$ whose modulus is equal to 1 is a rational function:

$$S_{11}(k) = \prod_j^{N_{\mathrm{bnd}}} (k + \mathrm{i}\kappa_j)/(k - \mathrm{i}\kappa_j) \,. \tag{2.3.20}$$

Examples of transparent potentials with various numbers of levels are presented in Figs. 2.9, 3.5–7.

2.3.4 The Rational Reflection Coefficient (no Bound States)

Since the absolute value of the reflection coefficient may be smaller than unity, we choose it in the form of a ratio of polynomials $p_m(k)$ and $p_n(k)$ with

$$S_{12}(k) = r p_m(k)/p_n(k) \,, \tag{2.3.21}$$

where the constant r is singled out to equate the coefficients of the higher orders in $p_m(k)$ and $p_n(k)$ to unity. For the potential to be real we assume

$$r = (i)^{m-n} r_0 \,,$$

where $\mathrm{Im}\{r_0\} = 0$, and the zeros of $p_m(k)$ and $p_n(k)$ must be situated symmetrically with respect to the imaginary axis in the complex k plane. The constant r_0 must be sufficiently small for $S_{12}(k)$ to be less than 1.

Now we denote by α_j^+ the zeros of the polynomial p_n in the upper k half-plane. We close the integration contour in (1.4.10) for Q with a semicircle of infinite radius with $\mathrm{Im}\{k\} > 0$ [we recall that owing to the absence of bound

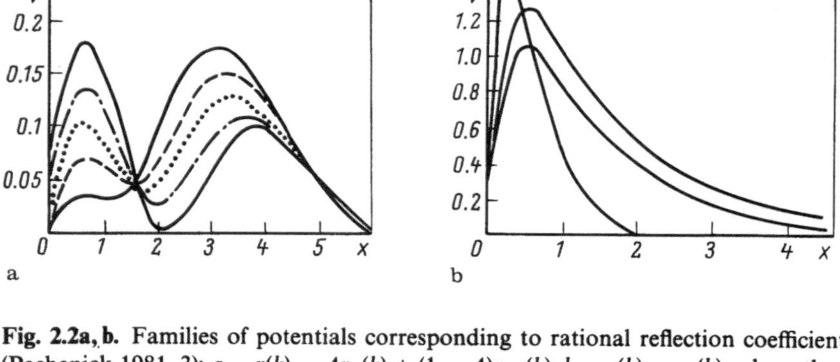

Fig. 2.2a, b. Families of potentials corresponding to rational reflection coefficients $r(k)$ (Pechenick 1981–3): a – $r(k) = Ar_3(k) + (1 - A)r_{10}(k)$; b – $r(k) = r_3(k)$, where the rational functions $r_3(k)$ and $r_{10}(k)$ have 3 and 10 poles, respectively, and no zeros; different curves correspond to different values of the parameter A which are chosen on the interval [0, 1]

states no sum exists in (1.4.10)]. Due to the exponentially decreasing factor the integral along this semicircle equals zero for $x > 0$. As a result, when $x > 0$, $Q(x)$ is written in the form of a sum of residues at the poles of S_{12} in the upper k half-plane, the number of which will be denoted by n_+:

$$Q_1(x + x') = \sum_{j=1}^{n_+} \rho_j \exp[i\alpha_j^+ (x + x')] , \tag{2.3.22}$$

where ρ_j coincides (within the factor i) with the residues of $S_{12}(k)$ at the points $k = \alpha_j^+$. From equation (1.4.9) with a degenerate kernel we find $K_1(x, x')$ that yields (Fig. 2.2):

$$V(x) = \frac{\hbar^2}{m} \frac{d^2}{dx^2} \ln \det(I + Z_1(x)) ; \quad x > 0 , \tag{2.3.23}$$

where the matrix Z_1 has the elements $Z_{1j}^{(1)} = i\rho_1 \exp(i(\alpha_1^+ + \alpha_j^+)x)/(\alpha_1^+ + \alpha_j^+)$, while for $f^+(k, r)$ we obtain formulas of the same type as for transparent potentials. For $x < 0$ the potential can be found in a similar way with the aid of Q_2 and K_2.

2.3.5 The Finite-difference Approach

An analogous technique for constructing solutions can also be applied in problems with a discrete space variable, although in this case there are no integral equations of the inverse problem, which could be reduced to algebraic ones by choosing degenerate kernels. Here, from the very beginning, sets of algebraic equations replace the Gelfand–Levitan–Marchenko equations. However, here also separable kernels $Q(m, n)$ can essentially simplify the solution, significantly decreasing the number of such equations. This is done by following the same scheme as in the case of a continuous coordinate. To avoid repetition, we shall only give one example illustrating the peculiarities of this approach.

Let the spectral function $\rho(E)$ coincide with the unperturbed $\mathring{\rho}(E)$ at $E > 0$. Then, in Q^{GL} only the contribution of bound states remains. It would seem that the most simple form of Q^{GL} is obtained when it is single-term, as discussed in previous sections. However, the unique characteristic of the difference problem does not permit us to reduce Q^{GL}. In the case of a discrete coordinate the spectrum is limited from above and, instead of the δ-function, the finite Kronecker symbol always occurs in the completeness condition. This imposes more rigid constraints on the spectral parameters. For example, when transforming the normalization constants of bound states, $c_j^2 - \mathring{c}_j^2 = \delta c_j^2$, one must be careful not to violate

$$\sum_j^N \delta c_j^2 = 0 ,$$

which follows from the completeness condition. Therefore, rather than changing only one normalizing factor in order to compensate for the corresponding deviation, at least one more factor c^2 for another level in the sum $\sum_j^N c_j^2$ should be changed. If in this case the position of the levels in V and $\mathring{V}$ coincides, then for Q^{GL} we obtain

$$Q^{\mathrm{GL}}(m, n) = \sum_{j=1}^{2} \delta c_j^2 \mathring{\varphi}(E_j, m) \mathring{\varphi}(E_j, n) .$$

With this Q^{GL} the set of algebraic equations (1.3.5) which is infinite if the interaction region is infinite reduces to two equations for $\varphi(E_j, m)$ giving (compare (2.3.19′)):

$$\varphi(E_i, m) = \sum_{j=1}^{2} \mathring{\varphi}(E_j, m)[P^{-1}(m)]_{ji} ,$$

where

$$P_{ij}(m) = \delta_{ij} + \delta c_i^2 \sum_{p=1}^{m-1} \Delta \mathring{\varphi}(E_i, p) \mathring{\varphi}(E_j, p) .$$

These functions determine

$$K(m, n) = - \sum_{i=1}^{2} \delta c_i^2 \varphi(E_i, m) \mathring{\varphi}(E_i, n) ,$$

with which we can find φ for any allowed values of E (compare (2.3.19)):

$$\varphi(E, m) = \mathring{\varphi}(E, m) - \sum_{i,j=1}^{2} \delta c_i^2 \mathring{\varphi}(E_i, m) P_{ij}^{-1}(m) \sum_{p=1}^{m-1} \Delta \mathring{\varphi}(E_j', p) \mathring{\varphi}(E, p) ,$$

and the potential from (1.3.7a):

$$V(n) = \frac{\hbar^2}{2m\Delta}[K(n+1, n) - K(n, n-1)] .$$

2.4 Potentials of a Finite Range and Infinitely Deep Wells. *R*-matrix Models

For forces differing from zero on the interval $[0, a]$, $a < \infty$, it is much more economical to deal with a discrete set of functions complete on the finite interval $[0, a]$ than with a continuous spectrum of wave functions on the semi-axis or on the entire axis. Indeed, all physical properties of a system in the case of finite range forces are determined by solutions on $[0, a]$ while free motion in the external region is readily taken into account. We shall make use of eigenfunctions of the Schrödinger operator which correspond to homogeneous boundary conditions at 0 and a (*R*-resonance states).

The form of the kernel Q of the inverse-problem integral equation in *R*-matrix theory (2.3.16)

$$Q(r, r') = \sum_{\lambda}^{\infty} \gamma_\lambda^2 \mathring{\varphi}(\mathring{E}_\lambda, r) \mathring{\varphi}(E_\lambda, r') - \sum_{\lambda}^{\infty} \mathring{\gamma}_\lambda^2 \mathring{\varphi}(\mathring{E}_\lambda, r) \mathring{\varphi}(\mathring{E}_\lambda, r') \tag{2.4.1}$$

is especially adapted for the construction of exactly solvable models. In (2.4.1) there are no integrals, and to render Q degenerate one must only restrict the summation to a finite number of terms. This can be achieved by choosing a set $\{E_\lambda, \gamma_\lambda^2\}$ in which only a few resonance parameters differ from the corresponding values $\{\mathring{E}_\lambda, \mathring{\gamma}_\lambda^2\}$ for free motion. Then, in (2.4.1) those terms in which both $E_\lambda = \mathring{E}_\lambda$ and $\gamma_\lambda^2 = \mathring{\gamma}_\lambda^2$ coincide, cancel out in both sums. As a result, a finite sum of terms with a factorized dependence on r and r' will remain in $Q(r, r')$.

The simplest (single-term) form of Q is obtained if the spectrum $\{E_\lambda\}$ is chosen to totally coincide with $\{\mathring{E}_\lambda\}$ and when only a single amplitude of reduced width γ_ν from the set $\{\gamma_\lambda\}$ is set to be different from $\mathring{\gamma}_\nu$:

$$Q(r, r') = \delta\gamma_\nu^2 \mathring{\varphi}(\mathring{E}_\nu, r)\mathring{\varphi}(\mathring{E}_\nu, r') \; ; \quad \delta\gamma_\nu^2 = \gamma_\nu^2 - \mathring{\gamma}_\nu^2 \; . \tag{2.4.2}$$

The usual procedure (2.2.1–7) for solving the integral equation of the R-matrix inverse problem (1.3.5') with $Q(r, r')$ from (2.4.2) yields

$$K(r, r') = -\delta\gamma_\nu^2 \mathring{\varphi}(\mathring{E}_\nu, r)\mathring{\varphi}(\mathring{E}_\nu, r')p^{-1}(r) \; ;$$

$$p(r) = 1 + \delta\gamma_\nu^2 \int_r^a \mathring{\varphi}^2(\mathring{E}_\nu, t)dt \; ; \tag{2.4.3}$$

$$V(r) = \mathring{V}(r) + \frac{2\hbar^2}{m} \delta\gamma_\nu^2 \mathring{\varphi}(\mathring{E}_\nu, r)p^{-1}(r) + \frac{\hbar^2}{m}(\delta\gamma_\nu^2)^2 \mathring{\varphi}^4(\mathring{E}_\nu, r)p^{-2}(r) \; ; \tag{2.4.4}$$

$$\varphi(E, r) = \mathring{\varphi}(E, r) - \delta\gamma_\nu^2 \mathring{\varphi}(\mathring{E}_\nu, r)p^{-1}(r) \int_r^a \mathring{\varphi}(\mathring{E}_\nu, t)\mathring{\varphi}(E, t)dt \; ; \tag{2.4.5}$$

$$\varphi(\mathring{E}_\nu, r) = \mathring{\varphi}(\mathring{E}_\nu, r)p^{-1}(r) \; . \tag{2.4.6}$$

We can now see (Fig. 2.3a; Poschel, Trubovitz 1987) how starting with an infinite rectangular potential well $\mathring{V}(r)$ the variation of a single spectral parameter γ_1 causes the deformation of the bottom of $V(r)$. Here γ_1 is chosen as the derivative $u_1'(a) = \gamma_1 \varphi_1'(r = a) = \gamma_1$ of the normalized ground state wave function $\psi_1 = u_1$ at the right-hand infinite wall. It is clear that to compensate the increase of the inclination $|\gamma_1|$ of u_1 at $r = a$ (Fig. 2.3b) and to not allow the increase of $\int u_1^2(r)dr$, the potential must have the form shown in Fig. 2.3a. The increasing attraction at the right-hand side in Fig. 2.3a makes the local curvature of the wave function greater so that it begins to decrease earlier when moving to the left. But, further, the function must not have a node until $r = 0$. So the repulsion at the left provides a monotonic decrease $u_1(r \to 0) \to 0$. At the same time the competition between the attraction and repulsion keeps all energy levels at the previous positions. If we should decrease $|\gamma_1|$, the potential perturbation would be of opposite sign.

It is interesting to compare Fig. 2.3a with the picture of perturbations of the harmonic oscillator potential when the normalizing constant of the lowest energy state is varied (Fig. 2.3d, Sukumar 1985). Pictures for spectral-equivalent potentials with an initial Coulomb interaction are given by *Fernandez* (1984).

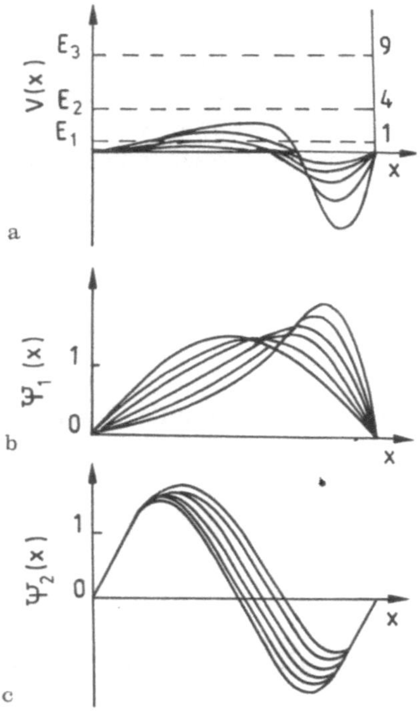

Fig. 2.3. Deformations of $V(r)$ (**a**) and wave functions (**b,c**) for the lower states in the infinite rectangular well caused by increase of γ_1 (Poshel, Trubowitz 1987)

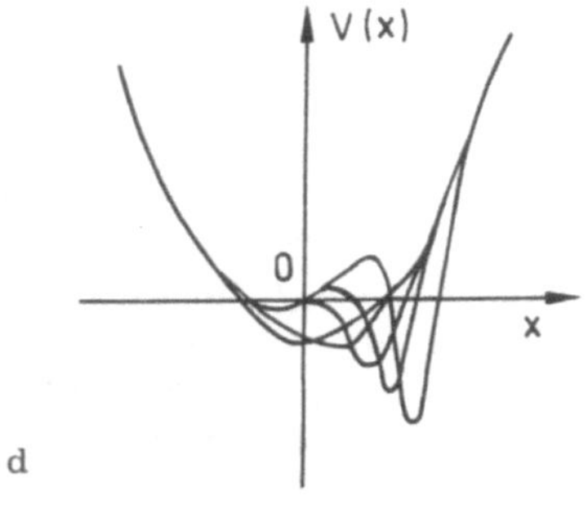

Fig. 2.3d. Same as in Fig. 2.3a but for a harmonic oscillator potential well with varying normalizing parameter of the ground state (Sukumar 1985)

Analogous speculations explain the deformation of $V(r)$, $\psi_1(x)$, $\psi_2(x)$ caused by the increase of a single spectral parameter $|\gamma_2|$ shown in Fig. 2.4 (Poshel, Trubowitz 1987).

Although the normalization constant of only a single ν-state has changed in the spectral data, the functions of all other states have been deformed having retained their $\gamma^2_{\lambda \neq \nu}$ (see Fig. 2.3c; Fig. 2.4c.)

If in the set $\{E_\lambda, \gamma^2_\lambda\}$ only one E_ν is taken to be different from its unperturbed value $\mathring{E}_\nu$ and all $\gamma^2_\lambda = \mathring{\gamma}^2_\lambda$, then two factorized terms will remain in $Q(r, r')$. In the general multi-term case it is necessary to solve sets of algebraic equations similar to (2.3.11).

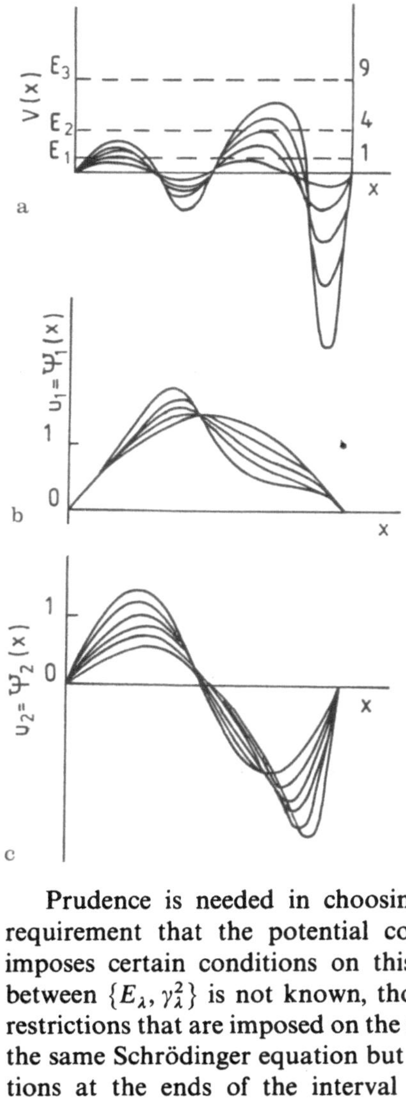

Fig. 2.4a–c. Same as Fig. 2.3a–c but with γ_2 increased (Poshel, Trubowitz 1987)

Prudence is needed in choosing the parameters $\{E_\lambda, \gamma_\lambda^2\}$, however. The requirement that the potential corresponding to the set $\{E_\lambda, \gamma_\lambda^2\}$ be local imposes certain conditions on this set. The explicit form of these relations between $\{E_\lambda, \gamma_\lambda^2\}$ is not known, though *Marchenko* (1977) has formulated the restrictions that are imposed on the set of two spectra $\{E_\lambda, E_\mu\}$ corresponding to the same Schrödinger equation but to differing homogeneous boundary conditions at the ends of the interval $[0, a]$. Specifically, the sequences E_λ and $E_\mu(\lambda, \mu = 1, 2, \ldots)$ corresponding to the boundary conditions $u(0) = u_\lambda(\pi) = 0$ and $u_\mu(0) = u'_\mu(\pi) = 0$ must be alternating:

$$-\infty < E_{\lambda=1} < E_{\mu=1} < E_{\lambda=2} < E_{\mu=2} < \ldots$$

and must satisfy the asymptotic conditions

$$E_\lambda = \lambda^2 - 2A + \alpha/\lambda\,; \quad E_\mu = (\mu - \tfrac{1}{2})^2 - 2A + \beta_\mu\,,$$

$$\sum_\lambda^\infty \alpha_\lambda^2 < \infty\,; \quad \sum_\mu^\infty \beta_\mu^2 < \infty$$

where A is an arbitrary real number; (Marchenko 1977). According to the theorem on two spectra by *Levitan* and *Gasymov* (1964) γ_λ^2 are also expressed in terms of $\{E_\lambda, E_\mu\}$.

Direct substitution of the obtained functions $\varphi(E, r)$ and the potential $V(r)$ into the Schrödinger equation, however, can also serve as a check of the correct choice of $\{E_\lambda, \gamma_\lambda^2\}$. It is also necessary to verify whether the required boundary conditions are satisfied for $\varphi(E, r)$; $u_\lambda(r) = \gamma_\lambda \varphi(E_\lambda, r)$ at the points $r = 0$ and $r = a$.

Now we would like to say a few words about the influence of energy shifts on the potential shape. In the case of a symmetrical potential its shape is completely determined by the positions of the energy levels (Sect. 1.4.3).

Can we predict the potential deformation which shifts the ground state level toward the E_2-level without perturbation of the positions of all other levels $E_{\lambda > 1}$? It is natural to suppose, in accordance with perturbation theory, that levels are most sensitive to variation of the potential form in the regions where the corresponding wave functions have maxima $|\psi_\lambda(x)|$. Indeed, Fig. 2.5a shows that there is a repulsive barrier in the middle of the interaction interval where $|\psi_1(x)|$ has a maximum, which pushes E_1 up. The attractive parts of the perturbation act on the ground state less effectively because there $\psi_1(x)$ is near to zero. For all other levels the repulsive and attractive parts of the perturbation counteract each other so that $E_{\lambda > 1}$ remain at the same positions. Likewise, to shift E_1 up in the harmonic oscillator well, we need potential deformations such as those shown in Fig. 2.6 (Sukumar 1985).

To shift the ground state down, we need a perturbation of sign opposite to that in Fig. 2.5a.

Now it is easy to predict how to raise only the second level. We need the potential deformation shown in Fig. 2.7 (Poshel, Trubowitz 1987) with two

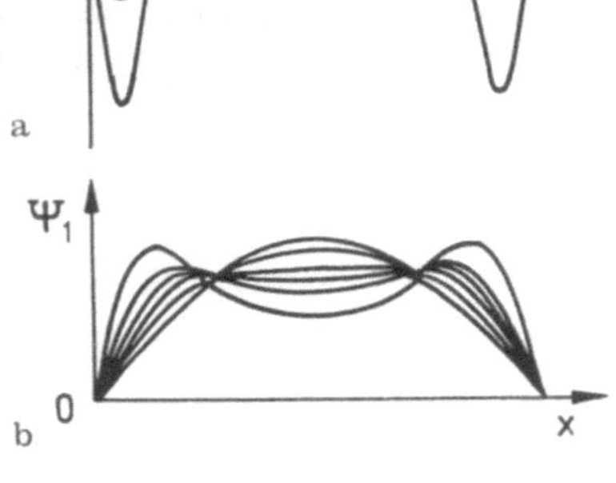

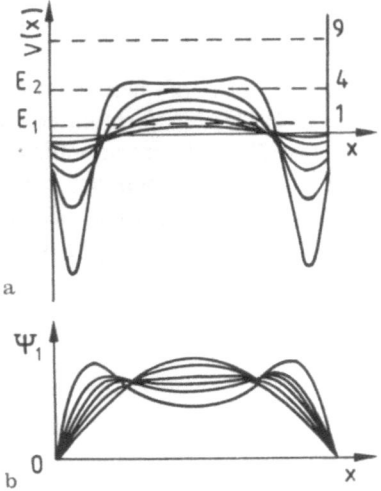

Fig. 2.5a,b. Deformation of the bottom (**a**) of an infinite rectangular potential well when the ground state energy is gradually shifted up to E_2 (Poshel, Trubowitz 1987); (**b**) deformation of $\psi_1(x)$

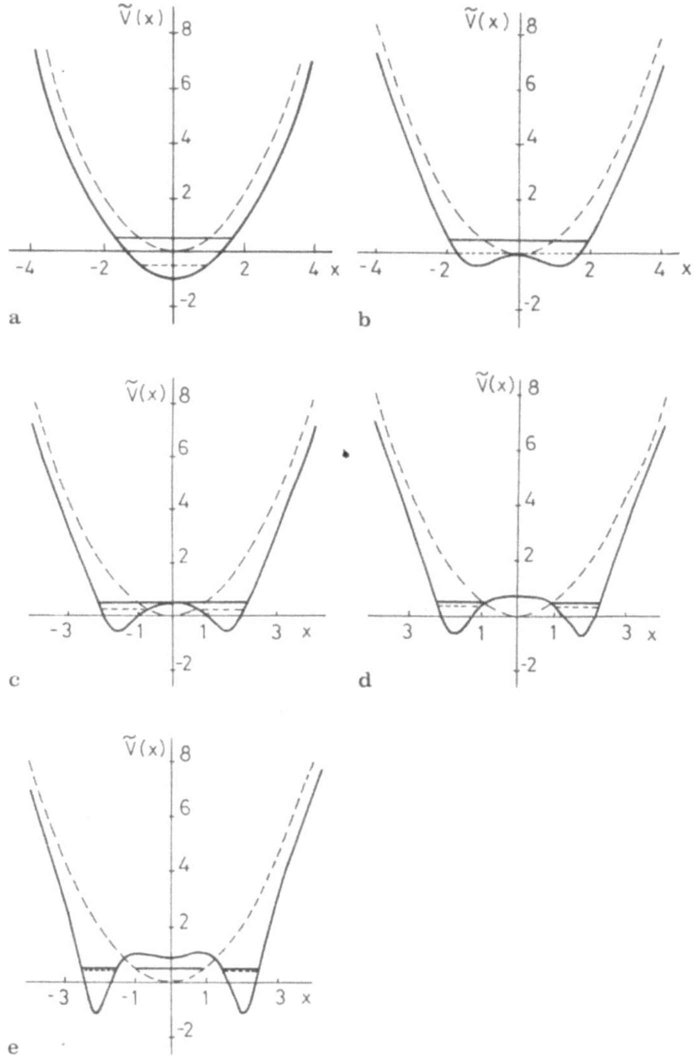

Fig. 2.6a–e. Same as Fig. 2.5a, but for the harmonic oscillator potential well (Sukumar 1985)

repulsive hills where $|\psi_2(x)|$ has maxima. The three attractive zones near the nodes of $\psi_2(x)$ have a weak effect on E_2, but counteract the repulsion of all other levels. A deformation of the opposite sign will pull E_2 down. More surprizing is the approximate additivity of potential deformations in response to variations of two spectral parameters, as shown in Fig. 2.8.

It is interesting to compare the energy levels in infinite and finite wells. Figure 2.9 shows them for finite transparent potentials (nonreflecting, "two- and

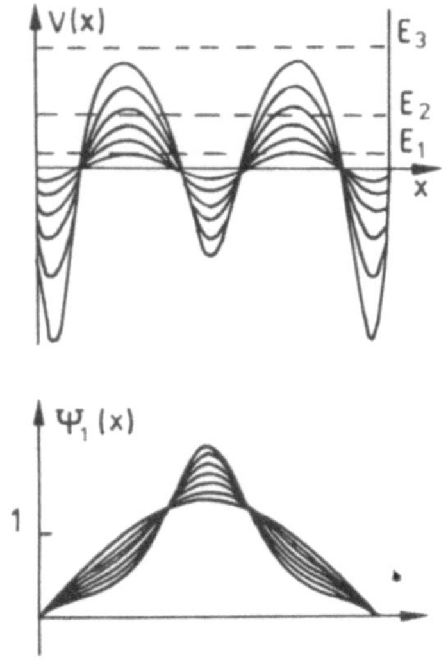

Fig. 2.7. Same as Fig. 2.5 but for $E_2 \to E_3$

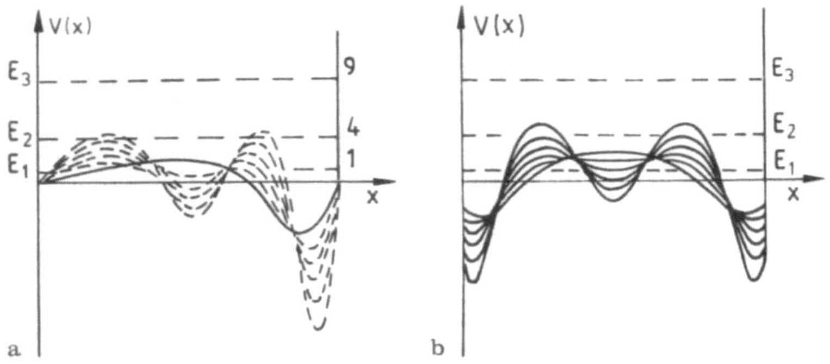

Fig. 2.8a, b. "Addition" of potential perturbations (Poshel, Trubovitz 1987): **(a)** Fixed γ_1 corresponding to maximum deformation of the potential (*solid line*) as shown in Fig. 2.3a, and added perturbations from Fig. 2.4 (dashed lines), with increasing γ_2 **(b)** E_1 is fixed at a level shifted up due to one of the deformations shown in Fig. 2.5a, and E_2 is increasing, to the third level causing the additive accumulation of perturbations from Figs. 2.5a, 7

three-soliton" wells). The upper level is shifted down by two **(a)** and three **(b)** local potential deepenings and by broadening of the whole well. The ground state level is pushed up by raising the central part of the potential bottom. Thus, this is also an example of the additivity of perturbations.

If we don't restrict ourselves to symmetrical perturbative potentials when shifting the energy levels, it is necessary to also determine the normalizing

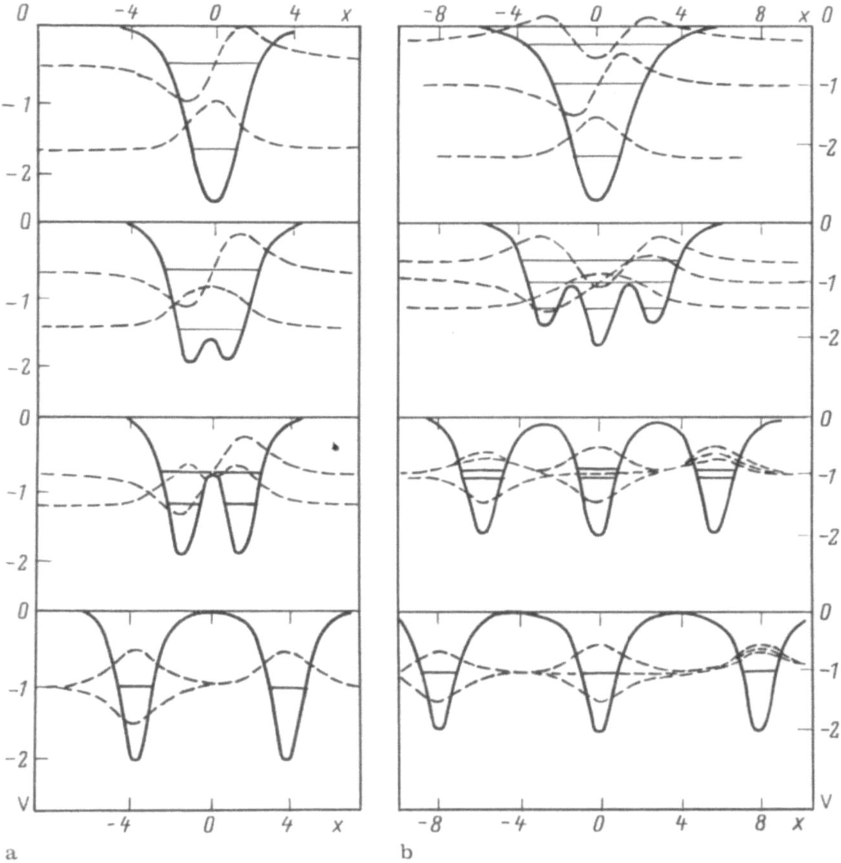

Fig. 2.9. The change in form of Bargmann potentials with two (**a**) and three bound states (**b**) as the levels approach each other: the *dashed curves* represent the wave functions of bound states (Kwong et al 1980). The potentials separate into wells which move from each other the farther, the smaller the splitting between the levels. Unlike the case with infinite walls at $x = 0$; $x = a$ (Figs. 2.5a,6b), the possibility to separate the partial deepenings is used here to isolate wave fragments.

constants. In Fig. 2.10 (Zakhariev et al 1990) the ground state energy E_1 is shifted to E_2, with γ_1 related to k_1 as in an unperturbed state: $\gamma_1 = (2/\pi)^{\frac{1}{2}}/k_1 \xrightarrow[k_1 \to k_2]{} \gamma_2$. The functions $\psi_1(x)$, $\psi_2(x)$ become very much alike in the right-hand part of the well. Near the node of $\psi_2(x)$ the barrier of the perturbative potential (Fig. 2.10) also decreases $\psi_1(x)$. To the left of the node the absolute values of the functions approximately coincide. Even if the normalizing constants γ_ν, $\gamma_{\nu+1}$ of approaching states are determined to be very different, both wave functions $\psi_1(x)$ and $\psi_2(x)$ are nevertheless approximately equal to within a constant (Fig.

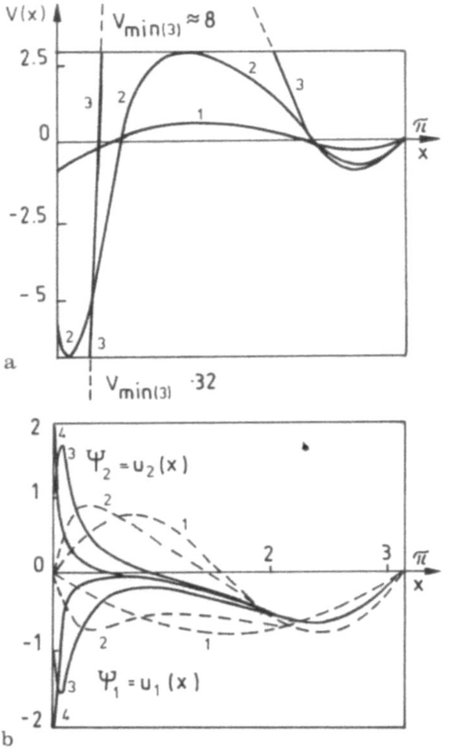

Fig. 2.10a. Perturbation of the infinite rectangular well when $E_1 \to E_2$: 1) $k_1 = 1.1$; 2) $k_1 = 1.5$; 3) $k_1 = 1.8$. **b** The moduli of functions $\psi_1(x)$ and $\psi_2(x)$ approach one another when $E_1 \to E_2$: 1) $k_1 = 1$; 2) $k_1 = 1.7$; 3) $k_1 = 1.9$; 4) $k_1 = 1.99$.

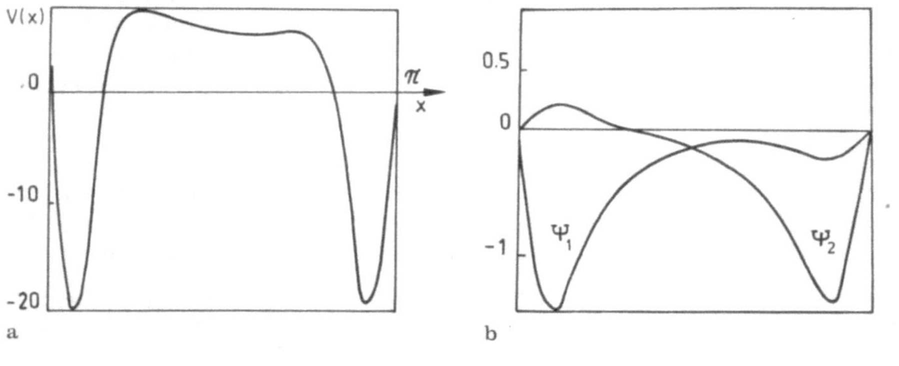

Fig. 2.11. Perturbation potentials (**a**) and wave functions (**b**): $\psi_1(x)$ and $\psi_2(x)$ for the second level lowered to $E_1(k_2 = 1.2)$; $\gamma_2 = 6\ (2/\pi)^{\frac{1}{2}}/k_2$

2.11, Zakhariev et al 1990) which may be different to the right and to the left of the nodes in order to compensate for the difference in the normalizing integrals. Strong violation of the constancy of the ratio $\psi_\nu(x)/\psi_{\nu+1}(x)$ occurs only under the barrier where the functions are small.

A useful simple rule for single-channel wave function behavior under a potential barrier can be given. There cannot be more than one node in a standing wave, positive maxima and negative minima are forbidden, as well as more than one positive minimum or negative maximum.

To demonstrate how the relations between the spectral parameters $\{E_\lambda, \gamma_\lambda^2\}$ might be violated we shall take for all the values of λ $E_\lambda = \mathring{E}_\lambda, \gamma_\lambda^2 = \mathring{\gamma}_\lambda^2$ and add a new R-resonance at the energy E_μ with a reduced width γ_μ^2. (E_μ differs from all $E_\lambda = \mathring{E}_\lambda$). Then Q will have nearly the same form as in (2.4.2), only with the substitution $\delta\gamma_\nu^2 \to \gamma_\mu^2$ and $\mathring{E}_\nu \to E_\mu$. For $V(r)$, $\varphi(E, r)$, and $\varphi(E_\mu, r)$ formulas similar to (2.4.4–6) are obtained. The solutions satisfy the Schrödinger equation with the potential $V(r)$. However, the boundary conditions $\varphi(0) = 0$ are violated because $\mathring{\varphi}(E_\mu, 0)$ is not equal to zero for E_μ not coinciding with any one of the $\mathring{E}_\lambda$($\mathring{\varphi}$ becomes zero at both ends of the interval $[0, a]$ only at the eigenvalues $\mathring{E}_\lambda$). An attempt to exclude (instead of adding) one of the nonperturbed R-resonances gives rise to a similar difficulty.

It is interesting that unlike in the R-matrix theory, in another case of a purely discrete spectrum, that is, bound states in an infinitely deep well, it is possible to not only shift levels of the potential $\mathring{V}$ and the respective normalization constants but also to add new levels and to exclude existing ones (Sukumar 1985). This is allowed when the width of the well is not limited, as in an infinite square well.

We now turn to infinitely deep potential wells on the semi-axis $r \geqslant 0$. For the initial potential $\mathring{V}(r)$ we shall take a well with a purely discrete spectrum of bound states $\mathring{E}_\lambda$ and normalizing constants $\mathring{c}_\lambda^2$. Let the exact solutions of the Schrödinger equation with $\mathring{V}$ be known.

The kernel $Q(r, r')$ for the new potential $V(r)$ with the spectral parameters $\{E_\lambda, c_\lambda\}$ has nearly the same form as (2.4.1), as in the case of R-matrix theory (Gelfand, Levitan 1951; Zhigunov et al 1974; Zakhariev et al 1980):

$$Q(r, r') = \sum_{\lambda=1}^{\infty} c_\lambda^2 \mathring{\varphi}(E_\lambda, r)\mathring{\varphi}(E_\lambda, r') - \sum_{\lambda=1}^{\infty} \mathring{c}_\lambda^2 \mathring{\varphi}(\mathring{E}_\lambda, r)\mathring{\varphi}(\mathring{E}_\lambda, r') . \qquad (2.4.7)$$

The difference between (2.4.1) and (2.4.7) lies only in the substitution $\gamma_\lambda, \mathring{\gamma}_\lambda \to c_\lambda, \mathring{c}_\lambda$, but one must also remember that the boundary conditions for φ, $\mathring{\varphi}$ in (2.4.7) are defined not at $r = a$, but at the origin of the coordinate system.

By using the example of a linear well $\mathring{V}(r) = r$, *Grosse* and *Martin* (1979) showed how the form of a potential $\mathring{V} + \Delta V = V(r)$ changes due to the shift of a single (Fig. 2.12a) or of two (Fig. 2.12b) lower levels. In case *a* in (2.4.7) for Q of the two infinite sums only two terms remain, one from each sum. A shift of two levels can be done in a four-term Q. From Fig. 2.12 it is seen that the first levels are lowered when the potential becomes deeper, i.e., they are dependent on the negative part of ΔV in the region of small r, while the positive part of $\Delta V(r)$ holds

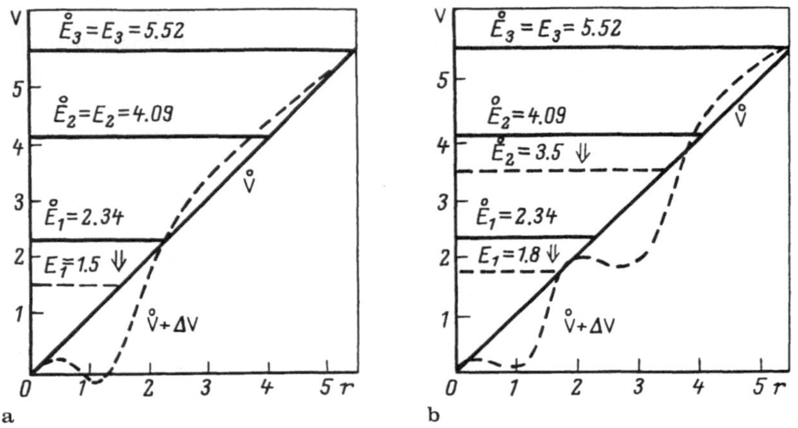

Fig. 2.12a,b. Shift of levels in a linear potential well $\mathring{V}$: **a** shift of a single level; **b** shift of two levels; *dashed line* shows the potential $\mathring{V} + \Delta V$ (Grosse, Martin 1979)

the remaining levels in their original positions. On the other hand, the positive part of $V(r)$ influences the lower states weakly, since it is situated beyond the "turning points" $r_{1,2}[V(r_{1,2}) = E_{1,2}]$ where the wave functions $\psi(E_1, r)$ and $\psi(E_2, r)$ are exponentially small.

The procedure for shifting levels $\mathring{E}_\lambda \to E_\lambda$, and changing normalization constants $\mathring{c}_\lambda^2 \to c_\lambda^2$, may be applied for approximate reconstruction of potentials from the set $\{E_\lambda, c_\lambda\}$. As will be shown in Chap. 3, changing the spectral parameters $[\mathring{E}_\lambda, \mathring{c}_\lambda\}$ $(\lambda = 1, \ldots, N)$ of the initial potential $\mathring{V}(r)$ to the values $\{E_\lambda, c_\lambda\}$ of the desired $V(r)$ it is possible to find $V^N = \mathring{V} + \Delta V^N$ whose form will tend towards V with the increasing N. The proof of such convergence was provided by *Adamyan* (1982).

Exclusion of a level. Such an operation, unlike the attempt to discard an R-resonance, can be done. We eliminate one term from the first sum in (2.4.7) and set the parameters $\{E_\lambda, c_\lambda\}$ equal to those for the unperturbed potential $\mathring{V}$ in the remaining sums. As a result of the cancelation of the summands in the two sums in Q only one term from the second sum which does not have a counterpart in the first sum will remain. With such a simple Q in a similar manner to (2.4.4) we find the potential $[p(r) = 1 - \mathring{c}_v^2 \int_0^r \mathring{\varphi}^2(\mathring{E}_v, t)dt]$:

$$V(r) = \mathring{V}(r) - \frac{2\hbar^2}{m}\mathring{c}_v^2 \mathring{\varphi}(\mathring{E}_v, r)p^{-1}(r) + \frac{\hbar^2}{m}c_v^4 \mathring{\varphi}^4(\mathring{E}_v, r)p^{-2}(r)\,, \tag{2.4.8}$$

having the same spectrum as $\mathring{V}$ with the exception of a single state at $\lambda = v$.

Abraham et al (1980) have calculated the correction to the oscillatory potential $\mathring{V} = x^2/2$ to exclude the ground state:

$$\Delta V(x) = \frac{4}{\sqrt{\pi}}\frac{\mathrm{e}^{-x^2}}{\operatorname{erfc}(x)}\left(\frac{\mathrm{e}^{-x^2}}{\sqrt{\pi}\operatorname{erfc}(x)} - x\right), \tag{2.4.9}$$

where

$$\mathrm{erfc}(x) = \frac{2}{\pi} \int_x^{\infty} \mathrm{e}^{-t^2} dt .$$

The eigenfunctions of the new potential $\mathring{V} + \Delta V$ are expressed through those of the oscillatory potential,

$$\mathring{\psi}_n(x) = \mathrm{e}^{-x^2/2} H_n(x)(\sqrt{\pi}\, 2^n n!)$$

by

$$\psi_n(x) = \mathring{\psi}_n(x) - (2/n\pi)^{\frac{1}{2}} \frac{\mathrm{e}^{-x^2}}{\mathrm{erfc}(x)} \mathring{\psi}_{n-1}(x) , \quad n = 1, 2 \ldots \tag{2.4.10}$$

Addition of levels is performed in a similar manner (Sukumar 1985). Figure 2.6 shows the results of addition from below of a single level to the spectrum of the oscillator (initial potential is shown by the dashed line).

Adamyan (1981) has shown that perturbating corrections $\Delta V(r)$ which cause a change in only a finite number of levels do not affect the asymptotic behavior of the potential as $r \to \infty$ and, moreover, that they tend to zero as $r \to \infty$. Generalization to the $l > 0$ case of the formalism of Bargmann transformation of infinite potential wells has been considered by *Gostev* et al (1982).

2.5 Potentials allowing Exact Solutions for Variable Angular Momenta

All the models considered above are obtained within the formalism of the inverse problem assuming a fixed angular momentum l. A similar approach with a constant energy leads to a significant increase in the number of exact solutions. Another effective method of constructing potentials and their corresponding wave functions in a closed analytic form adopts the Crum–Krein transformations. The initial version of these transformations (with $l = \text{const}$) was considered by *Agranovich* and *Marchenko* (1963) and *Chadan* and *Sabatier* (1977). Later their modification was proposed for $E = \text{constant}$ problems (Zakhariev et al 1982) and recently this generalization was found by *Rudyak* et al (1984, 1987) and *Suzko* et al (1985). Here, the obtained potentials correspond to explicit solutions for a set of parameters l and E situated on lines in the (λ^2, E)-plane ($\lambda = l + \frac{1}{2}$). This new method, from which the Crum–Krein ($l = \text{const}$) and the Crum–Krein–Pivovarchik ($E = \text{const}$) transformations are derived as special cases, is discussed in this section. Also considered are the Lipperheide–Fidelday models which are similar to the Bargmann models.

2.5.1 Potentials from Spectral Data at Fixed Energy and Variable l

The algorithm for construction of potentials of the Bargmann type allowing exact solutions for $E = \text{const}$ and arbitrary l has much in common with the case of l = const. Let $\mathring{f}(\lambda, k)$ be the known Jost function of the initial problem. We shall construct a new Jost function $f^+(\lambda, k)$ by adding to $\mathring{f}^+(\lambda, k)$ a factor similar to (2.3.1) except with a rational dependence on the variable angular momentum λ instead of on k as in (2.3.1):

$$f^+(\lambda, k) = \mathring{f}^+(\lambda, k) \sum_j^N \frac{\lambda - \alpha_j}{\lambda - \beta_j}, \tag{2.5.1}$$

The function $f^+(\lambda, k)$ is chosen in accordance with the requirement

$$\lim_{\lambda \to \infty} f(\lambda, k)/\mathring{f}(\lambda, k) = 1, \quad |\arg \lambda| < \pi/2, \tag{2.5.2}$$

unlike the case of l = const where both $f_l^+(k)$ and $\mathring{f}_l(k)$ were supposed to tend to 1 as $|k| \to \infty$.

For the potential to be real the equality $\beta_i = -\alpha_j^*$ must be satisfied:

$$f^+(\lambda, k) = \mathring{f}^+(\lambda, k) \prod_{j=1}^N (\lambda - \alpha_j)/(\lambda + \alpha_j^*). \tag{2.5.1}$$

To provide for the analyticity of $f^+(\lambda, k)$ in the right-hand complex λ half-plane and continuity for $\mathrm{Re}\{\lambda\} \geqslant 0$, it is necessary to choose $\mathrm{Re}\{\alpha_j\} > 0$.

Substituting (2.5.1) into (1.4.33) for $N = 1$, we obtain

$$\begin{aligned} Q(r, r') &= M_1^2 \mathring{f}^+(\alpha_1, k, r)\mathring{f}^{+*}(\alpha_1, k_1, r')/(rr') \\ &+ \frac{\alpha_1^{*2} - \alpha_1^2}{rr'\mathring{f}^+(\alpha_1, k)} \begin{cases} \mathring{\varphi}(\alpha_1, k, r)\mathring{f}^+(\alpha_1, k, r') & \text{for } r < r'; \\ \mathring{f}^+(\alpha_1, k, r)\mathring{\varphi}(\alpha_1, k, r') & \text{for } r > r'. \end{cases} \end{aligned} \tag{2.5.3}$$

With this kernel, from (1.4.32) we find

$$\begin{aligned} K(r, r') = &-\left[\frac{\alpha_1^{*2} - \alpha_1^2}{\mathring{f}^+(\alpha_1, k)} \frac{\mathring{\varphi}(\alpha_1, k, r)}{r} + \frac{M_1^2 \mathring{f}^+(\alpha_1, k, r)}{r}\right] \\ &\times \frac{\mathring{f}^+(\alpha_1^*, k, r')}{r'} W^{-1}[\mathring{f}^+(-\alpha_1^*, k, r), \ \mathring{f}^+(\alpha_1, k, r)c_1^2/ \\ &\times (\alpha_1^{*2} - \alpha_1^2) + \mathring{\varphi}(\alpha_1, k, r)/\mathring{f}(\alpha_1, k)], \quad W[ab] = ab' - ba'. \end{aligned} \tag{2.5.4}$$

Of all the family of potentials corresponding to (2.5.4) only those whose normalization constant is in the form $M_1^2 = f^+(-\alpha_1, k) \ (\alpha_1^{*2} - \alpha_1^2)/2ik\mathring{f}^+(\alpha_1, k)$ are real

$$V(r) = -\frac{\hbar^2}{m}(\alpha_1^{*2} - \alpha_1^2)|\mathring{f}^+(\alpha_1^*, k, r)|^2 W^{-1}[\mathring{f}^+(\alpha_1^*, k, r), \ \mathring{f}^+(\alpha_1^*, k, r)]. \tag{2.5.5}$$

2.5.2 Newton–Sabatier Potentials

When the number N of coefficients c^2 differing from zero is finite in (1.4.39), then in the system (1.4.41) there remain N algebraic equations which are solvable in a closed form. If only one coefficient c_L differs from zero, we obtain from (1.4.41)

(Newton 1982; p. 650–658):

$$\varphi_L(k, r) = krj_L(kr)p^{-1}(r)\,; \quad p(r) = 1 + c_L^2 L_{LL}(r)\,;$$

$$\varphi_{l \neq L}(k, r) = krj_l(kr) - c_L^2 L_{lL} krj_L(kr)p^{-1}(r)\,. \tag{2.5.6}$$

With this φ we find K by substituting (2.5.6) into (1.4.40) with $c_{l \neq L} = 0$, from which we have

$$V(r) = -\frac{\hbar^2}{m} r^{-1} c_L^2 \frac{d}{dr}\{r^{-1}[krj_L(kr)]^2 p^{-1}(r)\}\,. \tag{2.5.7}$$

From (2.5.6) as $r \to \infty$, we find the expression for the phase shifts $\sin\delta_l = 0$,

$$\tan\delta_l = c_l^2\{[1 + \tfrac{1}{2}\pi c_L^2/(2L+1)]\,[(l-L)(l+L+1)]\}^{-1}\,. \tag{2.5.8}$$

For $N > 1$ we have the summation over the angular momenta for which $c_l \neq 0$:

$$\varphi_l(k, r) = \sum_{l'}^{N} krM_{ll'}^{-1}(r)j_l(kr)\,,$$

where

$$M_{ll'} = \delta_{ll'} + L_{ll'}(r)c_{l'}^2\,; \tag{2.5.9}$$

$$V(r) = -\frac{\hbar^2}{m} r^{-1} \frac{d}{dr}\left\{r^{-1}\sum_{l}^{N}(kr)^2 j_l(kr)\sum_{l'}^{N} M_{ll'}^{-1}(r)j_{l'}(kr)\right\}. \tag{2.5.10}$$

The solutions φ and the potentials V with Q in the more general form (1.4.40) are obtained in a similar way.

Many calculations on the approximate solution of the inverse problem with the aid of Newton–Sabatier potentials are due to *Coudray* (1979).

2.5.3 The Generalized Crum–Krein Transformations

In this chapter, in the search for exactly solvable models a central role was played by the equations of the inverse problem. It is also possible, however, to transform directly known solutions of the Schrödinger equation y_0 with a certain potential V_0 in such a way that other solutions y_1 which correspond to another potential V_1 are obtained. Such transformations were proposed by *Euler* (1780–94, *Imshenetsky* (1882), *Crum* (1955) and *Krein* (1951). They play an important part in the mathematical apparatus of the inverse problem. These transformations became especially interesting when it became recognized that with their aid it is possible to solve nonlinear equations. (In soliton theory they are more often attributed to Darboux.)

The generalization of these transformations to potentials increasing towards infinity was suggested by *Levi* (1988).

The Euler–Imshenetsky–Crum–Krein–Darboux transformations yield potentials and solutions at arbitrary energies on the whole axis or on $(0, \infty)$ with fixed l. *Pivovarchik* (1982) extended the idea by developing the method for constructing solutions at $E = \text{const}$ and different l (Zakhariev et al 1982). Both approaches represent special cases of a more general method applicable to variable E and l (Rudyak et al 1984, 1987; Suzko 1985, 86). *Abramov* (1984) considers the quasi-classical case and *Bogdanov* and *Demkov* (1982) formulate the corresponding inverse problem.

From the potential $V_0(r)$ and the solution $y_0(r)$ found at fixed E' and $\lambda' = l' + \frac{1}{2}$ we obtain after the transformation, solutions with a new potential $V_1(r)$ for arbitrary E and λ:

$$y_1(r) = r/[y_0(r)\sqrt{1 + \alpha r^2}|] . \tag{2.5.11}$$

$$a < r < b$$

$$z_1(r, E, \lambda) = \frac{r}{y_0(r)\sqrt{|1 + \alpha r^2|}} W\{y_0(r), \chi_0(r, E, \lambda)\} \tag{2.5.12}$$

where $\alpha = (E' - E)/(\lambda^2 - \lambda'^2)$ and $\sqrt{|1 + \alpha r^2|}$ and $y_0(r)$ are assumed not to be zero on the interval $a < r < b$; the new potential is assumed to be

$$V_1(r) = V_0(r) - 2\left[\frac{d^2}{dr^2}\ln y_0(r) + \frac{d \ln y_0(r)/dr}{r(1 + \alpha r^2)}\right] - \frac{3\alpha}{|1 + \alpha r^2|^2} . \tag{2.5.13}$$

From independent solutions φ_0, χ_0 one obtains z_1, y_1 that are also independent.

The family of the found potentials $V_1(r)$ depends on three parameters: α, λ', E'. By varying them we obtain different $V_1(r)$ and the respective solutions $z_1(r, E, \lambda)$ with λ and E situated on a line in the (λ^2, E)-plane passing through a point (λ'^2, E') at an angle α to the E axis.

Now let us consider some special cases.

I. $l = \text{const}$. Let $\alpha = \infty$; then $\lambda = \lambda' = \text{const}$, i.e., the line of the parameters l and E for which the new exact solutions are constructed is parallel to the E-axis in the (λ^2, E)-plane (Fig. 2.13). These are the ordinary Crum–Krein trans-

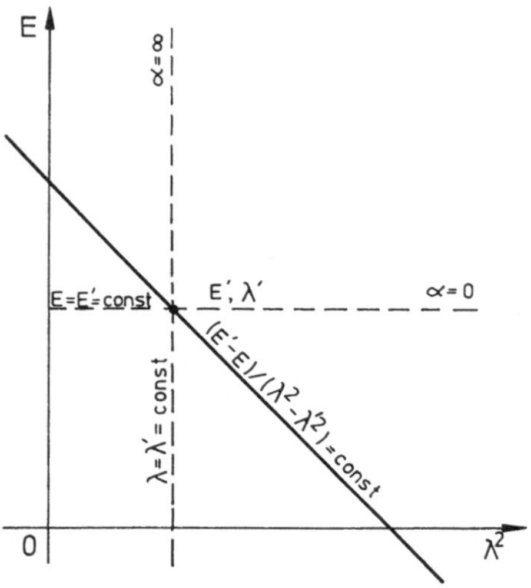

Fig. 2.13. Lines in the $[E, l(l+1)]$-plane corresponding to input data for Bargmann potentials in different approaches: Gelfand–Levitan–Marchenko (vertical $E = \text{const}$), Regge–Newton–Sabatier (horizontal $l = \text{const}$), generalized (with arbitrary slope) (Rudyak, Zakhariev 1987)

formations. By using them one can also construct the Bargmann solutions (Rudyak et al 1984, 1987; Suzko 1985, 1986). Thus, it is possible to construct the solutions of the radial Schrödinger equation with $l \neq 0$ from solutions with $l = 0$. The validity of formulas derived with these transformations is independent of the existence of the Gelfand–Levitan–Marchenko equations. Therefore, the Crum–Krein transformations are very powerful in proofs in the theory of the inverse problem. For a generalization to the multi-channel case we refer the reader to Chap. 5.

II. E = const. The case of $\alpha = 0$ corresponds to $E' = E$ = const. Equations (2.5.11,12) in the limit $\alpha \to 0$ transform into the Crum–Krein–Pivovarchik transformations:

$$y_1(r) = r/y_0(r) ; \tag{2.5.14}$$

$$z_1(r, \lambda) = r y_0^{-1}(r) W\{y_0(r), \chi_0(r, \lambda)\} . \tag{2.5.15}$$

III. Variables E and l, arbitrary α. Now let $E' = \kappa^2 < 0$, $l' = 0(\lambda' = \frac{1}{2})$; $E = E' - \alpha l(l+1)$; $V_0(r) \equiv 0$.

As $y_0(r)$ we take $\exp(\pm \kappa r)$. Then, in agreement with (2.5.11,13)

$$y_1(r) = r \exp(\mp \kappa r)/\sqrt{|1 + \alpha r^2|} ; \tag{2.5.16}$$

$$V_1(r) = \mp 2\kappa [r(1 + \alpha r^2)]^{-1} - \frac{3\alpha}{(1 + \alpha r^2)^2} . \tag{2.5.17}$$

From (2.5.12) we find two linearly independent solutions:

$$z_1(r, k, l) = r[j_l'(kr) \mp \kappa j_l(kr)]/\sqrt{|1 + \alpha r^2|} ;$$

$$p_1(r, k, l) = r[n_l'(kr) \mp \kappa n_l(kr)]/\sqrt{|1 + \alpha r^2|} ,$$

$$k = \sqrt{E' - \alpha l(l+1)} . \tag{2.5.18}$$

It is interesting that when $\alpha = 0$, the potential given by (2.5.17) transforms into the Coulomb potential whose solutions are the Bessel functions.

Now let $E' = 0$, $l' = 0\,(\lambda' = \frac{1}{2})$, $E = -\alpha l(l+1)$, $V_0(r) = 0$, and $y_0(r) = r$ be a partial solution of the equation $y_0''(r) = 0$. As the initial linearly independent solutions we shall take

$$\chi_0(r, E, l) = j_l(k = \sqrt{-\alpha l(l+1)}, r)$$

and

$$\eta(r, E, l) = n_l(k = \sqrt{-\alpha l(l+1)}, r).$$

In accordance with (2.5.11–13) we have

$$y_1(r) = 1\sqrt{|1 + \alpha r^2|} ; \tag{2.5.19}$$

$$V_1(r) = -\frac{2}{r^2(1 + \alpha r^2)} - \frac{3\alpha}{(1 + \alpha r^2)^2} \tag{2.5.20}$$

and two linearly independent solutions

$$z_1(r,k,l) = [rj_l'(kr) - j_l(kr)]/\sqrt{|1+\alpha r^2|}\,;$$
$$p_1(r,k,l) = [rn_l'(kr) - n_l(kr)]/\sqrt{|1+\alpha r^2|}\,,$$
$$k = \sqrt{-\alpha l(l+1)}\,. \tag{2.5.21}$$

The models for variable E and l can be generalized to the multi-channel case by using the Crum–Krein–Marchenko (Darboux) transformations.

2.5.4 Lipperheide–Fidelday Potentials

A new family of models with solutions in a closed analytic form was found by analogy with (2.5.5) where, instead of the Jost solutions $\mathring{f}(\lambda,k,r)$, one must substitute the functions $\mathring{\varphi}(\lambda,k,r)$ which are regular at zero [let $\mathring{V}(r)=0$] (Lipperheide et al 1981):

$$\mathring{\varphi}(\lambda,k,r) = (2/k)^{\lambda}\Gamma(\lambda+1)\sqrt{r}J_{\lambda}(kr) \xrightarrow[r\to 0]{} r^{\lambda+\frac{1}{2}}\,; \tag{2.5.22}$$

$$V(r) = \frac{\hbar^2}{m}\frac{d}{dr}\left[\frac{1}{r}\frac{|\mathring{\varphi}(\alpha,k,r)|^2}{\chi_{\alpha^*\alpha}}\right], \tag{2.5.23}$$

where

$$\chi_{\lambda\mu} = W[\mathring{\varphi}(\lambda,k,r), \mathring{\varphi}(\mu,k,r)]/(\lambda^2-\mu^2)$$

and J_λ is the Bessel function. The potential (2.5.23) is finite everywhere.

The regular solution φ of (2.5.23) is determined from

$$\varphi(\lambda,k,r) = \mathring{\varphi}(\lambda,k,r) - [\chi_{\lambda\alpha}(r)/\chi_{\alpha\alpha^*}(r)]\mathring{\varphi}(\alpha^*,k,r)\,. \tag{2.5.24}$$

From the asymptotic behavior of φ as $r\to\infty$ we find the scattering function ($\sigma_\alpha \equiv \exp[-i\pi(\alpha-\frac{1}{2})]$):

$$S(\lambda) = \left(\frac{\sigma_{\alpha^*}-\sigma_\alpha}{\alpha^{*2}-\alpha^2} - \frac{\sigma_\lambda-\sigma_\alpha}{\lambda^2-\alpha^2}\frac{\sigma_{\alpha^*}}{\sigma_\lambda}\right)\left(\frac{\sigma_{\alpha^*}-\sigma_\alpha}{\alpha^{*2}-\alpha^2} - \frac{\sigma_\lambda-\sigma_\alpha}{\lambda^2-\alpha^2}\right)^{-1}. \tag{2.5.25}$$

In the general case of a nonzero initial potential and of several complex parameters α_m, β_n ($m, n = 1, \ldots, N$), instead of (2.5.23–25), we have

$$V(r) = \mathring{V}(r) - \frac{\hbar^2}{mr}\frac{d}{dr}\left[r\frac{d}{dr}\ln\det\chi(r)\right], \tag{2.5.26}$$

where $\chi(r) = \|x_{mn}\| = \|\chi_{\alpha_m\beta_n}\|$ is an $N\times N$ matrix;

$$\varphi(\lambda,k,r) = \mathring{\varphi}(\lambda,k,r)\frac{\det\|x_{mn} - x_{\lambda_m}\mathring{\varphi}(\beta_n,k,r)/\mathring{\varphi}(\lambda,k,r)\|}{\det\chi(r)}\,; \tag{2.5.27}$$

$$S(\lambda) = \mathring{S}(\lambda)\frac{\det\left\|\dfrac{\mathring{\sigma}_{\beta_n}-\mathring{\sigma}_{\alpha_m}}{\beta_n^2-\alpha_m^2} - \dfrac{\mathring{\sigma}_\lambda-\mathring{\sigma}_{\alpha_m}}{\lambda^2-\alpha_m^2}\dfrac{\mathring{\sigma}_{\beta_n}}{\mathring{\sigma}_\lambda}\right\|}{\det\|(\mathring{\sigma}_{\beta_n}-\mathring{\sigma}_{\alpha_m})/(\beta_n^2-\alpha_m^2) - (\mathring{\sigma}_\lambda-\mathring{\sigma}_{\alpha_m})/(\lambda^2-\alpha_m^2)\|} \tag{2.5.28}$$

where $\mathring{\sigma}_\lambda \equiv \exp[-i\pi(\lambda-\frac{1}{2})]\mathring{S}(\lambda)$.

If $\mathring{V}(r)$ is a real-valued function and in addition the set $\{\beta_n\}$ is the complex conjugate of $\{\alpha_m\}$, then the potential $V(r)$ is also real and the S-function is unitary for real λ.

The variable λ enters into the exponents in the scattering functions (2.5.25,28) through σ_λ and $\mathring{\sigma}_\lambda$, so that the dependence of S upon λ is not rational, as in the case of Bargmann models. With (2.5.26,28) it is also possible to approximately solve the inverse problem. Examples are considered in Chap. 3.

2.6 Notes on the Literature

The first review of V^B potentials was written by *Bargmann* (1949) himself. The formalism of the Crum–Krein transformations is presented by *Agranovich* and *Marchenko* (1963). In the review by *Faddeev* (1963) the perturbations ΔV in $\mathring{V}$, which arise when the Jost functions or $\mathring{S}(k)$ are multiplied by factors rational in k, are mentioned. The one-dimensional inverse problem on the entire axis is presented by *Faddeev* (1976) with examples of explicit solutions.

Separate sections of *Newton* (1982) and *Chadan* and *Sabatier* (1977) are devoted to the Bargmann potentials.

The inverse problem for the linear Schrödinger equation served as a starting point for a new field of investigations in physics and mathematics: the development of methods to solve nonlinear equations (theory of solitons). Now it turned out that most exactly solvable models can be united by the new mathematical formalism (see, for example, Krichever 1989; Dubrovin 1981, 1988).

The flexibility and simplicity of the models considered in this chapter make them extremely convenient for demonstrating peculiarities in quantum systems. Another example is presented by the investigation of the transition from the quantum inverse problem to its quasi-classical approximation (Lipperheide et al 1978).

Solutions of the Bargmann type were also considered for the relativistic Dirac and Bethe-Salpeter equations as well as for the quasi-potential [references can be found in the review by *Zakhariev* et al (1982), see also *Bogdanov* and *Demkov* (1987) and *Leon* (1980)]. *Moses* (1983) and *Moses, Prosser* (1983) found a series of potentials allowing exact solutions and possessing unusual properties. The two different potentials

$$V_1(x) = \frac{\hbar^2}{m}\theta(x)\frac{4x^4 + 8\sqrt{2}x^3 + 12x^2 - 3}{(2\sqrt{2}x^3 + 6x^2 + 3\sqrt{2}x + 3)^2};$$

$$V_2(x) = \theta(-x)\frac{4\sqrt{2}(\sqrt{2} + 1)}{(2x - \sqrt{2} - 1)^2},$$

where θ is the Heaviside function, $\theta(x) = 0$ for $x < 0$ and $\theta(x) = 1$, when $x > 0$, have no bound states and yield the same reflection coefficient at all energies. Interestingly, they correspond to quite ambiguous forces derived from usual scattering data. Of course, they are long-range potentials.

Moses et al (1983) also considered potentials depending on the parity, $\hat{V}$

$$\hat{V}\psi(x) \equiv V_1(x)\psi(x) + V_2(x)\psi(-x)\,,$$

which represent a simplest special case of nonlocal (quasilocal) potentials. If one chooses

$$V_1 = (-4/c)\delta(x) + 4/(2|x| + c)^2, \quad \text{and} \quad V_2 = 4/(2|x| + c)^2\,,$$

then such a potential has a bound state in the continuum.

New models can be obtained from the models discussed in this chapter by cutting off the potential and wave functions at a certain r_0 and joining the solutions with free waves at $r \geqslant r_0$.

Exact expressions for the propagators D describing the evolution of the packet

$$\psi(x, t) = \int_{-\infty}^{\infty} D(x, t; x', 0)\psi(x', 0)dx$$

in a reflectionless potential are given by *Crandal* and *Litt* (1987).

In addition to *Flugge* (1971) other literature is available on exact solutions obtained for the direct problem. *Natanson* (1979) reviewed the results on potentials found by transformation of hypergeometric equations. Difference analogues of hypergeometric functions were given by *Suslov* (1989). Proceeding from the modified Matier equation, exact Schrödinger solutions were obtained by *Spector* and *Aly* (1964, 1965), for instance , for $V(r) \sim \pm 1/r^4$.

The problem of scattering on an up-side-down biharmonic oscillator

$$V(x) = V_0 - m\omega_1^2 x^2/2$$

has been exactly solved by *Prakash* (1976). The potential with $\omega_1 = \omega_2$ is considered by *Wheeler* (1953). *Nieto* (1979) presented examples of calculation of the uncertainty $\Delta x \Delta p$ for a number of exactly solvable models, for oscillatory, rectangular well, Coulomb, Peshl-Teller, and Morse potentials. Many problems with δ-like potentials are considered in the book by *Demkov* and *Ostrovsky* (1975); see also *Albeverio* et al (1988) and *Pavlov* (1987).

The method of factorization (Infeld and Hull 1951), closely related to the construction of Bargmann additions to an initial V, was used to obtain supersymmetric nonrelativistic models (Gendenstein and Krive 1985; Haymaker and Rau 1986; Berezovoy and Pashnev 1987; Andrianov et al 1984; Sukumar 1985, 1988; Pashnev 1986; Cooper et al 1987). The exact solution for the combined oscillatory-Coulomb potential is given by *Petrosyan* et al (1986).

Bargmann-type potentials that are singular at zero are considered by *Kvitko* and *Pivovarchik* (1987). An analogy to the Bargmann potentials can also be obtained with nonlocal potentials (*Muzafarov* 1985, 1987). For a special form of the S-functions the inverse-problem equations also reduce to a set of algebraic equations (but bilinear), which provides us with a new class of simple, exactly solvable models.

A finite well with a linearly tilted bottom was considered by *Churchill* (1987). A class of Sturm–Liouville operators on the semi-axis, which are isospectral with the Lagger operator, is considered by *Skoblin* (1987).

Group-theoretical methods (the algebraic approach to solving equations with dynamical symmetry) reveal common aspects of Bargmann models, coherent states, and the interacting boson model generalized to states of the continuum spectrum (Alhassid et al 1983, 1986). Exact solutions for effective periodical potentials are considered by *Zaslavsky* and *Ulyanov* (1987).

Classes of potentials with an arbitrary finite number of exactly known bound states ("quasi-exactly" solvable problems) were found by *Turbiner* (1988) and *Ushveridze* (1989). Exactly solvable nonlinear equations with Grassman variables are considered by *Elpik* (1987).

Simple transformations which reduce the equations of motion for Coulomb and oscillator potentials to equations with a Morse potential were proposed by *Haymaker* and *Rau* (1986).

Bargmann-type potentials with $\mathring{V}(x) = cx$ were considered by *Calogero* and *Degasperis* (1978).

A useful collection of simple illustrations of wave-particle features can be found in "The Picture Book of Quantum Mechanics" by *Brandt* and *Dahmen* (1985).

2.7 Exercises

1. How will the form of an oscillatory potential on the semi-axis qualitatively change if one new level is added to its spectrum close to the ground state? Compare with the case of an oscillator on the entire axis.
2. How will the form of a finite rectangular potential well qualitatively change if the normalization constant or the position of some resonance is varied?
3. Obtain, within the Marchenko approach, the potential and solution corresponding to the single-term kernel $Q = M \exp[-\chi(r + r')]$, when there is only a single bound state $[S(\kappa) \equiv 1]$. Check that the solution satisfies the Schrödinger equation on the semi-axis with the found potential, but not the initial requirement: $S \equiv 1$. This is explained by the disagreement between the requirements that the potential should have a bound state and that $S \equiv 1$ which violates the Levinson theorem: $\delta(0) = \pi$ (Agranovich, Marchenko 1963, p. 134).
4. Check, by direct substitution, that solutions of the Bargmann type (2.2.4,16) satisfy the Schrödinger equation with the potentials (2.2.6,17), respectively.

Chapter 3

Approximate Solutions

"To my mind it is the need to harmonize conflicting requirements that makes the concoction of algorithms a fascinating task".

B. Parlett (1980)

3.1 General Remarks

Although the algorithms for approximate solutions of the Schrödinger equation are presented in complete detail in a number of monographs, practically no monographs devoted to approximate solutions of the quantum inverse problem are available. We hope to close this gap, at least to a certain extent, in this chapter.

An idea of the asymmetric development of quantum theory may be obtained with the aid of the scheme in Fig. 3.1. The vertical axis represents the number of numerical solutions of the direct and inverse problems per year (horizontal axis). At present no more than several hundred papers devoted to numerical solutions of the $S \to V$ problem which amounts to less than 1% of the number of solutions of the Schrödinger equation. Truly, $V \to S$ computations are often performed with the aim of finding the form of potentials, which is actually a substitution of the $S \to V$ procedure by multiple repetition of $V_{\text{trial}} \to S_{\text{trial}}$, that is, fitting trial potentials V_{trial} to achieve a satisfactory agreement between S_{trial} and S. The merits and demerits of solution of an inverse problem by using the direct one will be discussed in Sect. 3.5.

In any case, to become more comprehensive, quantum mechanics must arrive at an equilibrium between the two formalisms $V \rightleftharpoons S$ since the necessity

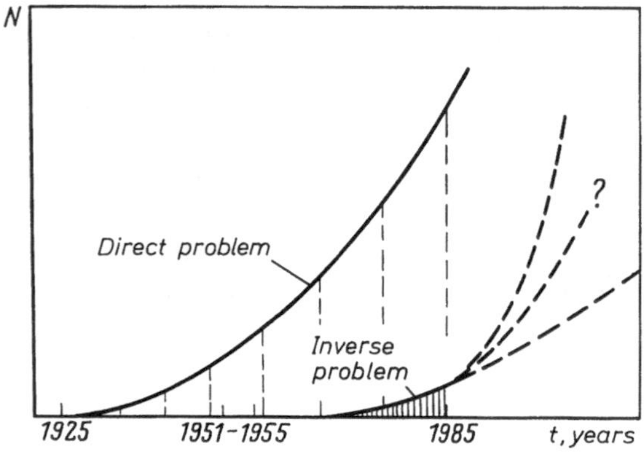

Fig. 3.1

for reconstructing V from S is as frequent a problem as that of finding a description of particle motion for a known V.

Although of the two $V \rightleftharpoons S$ operations only the inverse problem is essentially unstable, in performing computer calculations one must also beware of finding unstable solutions in the $V \to S$ problem. In both cases methods for eliminating uncertainties have much in common (Turchin et al 1970). Estimation of the accuracy of approximations and methods for regularization of solutions will be dealt with in Sect. 3.1; Sect. 3.2 is devoted to the reconstruction of interactions with the aid of exactly solvable models, to approximations of V by Bargmann potentials. For the first time, examples of methods of finding V from S are collected, permitting a demonstration of the quality of the reconstructed V.

Section 3.3 presents algorithms for reconstructing V both in the "stepwise" approximation and with the aid of the finite-difference approach, as well as the *method of finite elements* which has proved extremely efficient in computational mathematics for integrating differential equations. This method has only recently begun to be applied in quantum mechanics.

We shall not consider those approximate methods which are sufficiently well known from other books (perturbation theory, quasiclassical approaches, and methods of separable potentials). Only certain important points of the Bubnov–Galerkin method in scattering problems will be noted (Sect. 6.3). For further details on this method we direct the reader to *Zhigunov* and *Zakhariev* (1974).

3.2 Convergence of the Approximations, Stability and Regularization of Solutions

The modern mathematical theory of approximate methods is based on the technique of functional analysis (Krasnoselsky et al 1969; Mikhlin 1966, 1970) whose inclusion in quantum investigations is becoming more and more frequent. As computational techniques develop, the role of a strict analytic approach, that is, the study of convergence, of the stability of the applied methods, etc., will rapidly become more and more important.

Norm, Metric, Completeness. We begin by clarifying the notion of "proximity" between functions. It is convenient to consider functions as vectors in functional space (infinite-dimensional in the general case). By characterizing each vector ψ by its 'length' as $\|\psi\|$, it is possible to introduce the concept of "distance" between, for example, an exact function ψ and an approximate ψ^N as the modulus of the difference between them $\|\psi - \psi^N\|$. Naturally, one must always bear in mind the convention that a set of functions should form a linear normalized vector space. Thus, the proximity of functions is a relative concept: for example, a series of functions converging in one metric may diverge in another.

Sometimes it is possible to introduce a modulus corresponding to the desired physical quantity. Thus, the convergence $\psi^N \to \psi$ with respect to the "energy modulus" $\|\psi\| = [\int \psi^* H \psi \, dx]^{1/2}$ indicates that the corresponding energy values also converge $E^N \to E$. However, the problem of finding a modulus to characterize the measure of proximity between scattering parameters still remains open. The only case in which the proximity between wave functions of a continuous spectrum has been successfully related directly to the difference between scattering amplitudes is the zero-energy limit of the relative motion of colliding particles ($k = 0$).

As a rule, the function ψ is unknown and therefore by direct calculation of $\|\psi - \psi^N\|$ it is impossible to estimate the closeness between ψ^N and ψ. One may judge indirectly about the approach of ψ^N to ψ from how well ψ^N satisfies the Schrödinger equation. A measure of the divergence can be found by solving the *gap*

$$\sigma^N = (H - E)\psi^N \,. \tag{3.1.1}$$

A universal method of searching for ψ^N is the Bubnov–Galerkin method (Mikhlin 1966, 1970; Turchin et al 1970). In this method ψ^N is constructed in the form of a superposition of known auxiliary linearly independent functions χ_α from a complete set $\{\chi_\alpha\}$:

$$\psi^N = \sum_\alpha^N \Psi_\alpha^N \chi_\alpha \,, \tag{3.1.2}$$

where the coefficients Ψ_α^N are found by minimizing the gap σ^N, i.e., by making it orthogonal to the first N functions χ_α[1]:

$$(\sigma^N, \chi_\alpha) = 0 \,, \quad \alpha = 1, 2, \ldots, N \,. \tag{3.1.3}$$

In the one-dimensional case the scalar product (3.1.3) represents the integral $\int \chi_\alpha^*(r)\sigma^N(r)dr$. This method is applicable to problems involving an arbitrary number of dimensions and particles (Chaps 6, 7). Depending on the form of interaction and the basis functions χ_α, (3.1.3) may be algebraic, differential, integral, or integrodifferential. The methods of functional analysis allow treating all these strongly differing equations in a unique manner. The principal points in the solution of general problems can be grasped by studying the one-dimensional case.

Often convergence of the trial function $\psi^N = \sum_\alpha^N \Psi_\alpha^N \chi_\alpha$ to the exact solution ψ is confused with convergence of the expansion $\psi = \sum_\alpha^\infty \Psi_\alpha \chi_\alpha$. Actually, $\Psi_\alpha^N \neq \Psi_\alpha$. Thus, in bound states ψ belongs to the L_2 space of square-integrable functions, and for convergence of the expansion of ψ it is sufficient that the set $\{\chi_\alpha\}$ be complete in L_2, while the convergence $\psi^N \to \psi$ still requires proof. One

[1] We assume that χ_α exactly satisfy the boundary conditions, imposed on ψ. Otherwise, the gap must also contain the contribution of uncertainties due to the boundary conditions (see the generalized Bubnov–Galerkin method in Sect. 6.4).

may take advantage of the following theorem. Let the solution of the Schrödinger equation be reduced to an infinite set of algebraic equations for Ψ_α:

$$\Psi_\alpha + \sum_\beta^\infty a_{\alpha\beta}\Psi_\beta = j_\alpha \,. \tag{3.1.4}$$

At the same time Ψ_α^N are determined from the reduced set of N equations[2]

$$\Psi_\alpha^N + \sum_\beta^N a_{\alpha\beta}\Psi_\beta^N = j_\alpha \,. \tag{3.1.5}$$

The solution (3.1.5) tends towards (3.1.4) with respect to the space norm of number sequences l_2 if

$$\sum_{\alpha\beta}^\infty a_{\alpha\beta}^2 < \infty \,; \quad \sum_\alpha^\infty j_\alpha^2 < \infty \,; \tag{3.1.6}$$

and (3.1.4) has a unique solution in the l_2 space for any arbitrary $j_\alpha \in l_2$. The squared modulus of $\{\Psi_\alpha\}$ in l_2 is defined as $\sum_{\alpha=1}^\infty \Psi_\alpha^2$ while the convergence $\{\Psi_\alpha^N\} \to \{\Psi_\alpha\}$ is defined as $\lim_{N\to\infty} \sum_{\alpha=1}^\infty (\Psi_\alpha - \Psi_\alpha^N)^2 \to 0$ where $\Psi_{\alpha>N}^N = 0$.

It might seem that in this case in order to increase the accuracy of the solution it would be sufficient to increase the number of terms in the trial function $\psi^N = \sum_\alpha^N \Psi_\alpha^N \chi_\alpha$. However, sometimes this leads to exactly the opposite effect. The important point is that only the condition of indicated convergence is insufficient. On the one hand, as N increases the uncertainty related to the limited set of N functions, χ_α decreases, but on the other hand, the uncertainty due to computations using large matrices, such as calculating their reciprocals, increases. For a given N, the second tendency may turn out to be dominant.

The stability of a method relative to uncertainties depends on which functions $\{\chi_\alpha\}$ are chosen to be the basis. The sufficient condition for obtaining a result arbitrarily close to the truth when increasing N as well as the precision of $a_{\alpha\beta}$ and j_α and the accuracy of numerical solutions of finite sets of equations (3.1.5), is that the set $\{\chi_\alpha\}$ must be *strongly minimal.* A set $\{\chi_\alpha\}$ is said to be strongly minimal if the eigenvalues v_i^N of its Gram matrix in the energy metric $\{\int \chi_\alpha^* H \chi_\alpha dr\}$ have a lower positive constant limit as $N \to \infty$ (Mikhlin 1966, 1970).

The "conditionality" number (Faddeev, Faddeeva 1969) of the coefficient matrix of (3.1.5) serves as the measure of how many times larger the uncertainties of the solution Ψ_α^N of (3.1.5) in comparison with those of $a_{\alpha\beta}$ and j_α are. For symmetric matrices this number coincides with the Todd number, equal to $\max|v_i|/\min|v_i|$ where v_i are the eigenvalues of the matrix.

This discussion concerns mainly direct problems. For inverse problems the situation is complicated by the fact that their formulation is often ill-posed and

[2] Inhomogeneity in the equations for Ψ_α (on which j_α depend) may arise due to inclusion of incident waves in the asymptotic conditions for ψ.

hence, regularization of the procedure for finding V is necessary (Zhigunov et al 1983; Tikhonov, Arsenin 1979; Turchin et al 1970).

Correctness. Consider the operator A which transforms V from the set V into S from the set S($\mathrm{A}V = S$). We want to find V from S. Following *Hadamar* (1932), the problem is considered correct if for any $S \in \mathbf{S}$ the solution V; 1) exists in V; 2) is unique in V; 3) is stable in V, i.e., depends continuously on $S \in \mathrm{S}$. Infinitely small variations δV correspond to infinitely small δS. Violation of any of these conditions renders the problem non-correct.

Tikhonov (1979) introduced a new definition of correctness, permitting regularization of problems, which are noncorrect according to Hadamar.

Let $\tilde{V}$ be a subset in $\mathrm{V}(\tilde{V} \subset \mathrm{V})$, and let $\tilde{S}$ be its mapping in $\mathrm{S}: \tilde{S} = \mathrm{A}\tilde{V}$. According to Tikhonov the problem is correct if the exact solution

1) exists and belongs to $\tilde{V}$; 2) is unique in $\tilde{V}$ for any S from $\tilde{S}$; 3) is stable (continuous) with respect to $S \subset \tilde{S}$. Thus, according to Tikhonov correctness is achieved by reducing the size of the set in which V is sought. When $\tilde{V} = \mathrm{V}$; $\tilde{S} = \mathrm{S}$, Tikhonov's and Adamar's definitions coincide.

The formulation of the problem may be regularized in another way, by introducing a so-called quasisolution in place of the exact one. The quasisolution is an element $V \in \mathrm{V}$ for which the distance from $\mathrm{A}V$ to S is minimized (i.e., the closest element from S of the set $\mathrm{A}\tilde{V} = \tilde{S}$ to S, and S may lie outside of $\tilde{S}$). The use of quasisolutions eliminates difficulties related to requirements 1) and 3) of the Tikhonov definition: 1) may lead to an overdetermination of the problem, and 3) implies that S, together with its uncertainties, belongs to $\tilde{S}$, while sometimes the criteria of this are themselves unstable (Ivanov 1978).

Bearing in mind the complexity of regularization of inverse problems it is useful, at the beginning, to consider the solution of the most simple example of a noncorrect problem—computation of a derivative.

Let δy characterize the width of the corridor of errors for a certain function y (Fig. 3.2). For an arbitrarily small δy the derivative y' is not defined: all values of y' are allowed, since y can assume an arbitrarily slope at the given point without leaving the corridor. Nevertheless, in the finite-difference approximation y' is readily found due to the narrowing of the class of functions y. Indeed, the ends of the straight-line interval, the slope of which determines the finite-difference derivative

$$y'(n) = [y(n+1) - y(n)]/\Delta ,$$

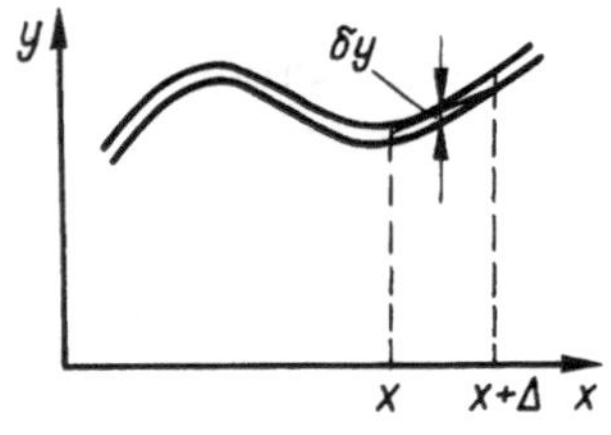

Fig. 3.2. Regularization of the noncorrectly formulated problem of defining the derivative of the function $y(x)$

are limited by the walls of the corridor (where Δ is the step). The accuracy of the calculation of y' can be increased infinitely if we are allowed to decrease δy. As it is often done in solving inverse problems, it is only necessary to decrease the regularization parameter (which is Δ in the example considered) in conjunction with δy. For each given δy there exists an optimal Δ. Obviously a step that is too small enhances the instability. The art of regularizing inverse problems lies in finding this optimum.

The subtle correlation between the choice of the round-off error and of the step Δ of the finite-difference approximation required for the numerical solution of differential equations is illustrated in Fig. 3.3 (May, Noye 1984). The absolute maximum error accumulated on a fixed interval of integration has a minimum at some optimal value of Δ (Fig. 3.3a). The smaller the Δ is, the better the finite-difference approximation, but the number of steps on the interval increases causing the round-off error to increase. The results of numerical solutions of an ordinary differential equation using Euler's method with different Δ-values on two different computers (CDC CYBER with 14 significant digits and a Hitachi PEACH minicomputer with 7 are compared in Fig. 3.3b).

3.3 Solution of Inverse Problems Using Bargmann Potentials

The variety and flexibility of exactly solvable models were demonstrated in Chap. 2. One can approximate scattering data (spectral data) for arbitrary one-particle, one-dimensional systems by choosing free parameters of S for Bargmann models. Since the same parameters define the explicit form of V^{B},

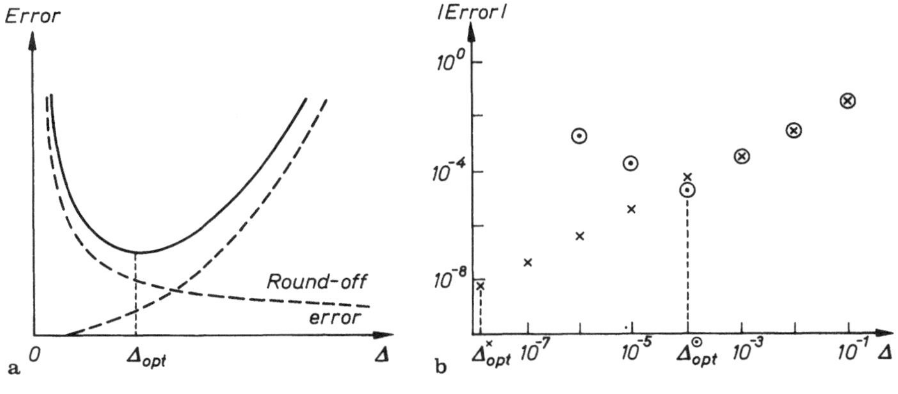

Fig. 3.3a. Qualitative behavior of the maximum error (*solid line*) as a function of the step value Δ for integration of an ordinary differential equation on a fixed interval **b** Numerically determined absolute errors and optimal Δ-values with two different values of round-off errors: points are calculated with 7 and stars with 14 significant digits

these potentials can be considered to be approximate solutions of the inverse problem.

Such calculations were performed for the most varied systems: inter-quark, nucleon–nucleon, α–α, nucleon–nuclear and other interactions [see the bibliography in *Zakhariev* et al (1982)]. However, practically no judgement on the quality of such solutions can be made without checking the method on examples involving known interactions.

Reconstruction of a Priori $V(r)$ from $S(k)$. The scattering parameters of the Bargmann model for approximating the S-function of a desired potential $V(r)$ can be fitted as follows (Malyarov et al 1973). The initial S-function is set equal to an approximation in the form of a rational function at $2n$ fixed points $k = k_s(s = 1, 2, \ldots, 2n)$:

$$S^{\mathrm{B}}(k_s) = \prod_{i=1}^{n} \frac{k_s - \alpha_i}{k_s - \beta_i} \equiv \frac{k_s^n + A_1 k_s^{n-1} + \ldots + A_n}{k_s^n + B_1 k_s^{n-1} + \ldots + B_n} = \mathrm{e}^{2\mathrm{i}\delta(k_s)} = S(k_s)\,. \tag{3.2.1}$$

This yields $4n$ linear algebraic equations for the imaginary and real parts of the coefficients A_j, B_j. From these coefficients we determine $\alpha_j(\beta_j)$ as zeros of the polynomial in the numerator (denominator) of the fraction in (3.2.1). The $S^{\mathrm{B}}(k)$ thus obtained immediately yields the potential $V^{\mathrm{B}}(r)$ as well as the wave functions ψ^{B}, since the exact expressions for V^{B}, ψ^{B} in the model are known through the set $\{\alpha_j, \beta_j\}$. A somewhat different method of determining $S^{\mathrm{B}}(k)$ has been discussed by *Lambert* et al (1975).

For computer calculations (Zakhariev et al 1983) potentials of various forms without bound states were chosen. The function $S(k)$ was found by numerical solution of the direct problem. The Bargmann S-function $S^{\mathrm{B}}(k)$ was chosen in the form

$$S^{\mathrm{B}}(k) = \prod_{j}^{N} \frac{(k - \alpha_j)(k + \alpha_j^*)(k + \beta_j)(k - \beta_j^*)}{(k - \beta_j)(k + \beta_j^*)(k + \alpha_j)(k - \alpha_j^*)}\,, \tag{3.2.2}$$

to take explicitly into account the fact that the zeros (poles) of the S-functions are situated symmetrically relative to the imaginary k-axis, and that to these zeros (poles) there correspond poles (zeros) lying symmetrically with respect to the real k-axis. The following potential corresponds to $S^{\mathrm{B}}(k)$ in the form of (3.2.2):

$$V^{\mathrm{B}}(r) = \frac{2\mathrm{i}\hbar^2}{m}\frac{d}{dr}\left\{\sum_{j}^{2N} \beta_i \frac{\alpha_j - \beta_j}{\alpha_j + \beta_j} f(\beta_j, r)\mathrm{e}^{\mathrm{i}\beta_j r}\right\}, \tag{3.2.3}$$

where $f(\beta_j, r) = \sum_i^{2N} (\mathbf{I} - 2P(r))_{ji}^{-1}\,\mathrm{e}^{\mathrm{i}\beta_j r}$; $\mathbf{I}$ is the unit matrix and

$$P_{ij} = -\frac{\beta_j}{\beta_j + \beta_i}\frac{\alpha_j - \beta_j}{\alpha_j + \beta_j}\mathrm{e}^{\mathrm{i}(\beta_j + \beta_i)r}\,. \tag{3.2.4}$$

Typical examples of comparison of the initial $V(r)$ and the reconstructed $V^{\mathrm{B}}(r)$ are presented in Fig. 3.4.

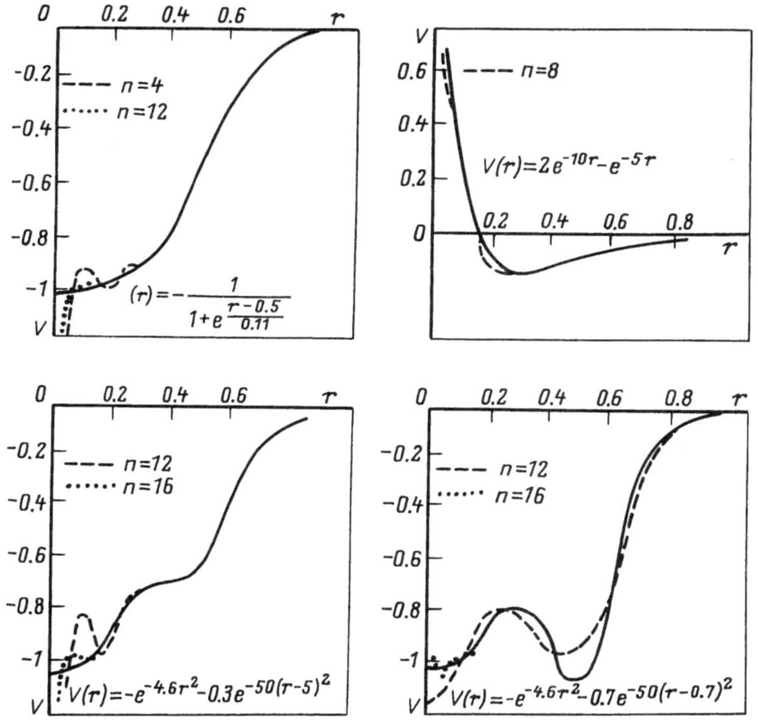

Fig. 3.4. Reconstruction of potentials by approximation of the scattering functions $S(k)$ with a model rational $S^B(k)$ with different numbers of poles n. *Solid lines*—initial potentials; *dotted* and *dashed lines*—determined approximate potentials. From (Zakhariev et al 1983)

Reconstruction of Infinitely Deep Potential Wells. A group of theorists from the Fermi Laboratory at Batavia (USA), in seeking a method for reconstruction of quark potentials, made a surprising observation. It turned out that when the inverse problem is solved for infinitely deep wells, the lower part of these wells is reproduced well with the aid of reflectionless Bargmann wells of finite depth (Thaker et al 1978; Schonefeld 1980). In fact, in spite of the essentially differing natures of the upper parts of the initial and the approximating V^B potentials V and V^B, the coincidence of the lower levels was sufficient for achieving $V^B(r) \approx V(x)$ in the same energy region. Figure 3.5 shows the approximate reconstruction of parabolic (harmonic oscillator), linear, and rectangular wells from the positions of the first eight lower levels (Schonefeld 1980). Indeed, to obtain such a good approximation, it is necessary to make a special choice of the position of the lower limit $E_{\min}^{\text{cont}}$ of the continuous spectrum for the transparent potential V_N^B. In Fig. 3.6 it is shown how for $N = 4$ the variation of $E_{\min}^{\text{cont}}$ influences the form of $V_N^B(x)$ when the positions of the levels of bound states are

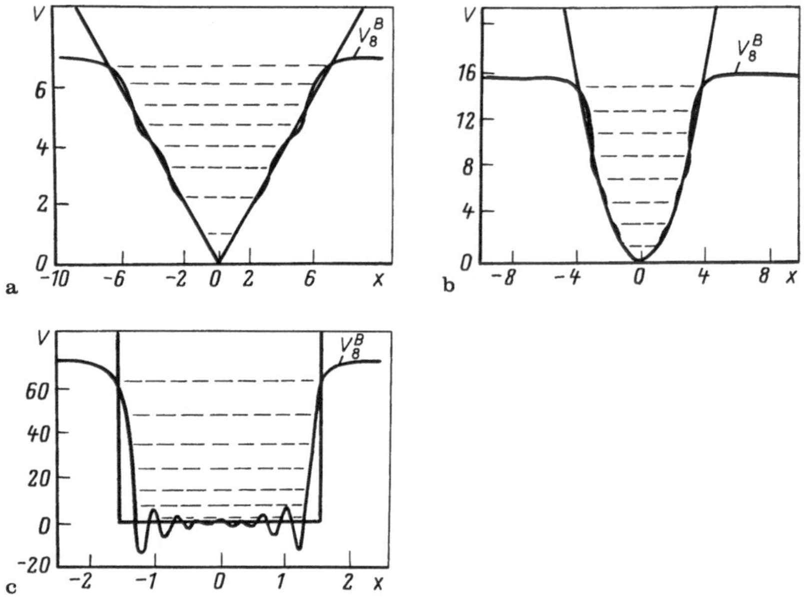

Fig. 3.5a–c. Reconstruction of linear (**a**), parabolic (**b**) and rectangular (**c**) infinite potential wells from eight lower levels using Bargmann potentials of finite depth. From (Schonefeld et al 1980)

fixed (Thaker et al 1978). The figure shows that value of $E_{\min}^{\text{cont}}$ at which oscillations in $V_N^{\text{B}}(x)$ are minimized. A curious trend can be observed: for different initial potentials and numbers of levels N in V^{B} the stabilizing parameter $E_{\min}^{\text{cont}}$ must be chosen to lie in the middle between the Nth and $(N+1)^{\text{th}}$ levels in V.

In solving the inverse problem *Thaker* et al (1978) made use of the fact that for reconstruction of symmetric potentials $V(x) = -V(x)$, it suffices to know only the position of the energy levels without knowledge of the normalizing constants. Figure 3.7 demonstrates how the quality of the approximation $V_N^{\text{B}}(x)$ of a parabolic well $V(x)$, becomes better with increasing N. The reconstruction of an infinite rectangular potential well turned out to be significantly worse than for the parabolic and linear wells. This is the usual consequence of sharp discontinuities in the shape of $V(x)$.

Numerous calculations similar to those of *Thaker* et al (1978, 1980), for potential wells of both finite and infinite depth have been done by *Asthana* and *Kamal* (1983). These authors also compare their reconstructed potentials with the results of *Thaker* et al (1978, 1980).

Schonefeld et al (1980) were the first to attempt to analyze the convergence of the sequence of approximations of transparent $V_N^{\text{B}}(x)$ to infinitely deep $V(x)$

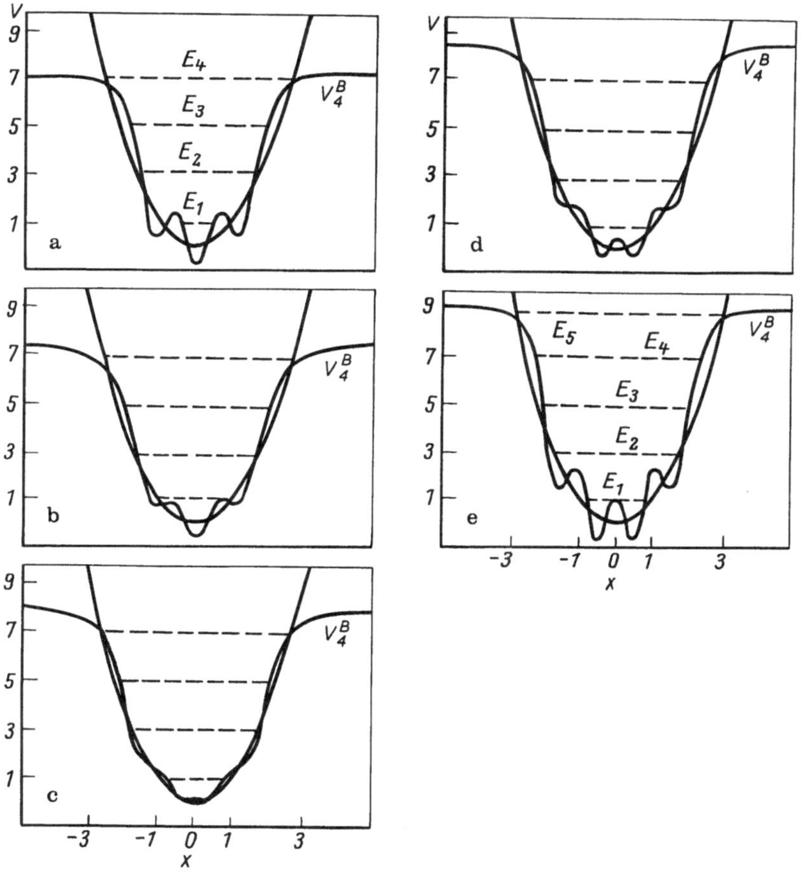

Fig. 3.6a–e. Dependence of a Bargmann potential V^B sharing four common levels with a parabolic well on the position of the edge E_{min}^{cont} of V^B. The edge of V^B rises smoothly from the position of the fourth level, E_4 (**a**) to E_5 (**e**); the choice of the middle position (**c**) regularizes the solution of the inverse problem (Thaker et al 1978)

with the enhancement of the number of levels N coinciding both in V and V_N^B. The deviation

$$\delta V(0) = V(0) - V_N^B(0)$$

was considered at only one point $x = 0$. In Fig. 3.7 it is shown that the relative uncertainty

$$\delta V(0)/(\Delta E_N),$$

where $\Delta E_N = E_N - E_1$ is the excitation energy of the uppermost N^{th} level in $V^B(x)$, tends towards zero as N increases. The absolute uncertainty $V(0)$

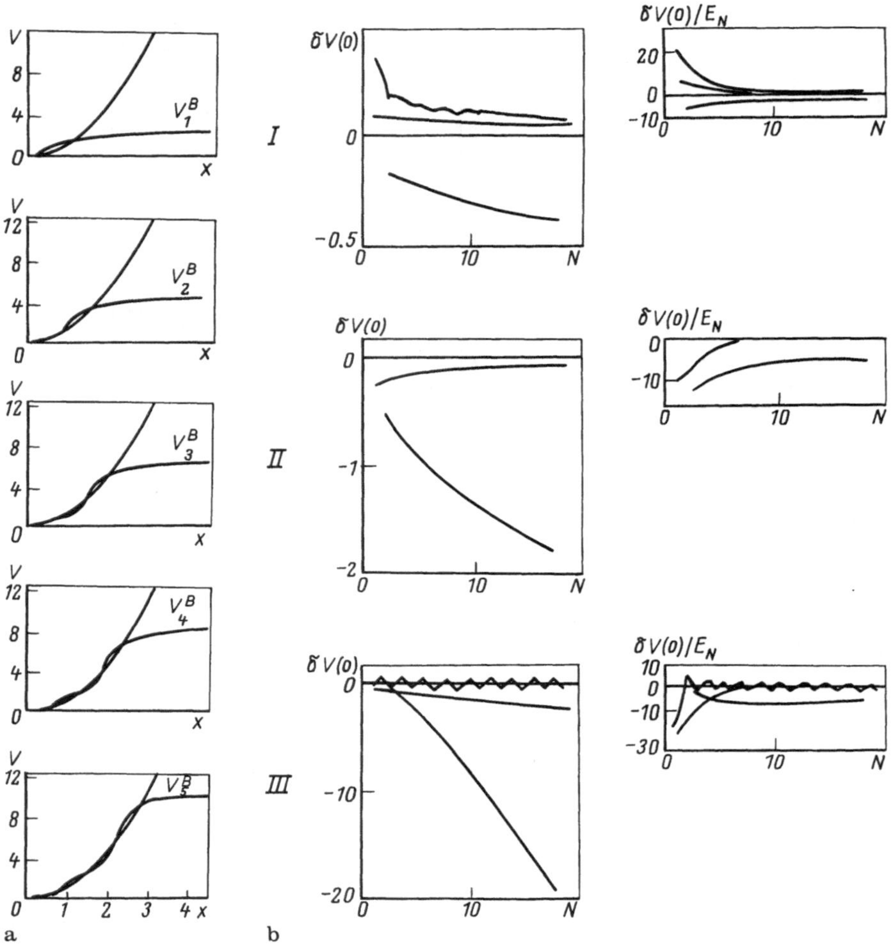

a b

Fig. 3.7a,b. Approximate reconstruction of $V(x)$ with the aid of reflectionless potentials of finite depth with $N = 1, 2, \ldots, 5$ lower levels. **a** reconstruction of a parabolic well; **b** dependence of the uncertainties of the approximation on the number of levels N for three positions of the edge of a well of finite depth for linear (I), parabolic (II), rectangular (III) wells. From (Thaker et al 1978)

increases when the choice of $E_{\min}^{\text{cont}}$ is wrong, but decreases for a harmonic oscillator potential and for a linear well when the stabilizing parameter is close to its optimal value. In the case of a rectangular well, even when $E_{\min}^{\text{cont}}$ is optimized, the uncertainty $V(0)$ oscillates as N varies, without damping, but with a small amplitude. *Stefanescu* (1982) later devoted a special study to the convergence of the approximation of this method.

Approximate reconstruction of $V(x)$ with a purely discrete spectrum can be performed by using another infinite well for the base potential $\mathring{V}(x)$. Constructing supplements to V of the Bargmann type $\Delta V(N)$ (Sect. 2.3) so as to make N lower levels coincide both in $\mathring{V} + \Delta V_N$ and V and to equate the respective normalizing constants, one may expect $\mathring{V}(x) + \Delta V_N(x)$ to approach $V(x)$ in shape as N increases. The conditions of convergence for this method was investigated by *Adamyan* (1981, 1982).

One example of such a solution to the inverse problem was devised by *McWilliams* (1979). He used a linear well $\mathring{V}(r) = \alpha r, \alpha > 0$ and attempted to reconstruct $V(r)$:

$$V(r) = \alpha r - \beta \exp(-r/r_0) .$$

In Fig. 3.8 it is shown how $\mathring{V}(r) + \Delta V_N(r)$ approaches $V(r)$ as N increases. This method was applied by the author for finding the interquark potential (see also Adamyan 1981, 1982).

The first model example of numerical reconstruction of the potential shape from spectral data in the framework of the R-matrix scattering theory using the Bargmann-type solutions was given by *Amirkhanov* et al (1989).

The initial potential well was chosen with an inclined linear bottom. It was reconstructed from its spectral R-matrix parameters $\{E_\nu, \gamma_\nu\}$ calculated numerically. Figures 3.9a–d demonstrate the shapes of the Bargmann-type $V_n^{\mathrm{B}}(r)$ potentials built of $n = 2$, 6, 12 and 18 pairs of spectral parameters $\{E_\nu, \gamma_\nu\}$ for the lowest R-matrix states.

The accuracy of inverse problem solutions does not increase monotonically with the number n and a fixed step Δ of the finite-difference approximation for numerical determination of input spectral parameters $\{E_\nu, \gamma_\nu\}$. Both the approximately reconstructed potentials deviate more from $V(r)$ with $n = 18$ than $n = 6$. Thus, more accurate numerical determination of $\{E_\nu, \gamma_\nu\}$ is required to get better results with $n = 18$ than with $n = 6$.

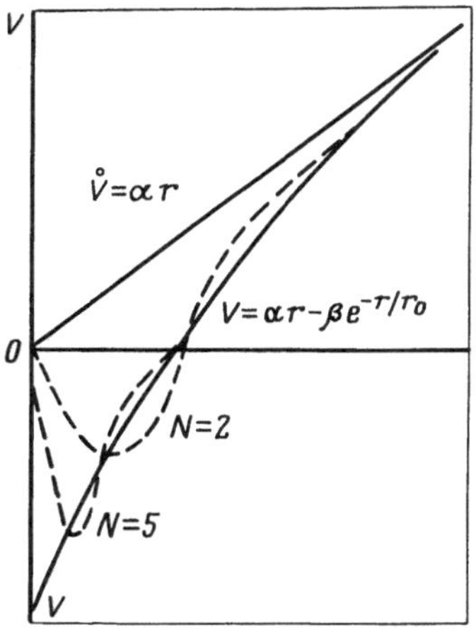

Fig. 3.8. Convergence of successive approximations: N lower levels of an auxiliary linear well $\mathring{V}(r)$ were shifted until they coincided with the levels of the desired potential $V(r)$ and the normalizing constants adjusted correspondingly. From (McWilliams 1979)

Methods of solution at fixed energy. The approach with $E = \text{const}$ has made a fundamental contribution to methods of approximate calculations. The reconstruction of a priori potentials of various forms, obtained by applying the well-known Newton–Sabatier method have already been reported (Chadan, Sabatier 1989). More detailed and improved calculations have been presented by *Coudray* (1979).

As one may see from Fig. 3.10, the quality of the reconstructed $V(x)$ becomes better at higher values of the fixed energy E. Especially subtle details of the behavior of $V(r)$ can be revealed only at $E = 400$ MeV. In addition, it is necessary to utilize scattering data for a sufficiently large number of angular momenta l (up to 30). Convergence to the form of the initial potential as the number of included partial phase shifts increases ($l < L = 1, 10, 20$), is shown in Fig. 3.11. It is remarkable that in practice the Newton–Sabatier works quite well, although in general the phase shifts at a fixed energy do not uniquely define the potential V. The ambiguity manifests itself mainly at small r and in the form of oscillations of the tail of V (in the class of rapidly falling potentials the solution is unique).

Munchov and *Scheid* (1980) proposed a modification of the Newton–Sabatier method for potentials with known behavior for $r > a$. Restriction of the region $[0, a]$ where the potential is to be reconstructed led to the numerical solution of

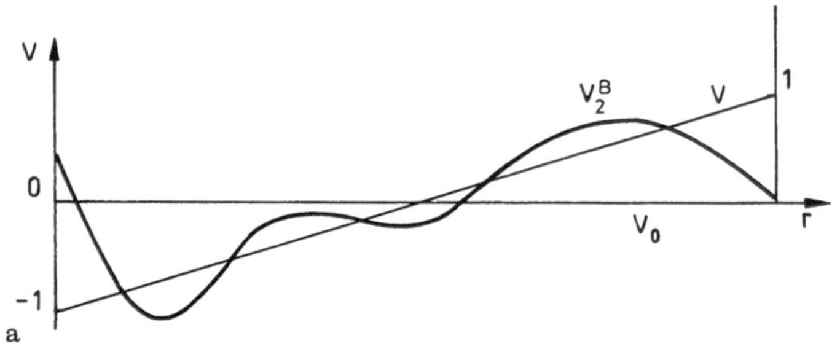

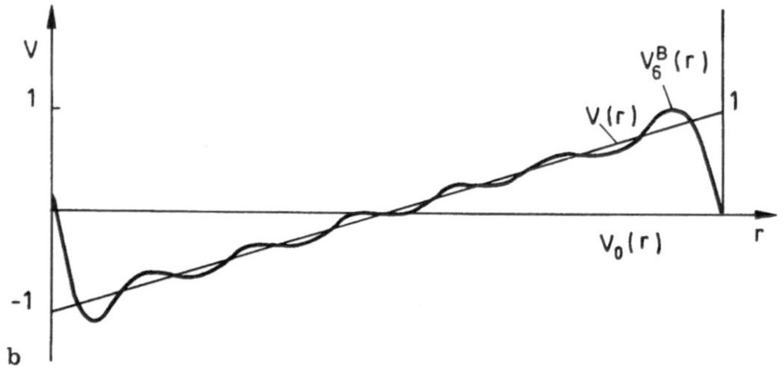

Fig. 3.9a,b (*continued*)

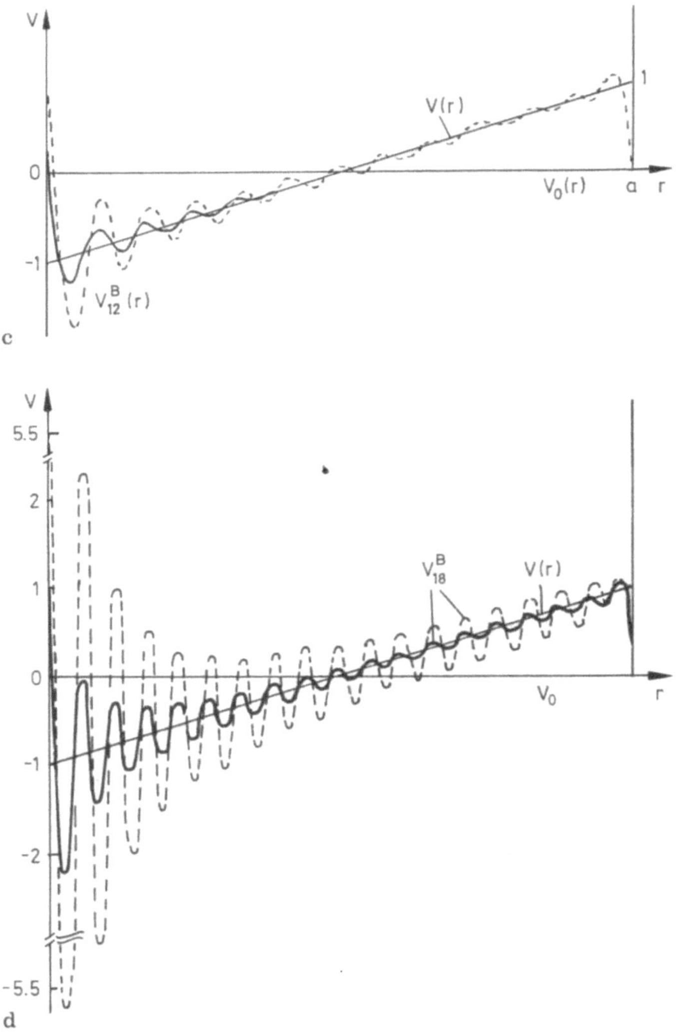

Fig. 3.9a–d. Reconstruction of the infinite well $V(r)$ with the inclined bottom in (**a**) 2-, (**b**) 6-, (**c**) 12-, (**d**) 18-level approximation. The input spectral parameters $\{E_\nu, \gamma_\nu\}$ were calculated numerically in the finite-difference approximation with steps Δ (*dashed lines*) and $\Delta/2$ (*solid wavy lines*). As a reference potential $V_0(r)$, the rectangular well was used. Bargmann potentials with n shifted levels E_ν and reduced widths γ_ν were used for the reconstructed potentials $V_n^B(r)$. Solid and dashed lines coincide in (**a**), (**b**)

the inverse problem becoming more stable. This made the requirements on the scattering data less stringent. It lowered the allowed energy and the number of initial phase shifts δ_l. Figure 3.12 demonstrates the resulting gain. The potential constructed in this new way for a well of rectangular form and finite depth is

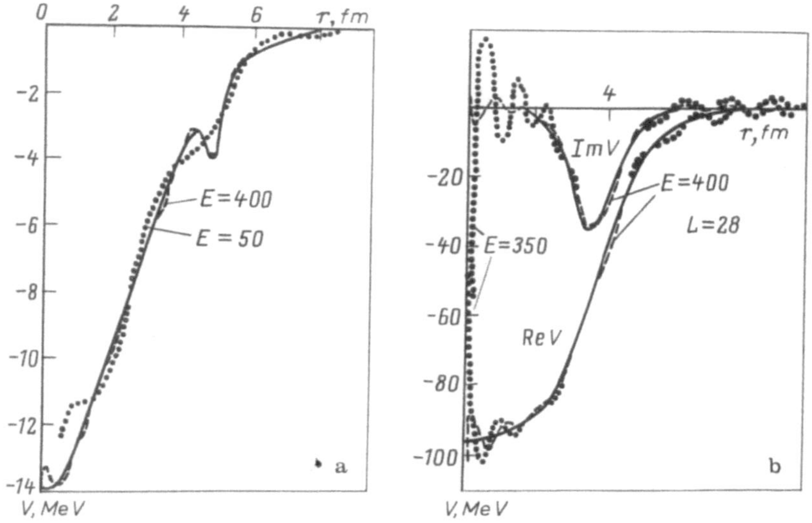

Fig. 3.10a,b. Reconstruction of a potential by the Newton–Sabatier method. The approximation gets better as the energy at which the initial scattering phases are taken increases. **a** $V(r)$ is real; **b** $V(r)$ is complex. From (Coudray 1979)

compared with the potential obtained from the previous calculations by *Coudray* (1979). Better agreement with the sought V is obtained even though the number of phases δ_l used is smaller ($l \leqslant 20$ instead of $l \leqslant 28$). In Fig. 3.12b the imaginary and real parts of the optical potential reconstructed from the phase shifts with $l \leqslant 13$ at $E = 14$ MeV are presented (Munchov, Scheid 1980). The corresponding results for the general Newton–Sabatier method yield strong deviations from the initial potential even when $E = 350$ MeV and $l \leqslant 28$ (Fig. 3.10b).

Following *Hron* and *Razavy* (1977), *Baldock* et al (1981) have analyzed alpha–alpha scattering. In test calculations with a known $V(r)$ they obtained quite good agreement at 20 MeV and for $l \leqslant 15$ (Fig. 3.13, dashed curve). For suppression of residual oscillations of the reconstructed potential they proposed to approximate it by a smooth curve at small r, thus further improving the quality of the solution (*dotted line* in Fig. 3.13). In this form the method already seems quite acceptable for practical applications.

The $E =$ const approach was used to find potentials for a number of nuclear systems by choosing parameters of the rational approximation of the S-function in the complex λ-plane (Malyarov et al 1975). The corresponding control calculations with a known $V(r)$ were not performed, however. *Lipperheide* et al (1981) consider it more practical to use their new exactly solvable models with the nonrational function $S(\lambda)$. Thus, for example, the new model potentials contain no nonphysical singularities, as may happen with V^{B} (Lipperheide et al

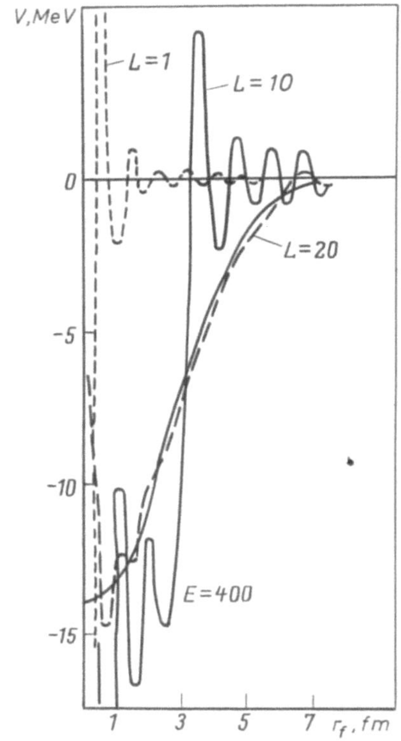

Fig. 3.11. Convergence of the approximations by the Newton–Sabatier method as the number of initial phase shifts L increases at a fixed energy (Coudray 1979)

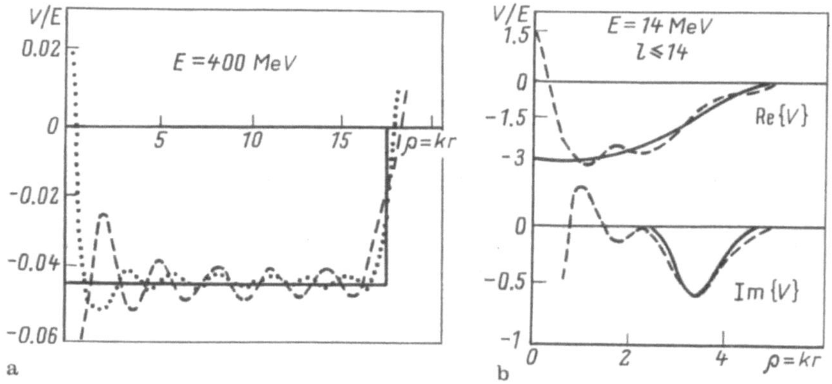

Fig. 3.12a,b. Reconstruction of potentials by the modified Newton–Sabatier method. The quality of approximation increases when the region in which the potential is reconstructed $r \leqslant a$ is decreased. **a** Reconstruction by Munchov et al of rectangular well from 20 phases (. . . . . . .) and calculations by the usual Newton–Sabatier method from 28 phases (– – –); **b** a complex potential is reconstructed from scattering data at the low energy $E = 14$ MeV, unattainable by the conventional method (Coudray 1979)

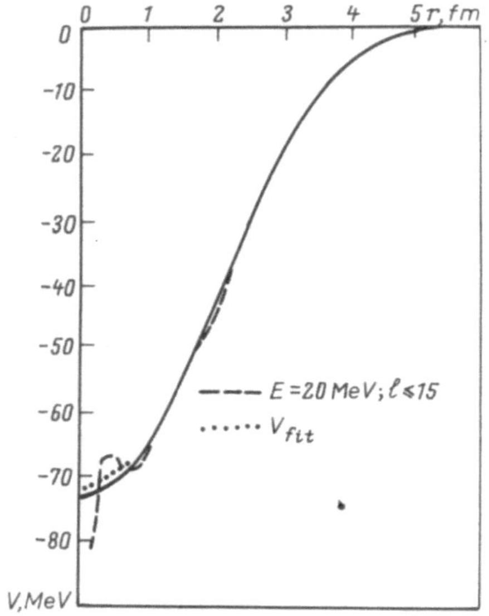

Fig. 3.13. Reconstruction of the alpha–alpha potential by the Munchov–Scheid method (Baldock et al 1981): *Dashed line* $E=20$ MeV, $l \leqslant 15$; *dotted line* smoothed out curve; yields an even better approximation

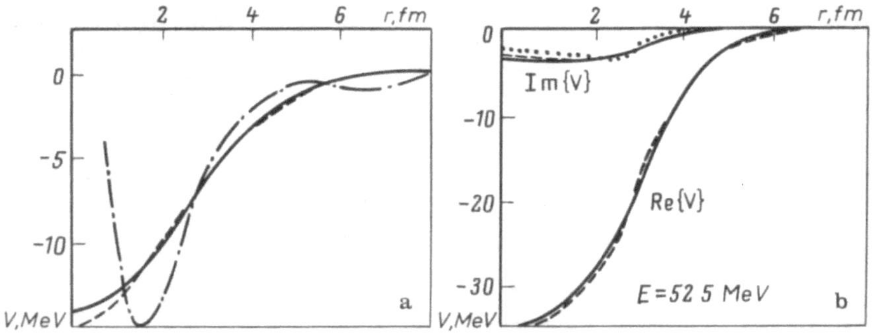

Fig. 3.14a,b. Reconstruction of potentials by the Lipperheide–Fidelday method (– – –). **a** comparison with the Newton–Sabatier method (Coudray 1979) (– . –) **b** complex $V(r)$, good agreement is obtained at relatively low energy by using only three pairs of parameters α_j, β_j (Lipperheide et al 1982)

1978). In Fig. 3.14 the results of the new method of reconstructing potentials are shown. Quite good agreement is obtained already for only three pairs of parameters α_j, β_j. Many similar results are shown by *Lipperheide* et al (1984).

Determination of the potential from (2.5.26) requires calculation of determinants. To avoid uncertainties due to the subtraction of large numbers, *Lipperheide* et al (1981) consider it better to make use of the following iteration procedure. First one finds the first correction $\Delta V_1(r)$ to the initial potential (which may be assumed to be equal to zero), corresponding to a single pair α_1, β_1 of λ values:

$$V_1(r) = \mathring{V}(r) + \Delta V_1(r) = \mathring{V}(r) - \frac{\hbar^2}{mr}\frac{d}{dr}\left[r\frac{d}{dr}\ln\mathring{\chi}(r)\right], \tag{3.2.5}$$

where

$$\mathring{\chi}(r) = W[\mathring{\varphi}_{\beta_1}, \mathring{\varphi}_{\alpha_1}]/(\beta_1^2 - \alpha_1^2)\,.$$

Then, the solutions are determined numerically with the new potential $V_1(r)$ utilized as the initial one. The second pair of λ values α_2, β_2 is introduced and the second correction $\Delta V_2(r)$ is found and so on.

In the next section, attention is drawn to the intrinsic possibility of developing methods for the reconstruction of potentials corresponding to widely used solution methods of the direct problem.

3.4 Approximation of Potentials by Steps, at Discrete Points, and by Splines; the Role of the Upper Part of the Spectrum

3.4.1 Piecewise-constant Potentials

The simplest examples of piecewise-constant (step-like) potentials—rectangular barriers and wells—serve as especially popular educational models, used in practically all textbooks of quantum mechanics.

Potentials composed of a sequence of rectangular steps (Fig. 3.15) are also convenient for approximate calculation of the one-dimensional Schrödinger equation. The applicability of this method does not decrease as the energy increases, which is a definite advantage over the finite-difference approximations.

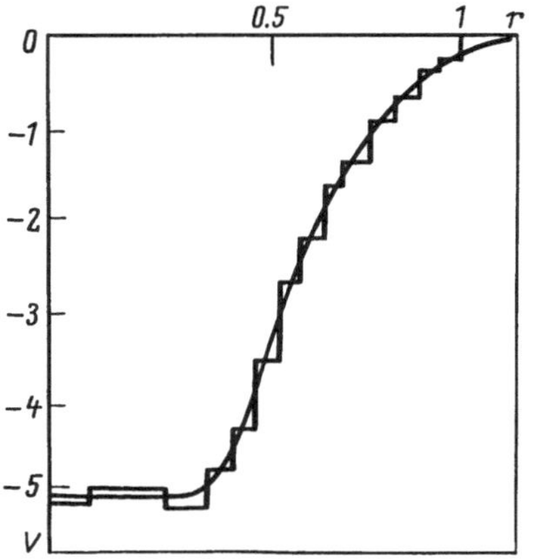

Fig. 3.15. Reconstruction of a potential in the piecewise-constant approximation. From (Zakhariev et al 1977)

The Direct Problem. An important merit of piecewise-constant potentials is the simple explicit form of the wave function on each of its steps. The general solution of the Schrödinger equation on the ith step $r_{i-1} \leqslant r \leqslant r_i$ where $V(r) = V_i = \text{const}$ is the sum of any two linearly-independent partial solutions with arbitrary coefficients A_i, B_i, for example:

$$\psi(r_{i-1} \leqslant r \leqslant r_i) = A_i \sin k_i r + B_i \cos k_i r \; ;$$
$$k_i^2 = 2m(E - V_i)/\hbar^2 \tag{3.4.1}$$

for $E \geqslant V_i$. Function ψ is expressed similarly when $E < V_i$, only the sine and cosine are replaced by increasing and decreasing exponentials (or by $\sinh \kappa_i r$ and $\cosh \kappa_i r$, where $\kappa_i = -ik_j$).

From the continuity condition for the wave function and its derivative we obtain a chain of recursion relations for the constants A_j and B_j:

$$A_i \sin k_i r_i + B_i \cos k_i r_i = A_{i+1} \sin k_{i+1} r_i + B_{i+1} \cos k_{i+1} r_i \; ;$$
$$A_i k_i \cos k_i r_i - B_i k_i \sin k_i r_i = A_{i+1} k_{i+1} \cos k_{i+1} r_i - B_{i+1} k_{i+1} \sin k_{i+1} r_i \, . \tag{3.4.2}$$

By applying these equations with different i in succession one solves the direct problem, i.e., one finds $\{A_j, B_j\}$. Various kinds of boundary conditions at one or at both edges of the integration interval are taken into account in the same way as in the case of the finite-difference Schrödinger equation.

The Inverse Problem. In this problem, besides A_j, B_j, it is also necessary to find V_j on each step. Unlike the elementary difference methods, here the values of the potential V_j enter into recursion relations nonlinearly: they occur in the arguments of the sine, cosine (exponentials, sinh, cosh) through $k_j = \sqrt{2m(E - V_j)}/\hbar$. This substantially complicates the search for V_j.

Now let us consider a single stage in the solution of the inverse problem in the case of a purely discrete spectrum; it will then be clear how the remaining stages are to be done and how to generalize the algorithm to systems with a continuous spectrum. We shall consider the known eigenfunctions u and their derivatives at the point $r_N = a$. Joining the solution $u_\nu(r)$ in the form (3.4.1) and its derivative $u'_\nu(r)$ on the Nth step $r_{N-1} \leqslant r \leqslant r_N = a$ together with $u_\nu(a)$ and $u'_\nu(a)$ we have, similarly to (3.4.2),

$$A_N^\nu \sin \sqrt{2m(E_\nu - V_N)} \frac{a}{\hbar} + B_N^\nu \cos \sqrt{2m(E_\nu - V_N)} \frac{a}{\hbar} = u_\nu(a) \; ;$$

$$A_N^\nu \sqrt{\frac{2m}{\hbar^2}(E_\nu - V_N)} \cos \sqrt{\frac{2m}{\hbar^2}(E_\nu - V_N)} a$$
$$- B_N^\nu \sqrt{\frac{2m}{\hbar^2}(E_\nu - V_N)} \sin \sqrt{\frac{2m}{\hbar^2}(E_\nu - V_N)} a = u'_\nu(a) \tag{3.4.3}$$

and the same for $E < V_j$. Hence we find A_N and B_N as functions of the parameter V_N. For determining V_N we shall take advantage, as usual, of the completeness relation, that is, that the functions $u_\nu (\nu = 1, 2, \ldots)$ are orthogonal at different points of the interval $[r_{N-1}, r_N]$:

$$\sum_{\nu}^{\infty} u_\nu(r) u_\nu(r' \neq r) = 0 \,. \tag{3.4.4}$$

We substitute into (3.4.4), instead of u_ν, their expressions in the form of (3.4.1) with $E = E_\nu, j = N$, where A_N^ν and B_N^ν are expressed explicitly through V_N from (3.4.2). As a result, we obtain a nonlinear equation for V_N.

In practice it is impossible to realize summation in (3.4.4) over an infinite number of ν values. To overcome this difficulty, it was proposed by *Zakhariev* et al (1977) to subtract from (3.4.4) a corresponding equality for the known functions $\mathring{u}_\nu(r)$ of free motion $[\mathring{V}(r) \equiv 0]$ which satisfies the same boundary conditions as u_ν. At large E_ν one may neglect the difference between the functions u_ν and $\mathring{u}_\nu$ and cut off summation over ν:

$$\sum_{\nu}^{\nu_{max}} [u_\nu(r) u_\nu(r') - \mathring{u}_\nu(r) \mathring{u}_\nu(r')] = 0 \,;$$

$$r_{N-1} \leqslant r \neq r' \leqslant r_N \,. \tag{3.4.5}$$

Since the sought parameter V_N enters into the arguments of many nonlinear functions, it is impossible to express it explicitly through known quantities. Therefore, we define V_N numerically as the value at which the sum in the left-hand side of (3.4.5) becomes zero (maybe only approximately). In a similar manner we find the other V_j.

Results of computer calculations characterizing the quality of reconstruction of the potential are presented in Fig. 3.15, where the smooth curve is the initial $V(r)$.

The piecewise-constant potential V_{step} with 20 steps was obtained from 50 pairs of R-matrix spectral parameters $\{E_\nu, \gamma_\nu\}$ the first 13 of which were found by solving the direct problem with $V(r)$, while the others were taken in the quasiclassical approximation with a sole parameter $\int V(r) dr$ determined from the asymptotic behavior of E_ν. The stability of the solution is achieved as in the other methods considered above, by a finite parametrization of the potential (in this case by restricting the number of steps N, i.e., by narrowing the class of functions among which V is sought). One can also apply, additionally, the *Tikhonov* et al (1979) regularization: Determine V_j from the minimum of the function

$$\mathscr{F}(V_j) + \alpha\{V_j - V_{j+1}\}^2 \,, \tag{3.4.6}$$

where $\mathscr{F}$ is the expression occurring in the left-hand side of (3.4.5). The additional term $\alpha\{V_j - V_{j+1}\}^2$ smooths out the form of V_{step}. The coefficient α is chosen to be sufficiently small for the suppression of oscillations in the solution and not to terribly violate the completeness relation defining V_j.

An example of the functional $\mathcal{F}$ for problems with a discrete and a continuous spectrum can be given by (ψ depends on V_j):

$$\mathcal{F}(V_j) = \sum_{\nu} \psi_\nu(r)\psi_\nu(r') + \frac{1}{2\pi} \int_0^{k_{\max}} \{\psi(k,r)\psi^*(k,r') - 4\sin kr \sin kr'\} dk \,. \tag{3.4.7}$$

In another modification of this model the potential is composed of pieces corresponding to exactly solvable models not necessarily with $V_j = \text{const}$. For instance, the last step can be replaced by some potential tail (Coulomb or exponential) with an exactly known solution.

The necessity of fitting the interaction strength at each step is a disadvantage of these methods when they are compared with the elementary finite-difference approach. On the other hand, the advantage of these methods is that the reconstructed wave function is determined at all values of j, at each step it has an explicit analytic form, and it retains its physical meaning as the energy increases. It is also easier to fit the interaction parameters sequentially (one at a time) than all together, as required when reconstructing potentials with the aid of the direct problem. This will be discussed further in Sect. 3.5.

Reconstruction of the Interaction in the Finite-difference Approach

It might have been expected that the development of exact finite-difference models of the inverse problem would also immediately solve the problem of approximate reconstruction of a potential $V(r)$ with continuous coordinate dependence. If the reconstructed $V^{\text{fd}}(r_n) \approx V(r)$, then there would only remain to interpolate $V(r_n)$ at intermediate points. It turns out, however, that many difficulties arise along this way. At present only the first examples have appeared of partially successful reconstruction of $V(r_n) \approx V(r)$ in cases where $E = \text{const}$ and $l = \text{const}$. A study of these cases yields an understanding of the discrete approximation in the inverse problem:

$E = \text{Const}$ Approach by Hooshyar, Razavy (1981). Substituting $\chi_l(r) = kr^{-(l+1)}\psi_l(r) \xrightarrow[r \to \infty]{} 1$ for the wave function $\psi_l(r)$ in the radial Schrödinger equation for a local spherically symmetric field $V(r)$ of finite range a,

$$-\psi_l''(k,r) + \frac{l(l+1)}{r^2}\psi_l(k,r) + u(r)\psi_l(k,r) = k^2\psi_l(k,r) \,, \tag{3.4.8}$$

where

$$k^2 = 2mE/\hbar^2 \,; \quad u(r) = \frac{2m}{\hbar^2} V(r) \,; \quad \lim_{r \to 0} \psi_l(kr) kr^{-(l+1)} = 1 \,,$$

we get:

$$\chi_l'' + \frac{2}{r}(-l+1)\chi_l' + (k^2 - u(r))\chi_l = 0 \,. \tag{3.4.9}$$

Now we shall pass to the difference analogue of (3.4.8) making use of the symmetric formula for the first-order difference derivative $\chi_l' \to (\chi_l(n+1) - \chi_l(n-1))/2\Delta$:

$$\chi_l(n+1) = A_n \cdot B_n(l)\chi_l(n) + c_n(l)\chi_l(n-1)\,;$$

$$n = 1, 2, \ldots, N; \quad N = a/\Delta;$$

$$A_n = 2 - \Delta^2 k^2 + \Delta^2 u(n)\,; \quad B_n(l) = nl(l+1+n)\,; \tag{3.4.10}$$

$$c_n(l) = (l+1-n)/(l+1+n)\,. \tag{3.4.11}$$

Note that $c_n(l) = 0$ for $l = n-1$. We rewrite (3.4.10) for $n = N$ in the form

$$\chi_l(N+1)/\chi_l(N) = A_N B_N(l) + c_N(l)[\chi_l(N-1)/\chi_l(N)]\,. \tag{3.4.12}$$

We shall take advantage of the fact that for $L = N-1$, in accordance with (3.4.11), $c_N(L) = 0$, so the second term in the right-hand side of (3.4.12) with unknown $\chi_l(N-1)$ vanishes, and one can determine A_N related to the potential,

$$u_n = u(n\Delta) = (A_n - 2 + \Delta^2 k^2)/\Delta^2 :$$

$$A_n = [z_N(L) + N]/[NB_N(L)]\,,$$

where $B_N(L)$ is a known quantity, and $z_N(L)$ is the logarithmic derivative of function χ at the edge of the interaction region determined from the scattering data of the problem with continuous r:

$$z_N(l) = N\frac{\chi_l(N) - \chi_l(N-1)}{\chi_l(N-1)} \approx r\chi_l'(r)/\chi_l(r)\Big|_{r=a}\,. \tag{3.4.13}$$

Similarly, assuming in (3.4.10) $n = N-1$,

$$\chi_l(N)/\chi_l(N-1) = A_{N-1}B_{N-1}(l) + c_{N-1}(l)[\chi_l(N-1)/\chi_l(N-2)]^{-1} \tag{3.4.14}$$

and taking into account that $c_{N-1}^{(L-1)} = 0$ for $l = L-1 = N-2$, we find the expression for A_{N-1} [i.e., $u(N-1)$] through $B_{N-1}(L-1)$ and $z_N(L-1)$ and so on. Several examples of approximate reconstruction of potentials have been presented by *Hooshyar and Razavy* (1981). Contrary to Newton–Sabatier (Chap. 2, p. 84) the *best* agreement is achieved at low energies (Fig. 3.16) when, however, it is difficult to measure many phase shifts δ_l (as $E \to 0$, it becomes more complicated). This method was modified for the multichannel case by *Hooshyar* and *Razavy* (1984).

The $l =$ const *Approach.* In this case the approximate solution of the inverse problem is hindered by the discrepancy between the spectral properties of the difference and differential Schrödinger operators which increases with energy. This discrepancy is evident in the analytically solvable example of an infinitely deep potential well: $V(r) = 0$ for $0 < r < a$; $V(r) = \infty$ for $r \leqslant 0$, $r \geqslant a$. The levels (or resonances of the R-matrix) are determined by

$$E_\nu = \nu^2\pi^2/a^2\,; \tag{3.4.15}$$

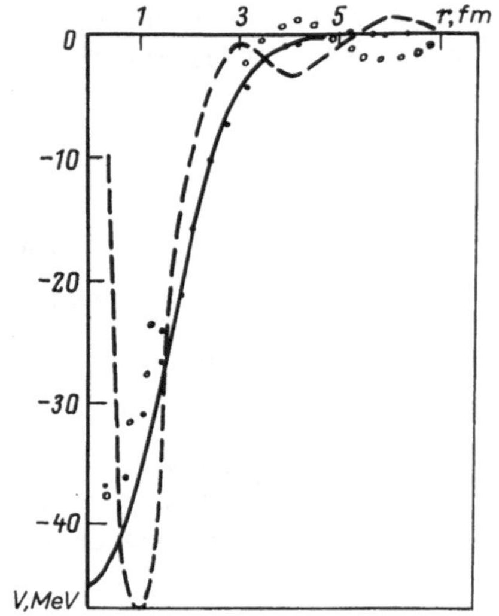

Fig. 3.16. Reconstruction of a potential in the finite-difference approximation at a fixed energy of 30 MeV (○ ○ ○) and at 2 MeV and $N=20$ (. . .) (Hooshyar, Razavy 1981). For comparison the result is shown that was obtained by the Newton–Sabatier method with $N=28$ (– – –) (Coudray 1977) at 30 MeV

for the differential operator and

$$E_\nu^{\mathrm{fd}} = [1 - \cos(\pi\nu/N)]/\Delta^2 \xrightarrow[\nu \ll N = a/\Delta]{} \nu^2\pi^2/a \tag{3.4.15a}$$

for the finite-difference operator.

In both cases only the *lower* parts of the spectra are close to each other. On the other hand, in solving the inverse problem the *whole* spectrum gets involved, owing to the utilization of the completeness of the eigenstates. The discrepancy does not vanish as Δ decreases. Although the range of energies where the properties of the difference and the continuous problems are close to each other increases with $1/\Delta$, the relative weight of the "good" part of the spectrum of the difference Schrödinger operator does not increase: as Δ decreases, the upper limit of the whole spectrum, $2/\Delta^2$, rises. Thus, the last level E_N^{fd} turns out to be $\pi^2/2 \approx 2.5$ times (!) lower than the Nth level $E_N = N^2\pi^2/a^2$ of the differential operator (Fig. 3.17a). A similar situation occurs for potentials of arbitrary form. At the same time, the initial data, even in a relatively stable model of the inverse problem, must satisfy rigorous requirements (in the course of solution the uncertainties are increased by many times).

In the search for a way out of such a situation attention was drawn to the fact that the upper bound of the good part of the spectrum (where the properties of the difference and differential equations are close to each other) rise, as Δ^{-1} increases, towards the range of energies where the perturbation theory is applicable. This promised simplification of the formulas when dealing with higher energy states. Although in perturbation theory there occur matrix elements of the potential which are unknown in the inverse problem, the first

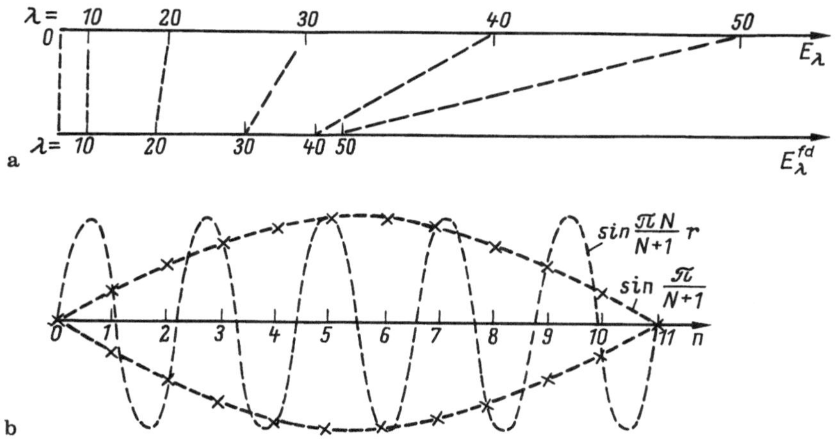

Fig. 3.17. Comparison of the spectra and wave functions of the differential and finite-difference problems: **a** spectra E_λ and E_λ^{fd} for $N = 50$; **b** eigenfunctions of the upper and lower states of the difference problem for $N = 11$ (× × ×) and the functions of the differential operator (– – –)

corrections

$$E_\nu^1 = \langle \mathring{\psi}_\nu | V | \mathring{\psi}_\nu \rangle$$

to the upper eigenvalues E_ν and E_ν^{fd} are insensitive to the details of the shape of the potential and depend only on its mean value $\bar{V}$, which is the same in both approaches

$$\bar{V}(n) \equiv \bar{V}(r) = \frac{2}{a} \int_0^a V(r) \sin^2 kr dr \, .$$

This opens up the possibility of approximate construction of the spectrum of the difference model E_ν^{fd} from E_ν at those energies where the potential can be considered as a small perturbation. Here, a measure of the precision is $\| V \| / d \leqslant 1$, the ratio of the norm of perturbation $\| V \|$ to the distance d between the levels of the unperturbed problem (3.4.15,15a). Consequently, the upper levels E_ν tend with the increase of ν to the eigenvalues of an infinite rectangular well with $V(r) = \bar{V} = \text{const}$ for $0 < r < a$. If the same were also valid for the difference equation, it would be easy to determine E_ν^{fd} from E_ν. But, as one can see from (3.4.15a), the gaps d between the levels E_ν^{fd} increase with ν only up to the middle ($\nu = N/2$) of the interval $[E_1^{fd}, E_N^{fd}]$, and at the very top of the spectrum are as small as for the first levels. This means that perturbation theory is applicable only in the middle part of the spectral interval $[E_1^{fd}, E_N^{fd}]$. In the special case of the non-perturbed problem (3.4.15a) the spectrum is exactly symmetric with respect to its middle point $E = 1/\Delta^2$. Having prevented the straightforward calculation for the transition $\{E_\nu\} \to \{E_\nu^{fd}\}$, the resemblance of the upper and the lower parts of the $\{E_\nu^{fd}\}$ spectrum turned out to be quite useful for the other approach. To show this, we shall first verify that for an arbitrary $V(n)$ the spectral symmetry ($\hbar^2 = 2m = 1$),

$$E_\nu^{\mathrm{fd}} - 1/\Delta^2 = 1/\Delta^2 - E_\nu^{\mathrm{fd}} ,$$

is weakly violated. Let us consider the difference equation (Case et al 1973):

$$[\psi(n+1) + \psi(n-1)]/2 = (1 - E\Delta^2)\exp[-\Delta^2 V(n)]\psi(n) , \qquad (3.4.16)$$

possessing a strictly symmetric spectrum. Indeed, if ψ is an eigenfunction of some E, then, upon performing the substitution of $-(1 - E\Delta^2)$ for $(1 - E\Delta^2)$ and of $(-1)^n\psi(n)$ for $\psi(n)$ (Fig. 3.15b), (3.4.16) is not violated. Equation (1.2.1a) differs from (3.4.16) only by the terms $V(n)\Delta^2$ which are second order in Δ. In addition, if (1.2.1a) is considered as a perturbed version of (3.4.16), then the difference between eigenvalues with the same ν in (1.2.1a) and (3.4.16) turns out to be symmetric in the first approximation with respect to $E = 1/\Delta^2$. As a result, we arrive at the prescription for constructing $\{E_\nu^{\mathrm{fd}}\}$ from the initial $\{E_\nu\}$: the lower E_ν^{fd} are set equal to E_ν, the next ones up to $\nu = N/2$ are found with the aid of $\bar{V}$ by applying perturbation theory, while the rest are determined by symmetry relative to $E = 1/\Delta^2$. For solving the model inverse problem it is also necessary to give the normalizing constants $\{c_\nu^{\mathrm{fd}}\}$ (or the reduced widths $\{\gamma_\nu^{\mathrm{fd}2}\}$ in the R-matrix approach) from the supposedly known values $\{c_\nu^2\}$ ($\{\gamma_\nu^2\}$).

This recalculation procedure for $\{E_\nu\} \to \{E_\nu^{\mathrm{fd}}\}$ is unsuitable for $\{c_\nu\} \to \{c_\nu^{\mathrm{fd}}\}$ ($\{\gamma_\nu\} \to \{\gamma_\nu^{\mathrm{fd}}\}$), since the perturbation $c_\nu(\gamma_\nu)$ is not expressed in a simple way through $\bar{V}$, as it was with the energy levels. One solution is to take advantage of the theorem of two spectra (Chap. 1, p. 36) (Melnikov, Zakhariev 1980). In the R-matrix it is possible to determine two spectral sets $\{E_\nu\}$ and $\{E'_\nu\}$ corresponding to two different choices of homogeneous boundary conditions for the eigenfunctions from the scattering function $S(k)$. We then perform the recalculation $\{E_\nu\} \to \{E_\nu^{\mathrm{fd}}\}$; $\{E'_\nu\} \to \{E_\nu'^{\mathrm{fd}}\}$, as shown above. Knowing $\{E_\nu^{\mathrm{fd}}\}, \{E_\nu'^{\mathrm{fd}}\}$ we determine with the aid of the theorem of two spectra

$$\gamma_\nu^{\mathrm{fd}2} = \prod_\mu (E_\nu^{\mathrm{fd}} - E_\mu'^{\mathrm{fd}}) \Big/ \prod_\mu{}' (E_\nu^{\mathrm{fd}} - E_\mu^{\mathrm{fd}}) , \qquad (3.4.17)$$

where the prime of the product just indicates the absence of the factor with $\mu = \nu$. Now, from $\{E_\nu^{\mathrm{fd}}, \gamma_\nu^{\mathrm{fd}}\}$ it is possible to reconstruct $V(n)$. In Fig. 3.18 the results of numerical calculations are presented. Even the potential with two minima has been reconstructed quite satisfactorily. However, an increase of the potential depth or of its range led to a rapid increase in the uncertainties. This is due to the conditions becoming less appropriate for the application of perturbation theory. On the other hand, the procedure for solution of the model inverse problem loses stability at small Δ. Thus, the search for more perfect algorithms in the finite-difference approximation must be continued.

The Method of Finite Elements

The method of finite elements combines the simplicity of difference schemes applied in solving the Schrödinger equation together with the merits of vari-

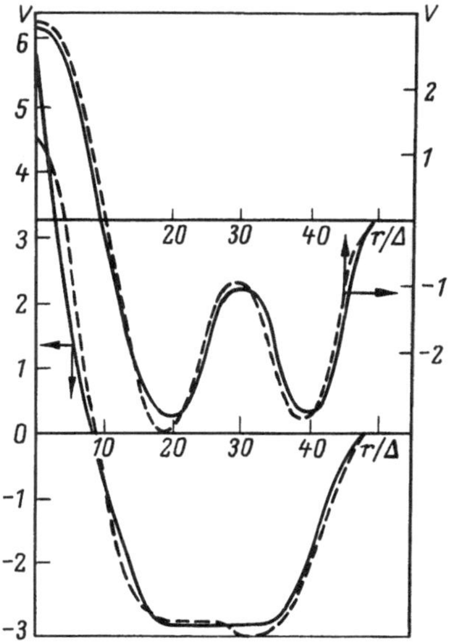

Fig. 3.18. Reconstruction of potentials in the finite-difference approximation from R-matrix resonance parameters (Melnikov, Zakhariev 1980)

ational approaches. Its strength is especially revealed in multi-dimensional problems, but its main features are best demonstrated in the case of a single variable. Until recently the method of finite elements was utilized only for solution of the direct problem. The possibility of reconstructing V within the framework of this formalism has been established only, in principle, with the aid of a device already applied in the step approximation.

The Direct Problem. In the simplest version of the method the wave-function is sought in the form of an expansion in the basis elements $\chi_i(r) \equiv |i\rangle$ depicted in Fig. 3.19:

$$\psi(r) = \sum_i c_i |i\rangle \, , \tag{3.4.18}$$

which signifies approximation of ψ by the broken line shown in Fig. 3.20. This is the approximation by so-called *linear splines*; quadratic and cubic splines lead to an even better approximation. Each coefficient c_i contains certain averaged information on $\psi(r)$ on the finite interval $r_{j-1} < r < r_{j+1}$. At the same time c_j is equal to the value of the trial function at point r_j: $c_j = \psi(r_j)$. The equations for c_j

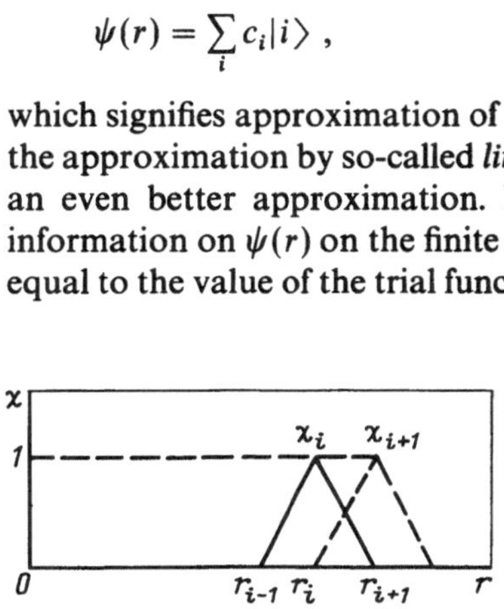

Fig. 3.19. Basis functions of the method of finite elements (linear splines)

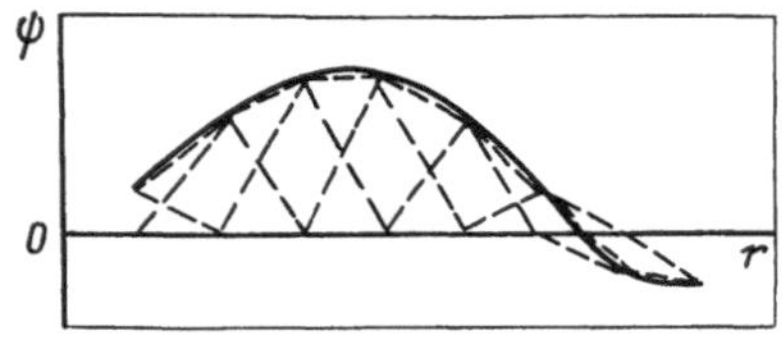

Fig. 3.20. Approximation of $\psi(r)$ (*solid line*) within the method of finite elements (*broken line*)

are reminiscent of the Schrödinger equation in the difference approximation, (1.2.1a).

We now substitute ψ in the form (3.4.18) into the Schrödinger equation, multiply on the left by $|j\rangle$ and integrate over r:

$$\sum_i \langle j|H-E|i\rangle c_i = 0\,. \tag{3.4.19}$$

Since the basis functions $|j\rangle$ overlap only with their immediate neighbors $|j\pm 1\rangle$, a set for the constants c_j of algebraic (difference) equations with a tridiagonal coefficient matrix is obtained:

$$\begin{aligned}&\langle j|H|j+1\rangle c_{j+1}+\langle j|H|j\rangle c_j+\langle j|H|j-1\rangle c_{j-1}\\&\quad = E[\langle j|j+1\rangle c_{j+1}+\langle j|j\rangle c_j+\langle j|j-1\rangle c_{j-1}]\,.\end{aligned} \tag{3.4.20}$$

Taking into account the explicit form of the functions $\chi_j(r)$ (Fig. 3.19):

$$|j\rangle = \begin{cases} 0 & \text{when } r\geqslant r_{j+1} \text{ and } r\leqslant r_{j-1}\,;\\ r/\Delta - j+1 & \text{when } r_{j-1}\leqslant r< r_j\,;\\ j+1-r/\Delta & \text{when } r_j\leqslant r<r_{j+1}\,,\end{cases} \tag{3.4.21}$$

where $r_j = j\Delta$. We shall rewrite (3.4.20) by calculating the matrix elements of d^2/dr^2 and of 1:

$$\begin{aligned}&-\frac{1}{\Delta^2}[c_{j+1}-2c_j+c_{j-1}]+V_{jj+1}c_{j+1}+V_{jj}c_j+V_{jj-1}c_{j-1}\\&= E\,[\tfrac{1}{6}c_{j+1}+\tfrac{2}{3}c_j+\tfrac{1}{6}c_{j-1}]\,,\end{aligned} \tag{3.4.22}$$

where

$$V_{ij} = \langle i|V|j\rangle/\Delta.$$

Unlike (1.2.1a) and (1.4.14), in the coefficients V_{ii} and $V_{ii\pm 1}$ in (3.4.22) the averaged influence of the interaction on the intervals (r_{j-1}, r_{i+1}), (r_i, r_{i+1}), (r_{i-1}, r_i), respectively, is taken into account. Now we shall obtain the completeness relation for the solutions of (3.4.22) satisfying the homogeneous boundary conditions

$$c_0 = 0\,;\quad c_{N+1}=0\,. \tag{3.4.23}$$

Such solutions exist at N energies $E = E_\nu$ and represent N-dimensional vectors $c^\nu \equiv \{c_1^\nu, c_2^\nu, \ldots, c_N^\nu\}$.

We now multiply (3.4.22) by the solution $c_i^{\nu'}$ at another energy E'_ν, and the corresponding equation for $c_j^{\nu'}$ by c_i^ν, then subtract the obtained equations from each other, and perform summation over i taking into account (3.4.23). As a result, we obtain the condition from which the orthogonality of c^ν and $c^{\nu'}$ for $\nu \neq \nu'$ follows (with the weight $\langle i|j\rangle$, where $\langle i|i\rangle = 2\Delta/3$; $\langle i|i\pm 1\rangle = \Delta/6$):

$$(E^\nu - E^{\nu'})\sum_{i,j}^{N} c_i^\nu \langle i|j\rangle c_j^{\nu'} = 0 . \tag{3.4.24}$$

Hence, together with the normalization of vectors c^ν we have:

$$\sum_{i,j}^{N} c_i^\nu \langle i|j\rangle c_j^{\nu'} = \delta_{\nu\nu'} . \tag{3.4.25}$$

From the condition of orthonormality (3.4.25) we obtain the completeness relation for c^ν. To this end we multiply (3.4.25) by $c_p^{\nu'}$, perform summation over ν' and, changing the order of summation, we find

$$\sum_{i}^{N} c_i^\nu \left\{\sum_{\nu'} c_p^{\nu'} \sum_j \langle i|j\rangle c_j^{\nu'}\right\} = c_p^\nu , \tag{3.4.26}$$

where the expression in braces is the Kronecker delta:

$$\sum_{\nu j}^{N} c_p^\nu \langle i|j\rangle c_j^\nu = \delta_{pi} , \tag{3.4.27}$$

or

$$\sum_{\nu}^{N} c_p^\nu c_i^\nu = K_{pi}^{-1} , \tag{3.4.27$'$}$$

where K is a tridiagonal matrix with elements $\langle i|j\rangle$. This completeness relation differs from (1.2.7) by the tridiagonal weight matrix $\| K_{ij} = \langle i|j\rangle \|$.

As in the case of the finite-difference Schrödinger equations (1.2.1a, 4.14), one can introduce in the region where no interaction is present the analogues of the solutions

$$c_n \sim \exp(\pm\, \mathrm{i}\theta n) ; \quad \sim \sin\theta n ; \quad \sim \cos\theta n .$$

Here the wave number θ is related to E in a way different from that in (1.3.15a, 17a). With the aid of these free waves the boundary (asymptotic) scattering conditions are formulated in the representation of the coefficients c_n just as readily as for $\psi(r)$. The R-matrix formalism in this approach is obtained with the aid of (3.4.27), as in the ordinary finite-difference case (1.2.13–19).

The inverse problem. We assume, as usual, that at two adjacent points at one of the ends of the interval $[1, N]$ E^ν and c_j^ν are known from spectral (scattering) data and from (3.4.23). We shall try to approximate the potential $V(r)$ by a broken line, just as for ψ. The matrix elements of V approximated by linear splines

$$V(r) = \frac{V(r_i) - V(r_{i-1})}{\Delta} r + V(r_i)(1 - i) + V(r_{i-1}) \cdot i \,,$$

$$r_{i-1} \leqslant r \leqslant r_i \tag{3.4.28}$$

are expressed linearly through the values of V at the points r_i.

The procedure of (1.2.11) is no longer suitable for solving the inverse problem, since in (3.4.22) three adjacent components c_j^ν occur with the factor E_ν which prevents one to get rid of the superfluous unknowns with the aid of the completeness relation (3.4.27). We can, however, make use of condition (3.4.27), as was done in approximating the step-like potential. Let $V_N = 0$ and let c_N^ν be known and $c_{N+1}^\nu = 0$. We shall express c_{j-1}^ν from (3.4.27) with $j = N$ through known quantities and the sought $V(r_{N-1})$. We substitute this expression for c_{N-1}^ν into (3.4.27) with $p = N - 1$ and $i = N$. We obtain a nonlinear equation for $V(r_{N-1})$:

$$\mathscr{F}(V(r_{N-1})) \equiv \sum_\nu^N c_{N-1}^\nu \left(\frac{\Delta}{6} c_{N-1}^\nu + \frac{2}{3} \Delta c_N^\nu + \frac{\Delta}{6} c_{N+1}^\nu \right) = 0 \,. \tag{3.4.29}$$

We find $V(r_{N-1})$ as the zero of function $\mathscr{F}(V(r_{N-1}))$. Repeating the same operations with a shift by one step into the interaction region, we successively find all the remaining $V(r_{i<N-1})$.

The formalism considered above represents one more finite-dimensional model of the inverse problem. As in the usual difference approach, the spectral properties of the operator in (3.4.20,22) differ significantly from those of the differential Schrödinger operator at high energies. Therefore, a prescription for transition from the spectral parameters of one operator to those of the other operator is needed.

Many of the examples of numerical solutions of the problem dealt with in this chapter reveal that much can be said on the form of interaction on the basis of the lower part of the spectral data. However, we still do not have algorithms for reconstructing potentials that would choose all the useful information on the forces out of the available measurements. A theory on the uncertainties in the $S \to V$ transformation has not yet been developed, but the first results of such a theory have already been obtained. Below we shall present estimates of these uncertainties found by *Marchenko* (1972) together with his collaborators *Lundin, Kozel* (1973), and in Sect. 3.5 we shall return to the inverse problem by applying repeated solution (direct problem) of the equation of motion.

Influence of the Upper Part of the Spectrum on the Shape of the Potential

One of the difficulties of the nonrelativistic inverse problem is that quantum mechanics cannot be applied at high energies, while for reconstruction of V usually data from an infinite energy interval are required. It is therefore interesting to estimate the uncertainty in $V(r)$ due to the absence of information on $S(E)$ at $E > E_{\max}$. No reasonable (finite) estimates can be obtained without restricting a priori the class of functions $V(r)$.

Assume $V(r)$ satisfies the condition

$$\int_r^\infty |V(r')|dr' \leqslant \alpha(r) ,$$

where $\alpha(r)$ is a continuous summable function on $[0, \infty]$ that does not increase there. It has been established that two such potentials $V_1(r)$ and $V_2(r)$ for which the scattering data coincide at $E < E_{max}$ differ by a quantity not greater than a certain function which decreases as E_{max} increases (Marchenko 1972):

$$|V_1 - V_2(r)| \leqslant \frac{10\hbar\left[\ln \dfrac{2m}{\hbar^2} E_{max} + 3\right]}{\sqrt{E_{max} \cdot 2m}} \alpha(r) \times \{9\alpha(r)\beta(\Delta r, r) + \gamma(\Delta r, r)\} + 4\Bigg/\sqrt{E_{max}\frac{2m}{\hbar^2}\left\{3\ln\frac{2mE_{max}}{\hbar^2} + 1\right\}} , \qquad (3.4.30)$$

where

$$\gamma(\Delta r, r) = \max_{j=1,2}\left\{\sup_{r<t<r+\Delta r} |V_j'(t)|\right\};$$

$$\beta(\Delta r, r) = \max_{j=1,2}\left\{\sup_{r<t<r+\Delta r} |V_j(t)|\right\};$$

$$\Delta r = 5\{\ln(2mE/\hbar^2) + 1\}/\sqrt{2mE}/\hbar ,$$

and E_{max} must be sufficiently large so that

$$5\hbar[\ln 2mE/\hbar^2 + 1]\alpha(r)/\sqrt{2mE} < 1 .$$

A similar limit was also obtained for the difference between the wave functions. Estimates such as (3.4.30) can be corrected under the assumption that $S_1(E) \to S_2(E)$ as $E \to \infty$. The quantity $|V_1 - V_2|$ decreases at a rate that rises with E_{max} for smooth V_j (Lundin, Kozel 1973). Thus, if $V_j(r)$ have n continuous and summable derivatives, in the estimate of $|V_1 - V_2|$ at sufficiently high E_{max} there appears a factor E_{max}^{-n+1}. Such estimates can also be given in the case when S_j are known with a given uncertainty at $E < E_{max}$.

Although the obtained formulas often yield a large upper limit for the uncertainties, in practice, it turns out to be possible by introducing additional assumptions concerning the potentials to determine $V(r)$ to good degree of accuracy even at small E_{max} (Fig. 3.4).

3.5 Reconstruction of Potentials by Multiple Solutions of the Direct Problem

The first attempts to approximate interactions by potentials of the most simple form (the rectangular well, the Saxon-Woods potential and others) were made by the "trial-and-error method". Two or three parameters were fitted to obtain the best description of the observable quantities, and after each fit the Schrödinger equation was once again solved. Essentially, this is the solution of the inverse problem by using the direct problem.

The algorithms of this approach to the inverse problem are continuously improving. At the beginning, this method was the only way to reconstruct potentials. But even after the appearance of the Gelfand–Levitan–Marchenko equations it still retains its important advantages: It requires no restrictions on the initial scattering data (such as $l = \text{const}$ or $E = \text{const}$)—*all* available characteristics of bound states and of the scattering (including cross sections) can be used. It is also convenient that the completeness relations need not be used in the calculations although they are necessary for proving the convergence of the approximations. However, the necessity of repeated solution of the equations of motion obviously makes this method less economical. It is also natural that difficulties related to instability, which are not seen when only a few interaction parameters are needed, manifest themselves when the class of functions among which the potential is sought is extended. Thus, the ill-posedness of the inverse problem cannot be overcome by solving it with the aid of the direct problem.

A number of investigations in this direction exist (Cooper et al 1989, iterative-perturbative approach with error estimations and references therein; Denisov 1977; Zhigunov et al 1983). *Wiesner* et al (1979) and *Christov* (1981) searched for the zero of the functional $\mathscr{F}(V_{\text{trial}}(x))$, by characterizing the discrepancy between the scattering data for the desired and the trial V_{trial} potentials by using the functional generalization of a simple procedure of the Newton method of tangents. In the Newton method the position of the zero of a function f is sought by successive iterations. At each step, at a defined point, a tangent to the function is constructed and its intersection with the horizontal axis serves as the next approximation to the point sought where $f = 0$. *Christov* (1981) investigated the convergence of these iterations in the inverse problem. He studied the continuous analogue of the Newton method in the solution of nonlinear functional equations.

Zhigunov et al (1983, 1985) investigated the nature of the deviation of the reconstructed potential from the desired one in real calculations: The influence of the incompleteness of the set of initial data and of uncertainties in their measurements on the ambiguity and instability of the solution of inverse problems.

3.6 Notes on the Literature

The sometimes bizarre effects of fixed precision arithmetic should be at the heart of work in . . . computations. Yet formal treatment of roundoff, though it seems to be necessary, rarely enlightens me . . . my heart always sinks when the subject of

> roundoff error is mentioned ... Fortunately there is an excellent little book by *Wilkinson* (1964) which supplies a thorough and readable treatment of the fundamentals.
>
> *B. Parlett* (1980)

The problem of regularization of the numerical solution of inverse (poorly formulated) problems was considered by *Tikhonov* and *Arsenin* (1979), *Ivanov* (1978), *Groetsch* (1984), and *Goncharsky* and *Bakushinsky* (1989). As a rule, the solution of direct and inverse problems finally reduces to the performance of elementary algebraic operations by a computer. One comparison of the efficiency and accuracy of these different methods is to estimate the total number of these operations, since the error in computation should be connected with the number of algebraic steps. Such an analysis has not yet been performed for the algorithms considered here, as far as we know.

References to numerous publications on the reconstruction of nucleon–nucleon potentials are given by *Coz* et al (1987) and *Gorbatov* et al (1979–88) and choices of various nuclear optical potentials may be found in *Atomic and Nuclear Data and Tables*, 1976.

On the generalization of the inversion procedure to the case of singular matrices see *Ben-Israel* and *Greville* (1974).

The inverse problem in the Born approximation is discussed by *Chadan* and *Sabatier* (1989), and by *Wu* and *Omura* (1962). In the direct problem a new textbook on the quantum-mechanical perturbation theory (diagram method) by *Kiselev* and *Lyapcev* appeared in 1989. The reconstruction of interactions in quasi-classical mechanics is also considered by *Chadan* and *Sabatier* (1989), as well as in the excellent article by *Wheeler* (1976). We also recommend the bibliography in *Funke* and *Zakhariev* (1987). It is remarkable that in the quasi-classical case for reconstruction of $V(r)$ lying lower than a certain energy E_1, the scattering data corresponding to $E < E_1$ are sufficient and a complete set is not required, contrary to the purely quantum case. The WKB formalism of the inverse problem represents a good physical example of application of the technique of *differentiation and integration of fractional order*.

The ordinary Schrödinger equation is sometimes used in relativistic problems except that the potential in it is substituted by a *quasipotential* and the energy is replaced by its relativistic analogue, expressed through the "rapidity". This is a variable canonically conjugate to the relativistic relative distance. (The "rapidity" parameter φ is related to velocity v by $v = \tanh \varphi$. Unlike the relativistic addition of velocities, in the case of rapidity the usual summation law is valid: $\varphi_{12} = \varphi_1 + \varphi_2$, so it is much easier to work with this quantity.) The inverse problem has been treated in this formalism as in the nonrelativistic case (Amirkhanov et al 1973, 1977; Malyarov et al 1979; Solovtsev 1984, 1987).

The inverse problem in tunneling was considered in the finite-difference approximation by *Hron* and *Razavy* (1981). A class of equivalent potentials corresponding to scattering data of nucleons on alpha particles was obtained by *Sofianos* (1985). The optical pion–helium potential was reconstructed by *Dea* et al (1985). The reconstruction of optical potentials from nucleon and α scattering data was considered by *Leeb* et al (1985), *Naidoo* et al (1984), *Fideldey* et al (1986).

The regularization of the numerical solution of the astrophysical inverse problem has been considered by the Soviet mathematicians *Goncharsky* et al (1987).

Chapter 4
The Levinson Theorem

4.1 General Remarks

In the development of any science the bulk of accumulated information gradually crystallizes into separate general laws or regularities which reflect especially deep relationships among the investigated objects. In quantum scattering theory the Levinson theorem, equally important for both the direct and inverse problems, is such a general law.[1]

In its basic form it establishes the relation between the difference $\delta_l(0) - \delta_l(\infty)$ of the phase shifts of partial waves of the relative motion of two colliding particles at zero and infinite energies E and the number of bound states m_l of these particles at a given l:

$$\delta_l(0) - \delta_l(\infty) = m_l \pi \ . \quad (4.1.1)$$

It is assumed that the particles have no spin and interact by a local spherically symmetric potential upon which restrictive conditions are imposed:

$$\int_0^\infty r|V(r)|dr < \infty \ ; \quad \int_0^\infty r^2|V(r)|dr < \infty \ . \quad (4.1.2)$$

The first condition in (4.1.2) signifies that the absolute value of the potential must not increase more rapidly than $1/r^{2-\varepsilon}$ where $\varepsilon > 0$ as $r \to 0$ and the second condition is that $|V(r)|$ decrease more rapidly than $1/r^3$ as $r \to \infty$.

Later it turned out to be possible to generalize the theorem to more complicated problems. In addition, various modifications have been devised to allow application of this theorem to cases where it cannot be applied directly. Several dozen studies on the Levinson theorem exist (see the references further): for nonlocal potentials, for charged particles, and for other cases in which forces exhibit singular behavior, and for potentials depending on energy and on the momentum operator. It is possible to relate $\delta_l(0) - \delta_l(\infty)$ to the number of Regge poles of a certain type. The theorem has been extended to multi-channel (Sect. 5.1), multi-dimensional (Sect. 6.2) and multi-particle systems (Sect. 7.5). Analogues of the Levinson theorem have been obtained for the finite-difference

---

[1] This chapter is essentially a revised and extended version of the review by *Beregy* et al (1973).

Schrödinger equation and even in classical mechanics. It has been established that the theorem represents a special case of a series of more general *sum rules.*

Levinson himself gave the proof of the theorem in the investigation of the uniqueness of the solutions of the inverse problem, and it became one of the necessary conditions imposed on scattering data for reconstruction of the local potential from these data. It can also serve as a useful landmark for the phenomenological choice of interaction parameters based on experimental results. For example, a potential well must be sufficiently deep to allow the existence of that number of levels that is indicated by the energy behavior of the scattering phases under the assumption that $\delta(\infty) = 0$ and that $\delta(k)$ is continuous.

Our discussion of various formulations of the Levinson theorem will permit us to consider and to correct certain facts in scattering theory. In this chapter, a general idea of the present state of the subject is given together with illustrations of the most interesting aspects of the problem using simple examples. The reader interested in additional details may find them in the references given below and in the bibliography of the review by *Beregy* et al (1973). The Levinson theorem for multi-channel, multi-dimensional and multi-particle problems is considered in Chaps. 5–7.

4.2 Simple Examples

Let the potential $V(r)$ in the Schrödinger equation with $l = 0$ satisfy (4.1.2). We shall define the following boundary conditions at $r = 0$ and $r \to \infty$:

$$\psi(k, 0) = 0 \; ; \quad \psi(k, r) \underset{r \to \infty}{\sim} A \sin(kr + \delta(k))/k \; . \tag{4.2.1}$$

The form of asymptotic behavior of ψ in (4.2.1) underlines that δ is the phase shift of the wave function $\psi(k, r)$ at large r with respect to the function $\mathring{\psi}(k, r) = A \sin(kr)/k$ of the free motion of a particle [with $V(r) \equiv 0$: $-\mathring{\psi}'' = k^2 \mathring{\psi}$].

We shall start by considering the behavior of $\delta(k)$ at high energies, which determines $\delta(\infty)$ in (4.1.1). Intuitively, it is clear that at energies much higher than $\max|V(r)|$ (if the potential is a limited function) one may neglect the influence of V on the motion of the particle. Consequently, as k increases, the difference between ψ and $\mathring{\psi}$ must decrease, asymptotically tending to zero together with the phase shift: $\delta(k \to \infty) \to 0$.

We shall explain this by using the example of waves scattered by a rectangular well of radius r_0 and depth V_0. Although the difference between the energy of a free particle $\mathring{E} = \hbar^2 k^2/2m$ and the effective kinetic energy of the particle in the field

$$E' = \hbar^2 k'^2/2m = \hbar^2 k^2/2m + V_0$$

remains constant as E increases ($E' - E = V_0$), the difference between the oscillation frequencies of the wave functions ψ and $\overset{\circ}{\psi}$ in the region $r \leqslant r_0$ causing the phase shift $\delta(k)$ to decrease as k increases. Indeed,

$$k'^2 - k^2 \equiv (k' + k)(k' - k) = 2mV_0/\hbar^2$$

and, consequently,

$$k' - k = 2mV_0/\hbar^2(k' + k) \xrightarrow[k \to \infty]{} 0 \ .$$

As the difference in frequencies vanishes, functions ψ and $\overset{\circ}{\psi}$ become identical, i.e., the phase shift vanishes: $\delta(\infty) = 0$. Cases of *singular potentials* for which $\delta(\infty) \neq 0$ are discussed in Sect. 4.2.

Now we shall consider the phase shift at $E = 0$. The wave function of free motion (Figs. 4.1,2)

$$\overset{\circ}{\psi} = A \sin(kr)/k$$

degenerates in this case into the straight line $\overset{\circ}{\psi} = Ar$ (zero oscillation frequency) passing through the origin, i.e., its phase is constant and equal to zero. For the same reason, the wave function ψ describing scattering on a rectangular well

$$\psi(0, r > r_0) = ar + b \ ,$$

also degenerates into a straight line outside of the range r_0 of action of the potential.[2] Both the functions Ar and $ar + b$, in the general case, intersect the r-axis at a finite distance Δr from each other: Ar crosses at the point $r = 0$ and $ar + b$ at $r = -b/a$. Compared with the infinite length of their "oscillation period" the value $\Delta r = b/a$ is negligible (the special case of $a = 0$ corresponding to an infinite scattering length will be considered later). Hence the phase shift must be a multiple of π. To verify the validity of (4.1.1), it only remains to show that this multiplicity equals the number of bound states in the well m.

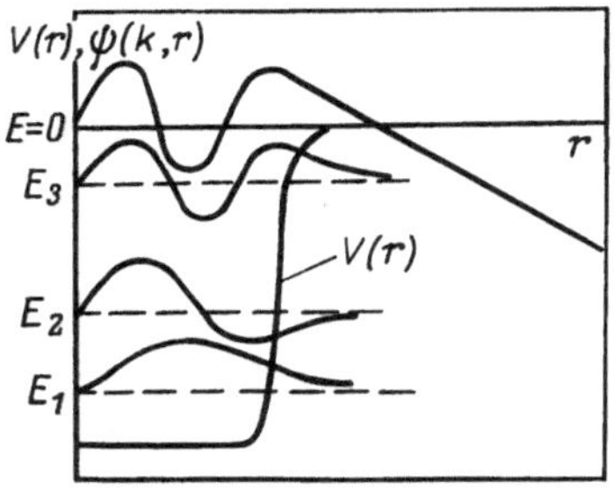

Fig. 4.1. Wave functions of bound states $\psi(k_j, r)$ and states belonging to the continuous spectrum $\psi(k, r)$ in the limiting case of $E=0$ for a particle in a potential well. In the region where $V=0$ the function $\psi(0, r)$ is a straight line inclined with respect to the r-axis

---

[2] At $k = 0$ the Schrödinger equation in the region where $V = 0$ assumes the form: $\psi'' = 0$, i.e., ψ depends linearly on r: $\psi = ar + b$.

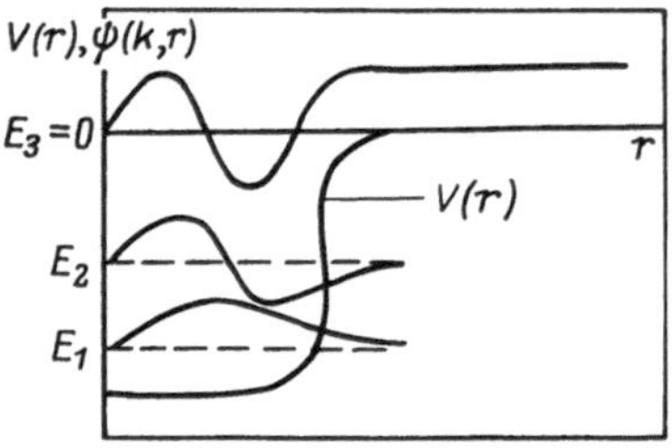

Fig. 4.2. Limiting case of a potential well when the energy of the upper level tends towards zero. At large r the wave function $\psi(0, r)$ becomes constant

The phase of $\psi(0, r)$ increases with r more rapidly than the phase of the eigenfunction of the lower energy m^{th} bound state possessing m nodes including the node at $r = 0$. This means that $\delta(0) > (m - 1)\pi$. At the same time the phase of $\psi(0, r)$ at the point $r = r_0$ cannot exceed $m\pi$, otherwise there would be more than m levels in the well. Thus, $m\pi$ is the only allowed value of $\delta(0)$. This along with the condition that $\delta(\infty) = 0$ represents the Levinson theorem (4.1.1).

Now we shall discuss the limiting case when a in $\psi(0, r > r_0) = ar + b$ becomes zero and the wave function is a constant when $r > r_0$ (infinite scattering length). Such a state is achieved if, for instance, one makes the binding energy of the $(m + 1)^{\text{th}}$ level go to zero while decreasing the depth of the well holding the $m + 1$ levels. The exponentially decreasing tail of the function ψ_{m+1} becomes a constant:

$$\lim_{\kappa_{m+1} \to 0} \left\{ \psi(\kappa_{m+1}, r > r_0) = \lim_{\kappa_{m+1} \to 0} M_{m+1} \exp(-\kappa_{m+1} r) \right\} = M_{m+1} = \text{const} .$$

Then $\psi(0, r)$ becomes non-normalized and the corresponding state is no longer bound. The phase of such a function is an odd multiple of the number $\pi/2$: it is as if the function $\psi(0, r > r_0)$ were a horizontal part of a sinusoidal curve stretched out to infinity and having zero frequency at its extremum. Such a "semi-bound" state, which is ready to transform into a truly bound state at any deepening of the well provides only half the contribution of a "normal" level to the Levinson relation:

$$\delta(0) - \delta(\infty) = (m + \tfrac{1}{2})\pi . \qquad (4.1.1')$$

This cannot be left without comment, however. For a partial wave with nonzero angular momentum $l \geqslant 1$, the $E = 0$ level gives the same contribution to the Levinson relation as all normal bound states. This is due to the existence of a centrifugal barrier which compels the function of such a state to decrease outside of the well with increasing r, as if confining it to the well. Thus, $\psi_{l>0}(0, r)$ turns out to be normalized, unlike $\psi_{l=0}(0, r)$.

Let us complement the above with illustrations of the dependence of phase shifts $\delta_l(k)$ on energy, angular momentum, and the depth of the potential well. Until now we have mainly dealt with $\delta(k)$ at fixed energies. It was shown that as the potential becomes deeper, $\delta(0)$ changes abruptly from 0 to π at the moment a new level arises (sometimes via the intermediate value of $\pi/2$). Now let us see

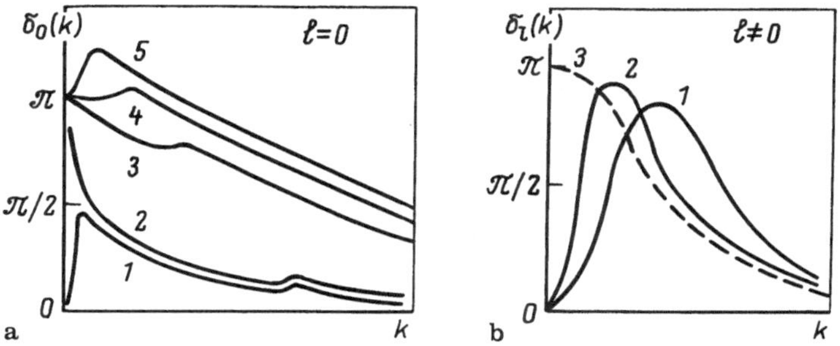

Fig. 4.3. Dependence of the phase shift on energy for rectangular potential wells of various depths. As the attraction becomes stronger ($1 \to 2 \to 3 \to 4 \to 5$), a level arises in the state $l = 0$ (**a**) and $l > 0$ (**b**). From (Newton 1982; McVoy 1967)

how this influences the energy dependence of $\delta_l(k)$. In Fig. 4.3 a series of curves which correspond to potentials $V(r)$ of various depths is plotted (Newton 1982). The curves are numbered in order of increasing attraction (the well becomes deeper). Interestingly, when $l > 0$, the appearance of a new level is preceded by the appearance of a resonance, i.e., of a quasistable state with $E > 0$ which is "confined" by the centrifugal barrier.

The notion of a phase function $\delta(k, r)$ for the potential $V(r)$ is closely related to the interaction range. First, we shall explain its meaning. At each fixed $r = r_0$ and energy $E = k^2\hbar^2/2m$ the quantity $\delta(k, r_0)$ is the usual phase shift of scattering in the "cut-off" potential $V_{r_0}(r)$ which is equal to the initial $V(r)$ at $r < r_0$ and to zero when $r \geqslant r_0$. For a fixed k the function $\delta(k, r)$ describes the dependence of the phase shift on the position of the cut-off point (cut-off of the tail) of the potential. It may be said that $\delta(k, r)$ characterizes the contribution to the phase shift of different parts of the interaction region. The theory of phase functions has been given in detail by *Babikov* (1988) and *Calogero* (1967).

In Fig. 4.4 the variation of the phase shift is represented as depending on the continuous increase in the tail region of the potential. At high energies $\delta(k, r)$ is close to zero for all r. As k decreases (in Fig. 4.4 the curves are numbered in the

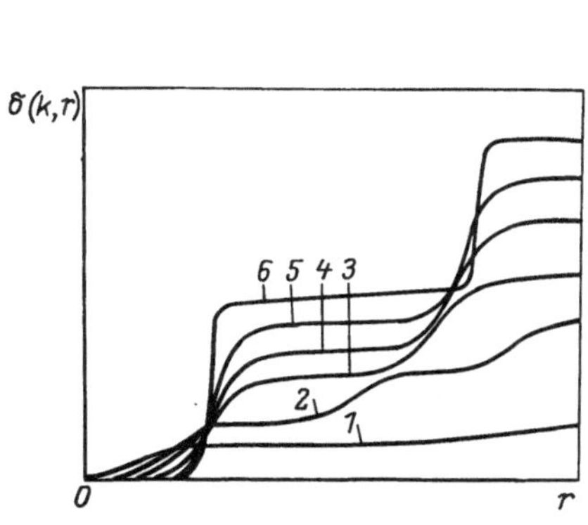

Fig. 4.4. Dependence of the phase functions $\delta(k,\ r)$ on r for scattering on the same potential at various energies: $E_1 > E_2 > E_3 > \ldots > E_6 \approx 0$. In the limit of $E=0$ a strictly step function $\delta(0,\ r)$ is obtained

order of decreasing E), the dependence of the phase function upon r acquires a more and more distinct step-like shape. At zero energy $\delta(k, r)$ is an exact step function. The height of each step is π, the jump occurs precisely at those r where a new level appears in the cut-off potential. At any $r = r_0$ the equality

$$\delta(0, r_0) - \delta(\infty, r_0) = n_{r_0}\pi$$

holds, where n_{r_0} is the number of bound states in the potential $V_{r_0}(r)$.

4.3 The Coulomb Potential and Other Singular Interactions

A special feature of the scattering of charged particles is the fact that the phase shift is unlimited. The phase difference between wave functions corresponding to motion in the field $V(r) \sim 1/r$ and to free motion increases logarithmically with r owing to the potential falling off slowly. In addition, in the case of Coulomb attraction there exists an infinite number of bound states. Therefore, the usual formulation of the Levinson theorem cannot be applied to such long-range forces.

Nevertheless, an analogous theorem was formulated by *Rein* (1969) and *Swan* (1968), although not for purely Coulomb interactions. Actually, for $V_{\text{Coul}}(r) = \text{const}/r$ an exact solution does exist. The case where a short-range $V(r)$ exists together with a Coulomb potential is of particular interest. The Schrödinger equation for the lth partial wave assumes the form ($\hbar = 2m = 1$):

$$\left[-\frac{d^2}{dr^2} + \frac{l(l+1)}{r^2} + \frac{\beta}{r} + V(r)\right]\psi_l(k, r) = k^2\psi_l(k, r)\,, \tag{4.3.1}$$

where $\beta = z_1 z_2 e^2$, z_1 and z_2 are the charges of the colliding particles in units of the elementary charge e. It turns out that relations such as (4.1.1) and (4.1.1′) also hold for phase shifts $\eta_l(k)$ characterizing the difference between scattering by the combination of fields $V(r) + \beta/r$ and scattering by $V_{\text{Coul}} = \beta/r$ only.

The partial wave $\psi_l(k, r)$ exhibits the asymptotic behavior:

$$\begin{aligned}\psi_l(k, r) \underset{r\to\infty}{\sim} &\exp[-\mathrm{i}(kr - l\pi/2 - \beta(\ln 2kr)/2k + \sigma_l)] \\ &- \exp[\mathrm{i}(kr - l\pi/2 - \beta(\ln 2kr)/2k + \sigma_l)]\, s_l(k)\,,\end{aligned} \tag{4.3.2}$$

where σ_l is the Coulomb phase:

$$\begin{aligned}&\sigma_l = \arg\Gamma(l + 1 + \mathrm{i}\beta/2k)\,; \\ &s_l(k) = \exp[2\mathrm{i}\eta_l(k)] = f_l(-k)/f_l^+(k)\,.\end{aligned} \tag{4.3.3}$$

Here $s_l(k)$ and its respective Jost function $f_l^+(k)$ characterize scattering in addition to Coulomb scattering (for purely Coulomb scattering $V_{\text{Coul}} = \beta/r : s_l \equiv 1, \eta_l \equiv 0$). Let us consider separately the cases of Coulomb repulsion ($\beta > 0$) and attraction ($\beta < 0$).

Repulsion. Let $\beta > 0$. Then bound states exist only because of the additional short-range potential $V(r)$. We shall denote by n_l the number of levels in the field $V(r) + \beta/r$ with the given l. For $\beta > 0$ the modified Levinson theorem assumes the form

$$\begin{cases} \eta_l(0) - \eta_l(\infty) = n_l\pi, \text{ if } l \geqslant 1 \text{ and when } l = 0, \text{ if } f_0(0) \neq 0\,; \\ \eta_0(0) - \eta_0(\infty) = (n_0 + \tfrac{1}{2})\pi, \text{ if } f_0(0) = 0 \text{ when } l = 0\,. \end{cases} \tag{4.3.4}$$

The last relation implies the existence of a level at $E = 0$. If one supposes that $V(r)$ satisfies the conditions

$$\lim_{r\to 0} r^2 V(r) = 0 \quad \text{and} \quad \lim_{r\to\infty} r^3 V(r) = 0\,, \tag{4.3.5}$$

one can show that $\eta_l(\infty) = 0$ (Swan 1968).

The transition from the ordinary phase shift δ to the shift relative to the solution with the reference potential made use of here turns out to be also useful for generalization of the Levinson theorem to other singular potentials, besides the Coulomb one, that do not satisfy (4.1.2).

Attraction. Let $\beta < 0$. In the potential $V(r) + \beta/r$ there exists an infinite number of bound states. They are all shifted with respect to the levels of the purely Coulomb case for which the following simple Bohr formula holds:

$$E_{nl} = -\beta^2/4n^2\,, \tag{4.3.6}$$

where n is the principal quantum number.

For $V(r) + \beta/r$ one can also write a relation similar to (4.3.6) for the positions of shifted levels. Except in this case it is necessary to substitute the integers n in the right-hand side of (4.3.6) by n'_l which are chosen especially for describing the discrete spectrum of the combined field, i.e.,

$$E_{nl} = -\beta^2/4n_l'^2\,. \tag{4.3.7}$$

In order to characterize the difference between the Bohr levels and the levels in the potential $V(r) + \beta/r$, we introduce the concept of a *quantum defect* for the n_l^{th} bound state:

$$\mu_l(n) = n - n'_l\,. \tag{4.3.8}$$

For a purely Coulomb field $\mu_l(n) = 0$.

Now we denote the limiting value of $\mu_l(n)$ as $n \to \infty$ by μ_l:

$$\mu_l = \lim_{n\to\infty} \mu_l(n)\,. \tag{4.3.9}$$

The analogue of the Levinson theorem for $\beta < 0$ is thus formulated with η_l and μ_l:

$$\eta_l(0) - \eta_l(\infty) = \mu_l\pi\,. \tag{4.3.10}$$

where $\eta_l(\infty) = 0$ if the conditions of (4.3.5) are satisfied.

The limiting quantum defect μ_l can be determined approximately from spectroscopic measurements for large n or by numerical solution of the Schrödinger equation (Swan 1968).

Now we shall consider other potentials exhibiting singular behavior as $r \to 0$. For example, let $V(r) = \beta/r^2$ be an additional component of the centrifugal potential $l(l+1)/r^2$ (Swan 1963):

$$l(l+1)/r^2 + \beta/r^2 = \nu(\nu+1)/r^2 . \tag{4.3.11}$$

where

$$\nu = [-1 + (1 + 4l + 4l^2 + 4\beta)^{\frac{1}{2}}]/2 . \tag{4.3.12}$$

As in the case of ordinary centrifugal forces, an energy independent phase shift is obtained:

$$\delta_l(k) = \pi(l - \nu)/2 . \tag{4.3.13}$$

Potentials such as $\beta\chi(r)/r^2$, where $\chi(0) = 1$ and $\lim_{r\to 0}\chi(r) = 0$, yield the same phase shift $\delta_l(\infty) = \pi(l-\nu)/2$.

For $V(r) = \beta/r^2$ and $V(r) = \beta\chi(r)/r^2$ the Levinson theorem is formulated in the usual way. However, when $\beta < -l(l+1)$ the wave function becomes infinite as $r \to 0$ and when $\beta < -\frac{1}{4} - l(l+1)$, "the particle falls onto the center" and the concept of phase shift is no longer applicable.

Potentials increasing more rapidly than $1/r^2$ as $r \to 0$ yield an infinite phase shift as $k \to \infty$. Thus, for potentials with a $1/r^{2+\varepsilon}$ dependence as $r \to 0$ and where $0 < \varepsilon < \infty$,

$$\delta(k \to \infty) = -k^{\varepsilon/(2+\varepsilon)} . \tag{4.3.14}$$

This result was obtained by *Swan* (1968), but a strict proof was not provided. Potentials with a hard core [$V(r) = \infty$ for $r < a$] give

$$\delta(k \to \infty) = -ka . \tag{4.3.15}$$

Although in the latter two cases (4.1.1) is not strictly applicable, the relation between the phase shift as $k \to 0$ and the number of levels m remains (Calogero, 1967):

$$\delta(0) = (m + \tau/2)\pi , \tag{4.3.16}$$

where τ is the multiplicity of the level at $k = 0$.

Pivovarchik (1987) has recently shown that for potentials of the form

$$V(r) = gr^{-m} + u(r) ,$$

where $u(r)$ satisfies the condition $\int |u(r)| e^{\varepsilon r} dr < \infty$, and $m > 2$ is an integer, one can obtain an analogue of the Levinson theorem for the difference between the phase shifts in the potentials $\mathring{V} = gr^{-m}$ and $V(r) = gr^{-m} + u(r)$ caused by the perturbation $u(r)$:

$$(\delta(k) - \delta_0(k))\Big|_0^\infty = \begin{cases} \pi n & \text{if } f^+(0) \neq 0 ; \\ \pi(n + \frac{1}{2}) & \text{if } f^+(0) = 0 . \end{cases} \tag{4.3.17}$$

The relation between Regge poles and scattering phases. This is best dealt with in the section devoted to singular potentials, since the very existence of these poles is due to the centrifugal energy term in the Schrödinger equation for partial waves which exhibits a singularity at $r = 0$:

$$-\frac{\hbar^2}{2m}\psi_l''(r) + \frac{l(l+1)}{r^2}\psi_l(r) + V(r)\psi_l(r) = E\psi_l(r) . \tag{4.3.18}$$

Let us denote by $N_l(E = 0)$ the number of Regge poles lying at $E = 0$ on the real l-axis to the right of the given angular momentum l. The quantity $N_l(E = 0)$ is equal to the number of bound states n_l (Ciaffaloni 1963).

We shall now explain this by following the behavior of an arbitrary eigenvalue $E_{\mu l}$ as l is continuously and monotonically increased in (4.3.18). As the centrifugal barrier $l(l + 1)/r^2$ increases, the depth of the effective potential well

$$l(l+1)/r^2 + V(r)$$

will decrease and at the same time the level $E_{\mu l}$ will rise until at a certain $l = l_\mu^{(0)}$ its binding energy will become equal to zero. A bound state corresponds to a Regge pole on $\mathrm{Re}\{l\}$ which will move to the right along $\mathrm{Re}\{l\}$ as $E_{\mu l}$ increases, i.e., part of the respective Regge trajectory (i.e., the curve traced by Regge poles in the complex l plane as energy varies), at $E \leqslant 0$ is situated on the real l axis. As the energy increases further the trajectory leaves $\mathrm{Re}\{l\}$ (Fig. 4.5). Thus, it is evident that each level from n_l has "its own" Regge trajectory which, as the energy decreases, necessarily passes through all the values $E_{\mu l}$ lying to the left of $\mathrm{Re}\{l\}$ ($\mathrm{Re}\{l\} > -\frac{1}{2}$) where the poles are related to the levels $E_{\mu l}$.

The Levinson theorem can be rewritten for Regge poles in the form (Ciaffaloni 1963):

$$\delta_l(0) - \delta_l(\infty) = \pi N_l(E = 0) . \tag{4.3.19}$$

Relation (4.3.19) is valid also for any arbitrary real $l > -\frac{1}{2}$, not only for integer l.

An equality similar to (4.3.19) has been obtained for Regge poles by *Lopez* and *Massida* (1968). They find that the number $n(k)$ of Regge poles lying in the right-hand side of the $\lambda = l + \frac{1}{2}$ plane at $E = \hbar^2k^2/2m$ is proportional to the phase $\omega(\lambda, k)$ of the Jost function for $\lambda = 0$. Lopez and Massida demonstrated that

$$\omega(0, k) = -\pi n(k) . \tag{4.3.20}$$

A similar theorem was also considered for the Coulomb interaction (Lopez, Massida 1970).

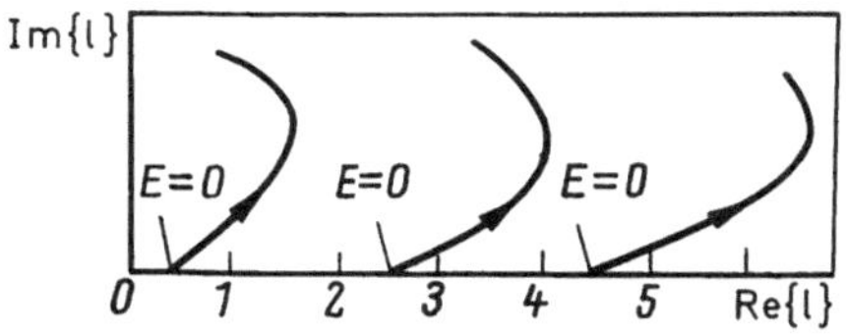

Fig. 4.5. Regge trajectories for scattering by a potential. The poles of the scattering function S in the complex angular momentum l plane first move along the $\mathrm{Re}\{l\}$ axis, as the energy increases up to $E=0$, the arrows on the trajectories point to the increase in energy.

4.4 Other Types of Interactions

We shall now consider velocity (that is, momentum) and energy dependent potentials, as well as cases of the finite-difference Schrödinger equation and propagation of waves along the whole axis $-\infty < x < \infty$, and nonlocal forces.

4.4.1 Potentials Depending on Velocity

For a phenomenological description of low-energy nucleon–nucleon interactions the dependence on differential momentum operator $\hat{p} = -i\hbar d/dr$ is sometimes introduced into the potential:

$$V(r, \hat{p}^2) = V_0(r) + \left[\frac{\hat{p}^2}{2m} V_1(r) + V_1(r)\frac{\hat{p}^2}{2m}\right] \Big/ 2 \,. \tag{4.4.1}$$

When $V_1(r) > 0$, the part of $V(r, \hat{p})$ dependent on $\hat{p}$ accounts for the repulsion of the nucleons at small distances (Beregy et al 1973). The use of $V(r, \hat{p})$ instead of the ordinary potentials $V(r)$ possessing infinite repulsive cores permits one to overcome the difficulties arising in solving multi-particle problems: Usually, because of the infinite increase in $V(r)$ the integrals for the matrix elements diverge.

Butera and *Girardello* (1967) considered superpositions of Yukawa potentials for the components V_0 and V_1 in (4.4.1):

$$V_j(r) = \frac{1}{r} \int_m^\infty \sigma_j(\mu) \exp(-\mu r)\, d\mu \,, \quad j = 0, 1 \,, \tag{4.4.2}$$

where the weighting factor σ satisfies the conditions

$$\int_m^\infty \sigma_0(\mu)\, d\mu < \infty \,; \quad \int_m^\infty \mu^j \sigma_1(\mu)\, d\mu < \infty \,, \quad j = 0, 1, 2 \,. \tag{4.4.3}$$

We omit the derivation of the Levinson theorem and only present the final result:

$$\delta_l(0) = \pi(n_l + \tau/2) \,, \tag{4.4.4}$$

where τ differs from zero only at $l = 0$ and is equal to the multiplicity of states at $E = 0$. *Butera* and *Girardello* (1967) included in the left-hand side of (4.4.4) the quantity $\lim_{k\to\infty} [\delta_l(k) - k_0]$, which must equal zero.

4.4.2 Potentials Depending on Energy

Calogero et al (1967) derived the Levinson theorem for potentials $V(E, r)$ where the phase shift at infinite energy $\delta(\infty)$ was equal to zero. At $E > 0$ the potentials satisfied

$$|V(E, r)| < M(E) r^{a-2} \,, \quad a > 0 \,, \tag{4.4.5}$$

where $\lim_{E\to\infty}[M(E)E^b] = 0$ for $a \neq 1$, $b = \frac{1}{2}$ for $a > 1$, and $b = \frac{a}{2}$ for $a < 1$, while for $a = 1$ $\lim_{E\to\infty}[M(E)E^{\frac{1}{2}} \ln E] = 0$. It was also necessary to restrict the rate of change of $V(E, r)$ with E for $E < 0$ so that the number of bound states n in $V(E, r)$ would not differ from n for the potential $V(r) = V(0, r)$ independent of energy:

$$\partial V(E, r)/\partial E \leqslant 1 , \quad E < 0 . \tag{4.4.6}$$

The Levinson relation has the familiar form

$$\delta(0) = n\pi . \tag{4.4.7}$$

A special example of an energy-dependent potential $V(r) + kQ(r)$ was investigated by *Pivovarchik* (1982). As usual, zeros of the Jost function correspond to bound states. However, here they are usually shifted from the imaginary $\mathrm{Im}\{k\}$ axis, contrary to the real $V(r)$ that is independent of E. As before, the Levinson theorem is satisfied, except that in counting the bound states, one must take into account their multiplicity (the shifted zeros of the Jost function may overlap).

4.4.3 The Finite-difference Schrödinger Equation

Guseinov (1976) derived an analogue of the Levinson theorem corresponding to equation (1.1.1a):

$$\frac{\ln s(+0) - \ln s(\pi - 0)}{2\mathrm{i}} = n\pi + \frac{2 - s(0) - s(\pi)}{4} . \tag{4.4.8}$$

Since the continuous spectrum of the finite-difference Schrödinger operator (1.1.1a, 1.12a) has an upper limit $E_{\max} = 2\hbar^2/m\Delta^2$ and, since the angular variable θ is utilized instead of the argument of the function k, in (4.4.8), $\ln s(\pi - 0)/2\mathrm{i}$ is obtained in place of $\delta(\infty)$. The fraction in the right-hand side of (4.4.8) corresponds to quasibound states at $E = 0$ and at $E = E_{\max}$. The number n includes all the discrete states situated both below and above the continuous spectrum. The existence of the latter is an interesting peculiarity of the difference problem.

4.4.4 Motion along the Axis

We shall now introduce the scattering matrix $S(k)$ composed of the transmission and reflection coefficients, $t = t^{\mathrm{l}} = t^{\mathrm{r}}$ and $r^{\mathrm{r}}, r^{\mathrm{l}}$, respectively, for the cases when the incident wave is present either to the left or to the right of the interaction region:

$$S = \begin{pmatrix} t & r^{\mathrm{r}} \\ r^{\mathrm{l}} & t \end{pmatrix} . \tag{4.4.9}$$

A unitary matrix 2×2 has two eigenvectors

$$S\mathbf{A}_s = \exp(2\mathrm{i}\delta_s)\mathbf{A}_s, \quad s = 1, 2 . \tag{4.4.10}$$

This means that if waves are incident on both sides of the interaction region with amplitudes equal to the components of the vector-column $\mathbf{A}_s$, then the amplitudes of the waves going out in different directions retain the same proportion [only a common phase factor $\exp(2i\delta_s)$ is added].

The Levinson theorem is formulated precisely for the eigenphase *shifts* $\delta_s(k)$:

$$\sum_{s=1,2} [\delta_s(0) - \delta_s(\infty)] = \pi n , \tag{4.4.11}$$

or, if a quasibound state exists at $k = 0$:

$$\sum_{s=1,2} [\delta_s(0) - \delta_s(\infty)] = \pi(n + \tfrac{1}{2}) , \tag{4.4.11'}$$

Here, we once more encounter the situation when in a one-dimensional one-channel problem features of the two-channel problem are observed. The Levinson theorem in multi-channel problems is obtained by a simple generalization of (4.4.11′), as it will be shown in Chap. 5.

4.4.5 Nonlocal Potentials

Local potentials represent a particular example of a wider class of interactions which are defined in the Schrödinger equation by the kernel of the integral operator $V(r, r')$:

$$-\frac{\hbar^2}{2m}\psi''(r) + \int V(r, r')\psi(r')dr' = E\psi(r) . \tag{4.4.12}$$

The equation of motion with the nonlocal potential $V(r, r')$ is integro-differential. It becomes an ordinary differential Schrödinger equation with a local $V(r)$ when the following choice is made:

$$V(r, r') = V(r)\delta(r - r') .$$

The interaction of complex systems (in the unified theory of nuclear reactions, for example) can be reduced to nonlocal potentials. They are used quite often in solutions of three-particle Faddeev equations, since the latter are significantly simplified when the nonlocal potentials are separable.

In nonlocal potentials of even finite range the existence of bound states embedded in continuum is possible. Such states arise when the inhomogeneous integral equation, following from the Schrödinger equation, and the corresponding homogeneous equation both have a solution. In this case the solution of the Schrödinger equation becomes ambiguous, and, besides the usual solution exhibiting the asymptotic behavior

$$\psi(k, r) \underset{r\to\infty}{\sim} \sin(kr + \delta(k)) ,$$

there also exists a solution for which $\psi(k, r) \underset{r\to\infty}{\rightarrow} 0$ (their arbitrary linear combination is also a solution).

Bound states with positive energy are unstable with respect to changes of the potential and may vanish under small variations of the interaction parameters. When such states exist, the Levinson theorem may be formulated in two different ways (Bolsterly 1969). If the phase shift is defined as a continuous function of energy, then the states embedded in the continuum must also be counted as belonging to the bound states (Beregy et al 1973):

$$\delta(0) - \delta(\infty) = \pi\left(m + m' + \frac{\tau}{2}\right), \tag{4.4.13}$$

where m and m' are the numbers of bound states with $E < 0$ and $E > 0$, respectively; τ is the multiplicity of quasibound states of zero energy. It is more consistent, however, to make use of the Levinson theorem in its conventional form [without m' in the right-hand part of (4.36)], but in this case the phase shift is determined with a jump by π at each bound state in the continuous spectrum as shown by the dashed curve in Fig. 4.6 (Bolsterly 1969). The latter case is joined continuously with the case of a resonance at $E > 0$ when its width tends towards zero. Such ambiguity in the definition of phases previously led to the erroneous idea that in this case the Levinson theorem is violated.

4.5 Notes on the Literature

As it was demonstrated by *Buslaev* and *Faddeev* (1960, 1962, 1966) (see also Merkuryev and Faddeev 1985), the Levinson relations represent a special case of the trace formulas:

$$\sum_{l=1}^{L} m_l = -\frac{1}{2\pi i}\ln\det s(0);$$

$$\sum_{l=1}^{L} m_l E_l^{\mu} = \frac{\mu}{2\pi i}\int_0^{\infty} dE E^{\mu-1}\left[\ln\det s(E) + 2i\sqrt{E}\frac{1}{4\pi}\int_0^{\infty} dr\, V(r) + 2i\frac{1}{\sqrt{E}}\sum_{l=0}^{\mu-1}\frac{Q_l}{E_l}\right], \quad \mu = 1, 2, \dots,$$

where Q_l is expressed through $V(n)$.

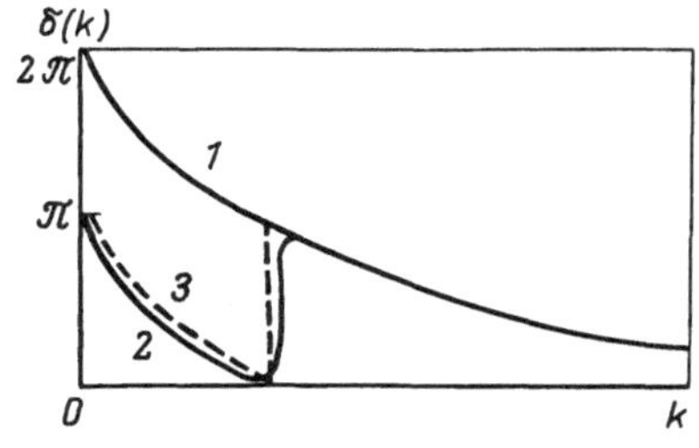

Fig. 4.6. Dependence of the phase shift on energy in the case of a nonlocal potential: *curves 1* and *3* correspond to differing definitions of the phase shift in the case of existence of a bound state embedded in the continuous spectrum; *curve 2* represents the phase shift when this bound state turns into a narrow resonance under a small variation of the potential; *curve 3* represents the limiting case for *2* when the width of the resonance is negligible.

Closely associated with the formalism presented in this chapter is the theory of time delays (Sect. 1.4, p. 26), discussed by *Martin* (1981). Here we shall only note that since the delay or advance $\Delta t(k)$ of a wave packet in an external field V relative to free motion is characterized by the derivative of the phase shift $\delta(k)$ with respect to k at the average energy of the packet, the integral of Δt over k is proportional to the number of bound states in V, according to the Levinson theorem.

A classical analogy of the Levinson theorem can be formulated with time delays. Unlike the quantum case, the respective sum rules are of a local nature, that is, they can be written for an arbitrary vicinity of any point.

The question may arise as to whether there exists an analogy of the Levinson theorem within the $E = \text{const}$ approach. Indeed, the completeness of the set of solutions with $E = \text{const}$ and l along the imaginary $\text{Im}\{l\}$ axis permits derivation of a formula such as the Levinson relation only for phase shifts with nonphysical values.

The Levinson theorem has been used for extending the range of application of the Born approximation (Stein, Green 1982).

A generalization of the Levinson theorem for the relativistic Dirac and Klein–Gordon equations is given by *Ma* et al (1985, 1986), and *Clemence* (1989).

A novel approach to the proof of the Levinson theorem has been proposed by *Iwinsky* et al (1986), and for the perturbed periodical potential by *Firsova* (1985).

The relationship between the forms assumed by the Levinson theorem in the one-, two-, and three-dimensional cases is discussed by *Gibson* (1987); see also the references given therein.

The criterion of existence of an infinite number of resonances (of a faster than exponential decrease of the potential at large distances) is given by *Hwang* (1987).

Levinson theorem for eigenphase shifts ("micro Levinson theorem") for noncentral forces was given by *Newton* (1989).

Part II. Multi-channel, Multi-dimensional, Multi-particle Problems

The one-dimensional and one-channel systems considered in Part I actually represent rare exceptions among real quantum objects. Although we have paid significant attention to these systems, we have borne always in mind their main role of serving as an auxiliary step in the solution of more complex and interesting physical systems.

Usually such systems are described by Schrödinger equations with partial derivatives and nonseparable variables. Generally they cannot be solved in a straightforward manner. If, however, one replaces ψ by its expansion $\psi = \sum_\alpha \psi_\alpha \chi_\alpha$ in a set of known functions χ_α, then sometimes it turns out to be possible to obtain for the coefficients ψ_α equations that are more simple than the initial equation for ψ. Such a method is widely applied in the theory of strong coupling of channels in nonrelativistic quantum mechanics (Zhigunov, Zakhariev 1974). It is remarkable that sets of multi-channel equations can often be considered as matrix generalizations of one-channel equations. Therefore, a close analogy between many properties of new complex objects and those of their one-dimensional prototypes can be established. We shall take advantage of this and in the course of presenting in Chap. 5 the simplest version of the formalism of the direct and inverse multi-channel problems we shall only draw special attention to qualitatively new aspects. This, in turn, will simplify the discussion in Chaps 6 and 7 on multi-dimensional and multi-particle systems based on the approximation of strong coupling of channels described in Chap. 5. Complications will be introduced gradually, when necessary; for example, systems of multi-channel equations with different excitation thresholds of individual channels will be first considered in Chap. 6.

Issues of the theory of many-body reactions are only briefly touched upon. At present only the first publications have appeared on the inverse problem for scattering processes involving rearrangement of particles, Chap. 7 (Zakhariev 1987, 1988; Dubovik et al 1989; Vinitsky et al 1989, 1990; Vinitsky, Suzko 1990).

Chapter 5
Multi-channel Equations

5.1 General Remarks

The term "channel" is used here in a narrow sense. If several one-dimensional Schrödinger equations are combined to form a single system, then each of these equations will be said to be related to one channel, while the components of the column of solutions will be called channel wave functions.[1] The unified technique of strong coupling of channels is used for describing a large variety of physical objects. Their complicated equations of motion reduce to sets of coupled equations.

In this chapter we shall only consider finite numbers of channels. An example of an infinite set of coupled equations is provided by the equations for the partial waves of a multi-dimensional problem with a nonsymmetric potential, discussed in Chap. 6.

A *continuum of channels* is sometimes considered in the course of describing many-body processes. The energies in all the channels are assumed to be equal. The case of channels with different excitation thresholds will be encountered in Chaps 6 and 7.

Elements of the multi-channel formalism (interaction matrices, matrix solution, and so on) are introduced in Sect. 5.2 and are used throughout the book. The matrix generalization of the one-channel inverse problem is discussed in Sect. 5.3.

5.2 Formalism of Multi-channel Coupling

The first notions of the theory of strong coupling of channels are already provided by the simplest example of the set of one-channel Schrödinger equations of the type (1.2.1),

$$-\frac{\hbar^2}{2m}\Psi''_{\alpha\beta}(k,x)+\sum_{\alpha'}^{M}V_{\alpha\alpha'}(x)\Psi_{\alpha'\beta}(k,x)=E\Psi_{\alpha\beta}(k,x)\,;$$

$$K^2=k^2 1=2mE/\hbar^2\,;\quad 1_{\alpha\beta}=\delta_{\alpha\beta} \tag{5.2.1}$$

---

[1]As channels are often called states in composite complexes differing in the arrangement of particles in the fragments (Chap. 7).

coupled by the $M \times M$-interaction matrix $V_{\alpha\alpha'}(x)$. Thus, it is possible to describe the one-dimensional motion of a particle with spin in a field capable of changing its direction of motion (in this case the channel index indicates the projection of spin).

To scalar wave functions, the Jost functions, the Green's functions, etc., there correspond, in the multi-channel case, their matrix analogues. The matrix form of the theoretical relations is compact so that often they even appear identical with their one-channel prototypes. This, however, may also lead to the possibility of missing the unique features specific to the multi-channel apparatus. For example, here one must be careful to observe the proper order in multiplying matrices.

For each of the coupled channels the law of conservation of flux is in general violated; wave fluxes can pass from one channel to another owing to *source* terms (coupling):

$$\sum_{\alpha' \neq \alpha}^{M} V_{\alpha\alpha'}(x)\Psi_{\alpha'}(k, x) \quad \text{in (5.2.1)} .$$

Only the *total* flux is conserved. In this case the notion of ordinary scattering phase shifts cannot be applied, instead *eigenphase shifts* are introduced. They are used, for example, to formulate the multi-channel Levinson theorem.

We shall often write the matrix formulas and formulas for separate components in parallel, so as to make it possible for the reader to focus attention either on the general relations between matrices as whole objects, or on the details of the internal matrix structure of the relations.

Equations of motion and wave functions. Compare the M-channel equations (5.2.1) with the Schrödinger equation (1.2.1). Instead of the single function $\Psi(k, x)$ in (1.2.1), we have M^2 functions $\Psi_{\alpha\beta}(k, x)$; $\alpha, \beta = 1, 2, \ldots, M$. The first index α of $\Psi_{\alpha\beta}$ in (5.2.1) determines the number of the channel component of the wave function. The second index β serves for indicating to which of the independent column solutions, corresponding to differing boundary conditions, a given component belongs. Naturally, no coupling with respect to β occurs in (5.2.1). We should note that here the energy is identical in all M channels. Thus, K is proportional to the unit matrix 1. The more general case of different energies will be considered in Chap. 6.

Instead of the scalar potential function $V(x)$ in (1.2.1), the dynamics of the system (5.2.1) is governed by the interaction matrix $V(x) \equiv \| V_{\alpha\beta}(x) \|$. The coupling of equations in (5.2.1) is realized by the nondiagonal elements $V_{\alpha\beta \neq \alpha}(x)$ (that is, $V_{\alpha\beta,\, \beta \neq \alpha}$). If they are set equal to zero, the system (5.2.1) breaks up into one-channel equations such as (1.2.1), which can be solved independently. To this end it is necessary to choose independent boundary conditions for the vector solutions of (5.2.1) in diagonal form. In this case the matrix of solutions $\Psi_{\alpha\beta}$ also becomes diagonal.

Boundary Conditions. Matrix Analogues $F^{\pm}$, Φ, Ψ of the Solutions $f^{\pm}$, φ, ψ

As in the case of a single channel a unique solution of the equations of motion (5.2.1) is defined by the boundary conditions. While (1.2.1) has two linearly independent solutions, the general solution of the set of M ordinary differential equations of the second order (5.2.1) is composed in the form of a linear combination of its $2M$ partial solutions corresponding to the set of $2M$ linearly independent boundary conditions.

Usually the interaction is assumed to vanish at large distances, $V_{\alpha\beta}(x) \xrightarrow[|x|\to\infty]{} 0$ (x is either a point on the whole axis or the radial variable). Asymptotically, the channels split. Thus, the solutions asymptotically describe free motion in each channel. Therefore, the generalized Jost solutions $F^{\pm}_{\alpha\beta}(k, x)$ satisfying the following conditions may, for instance, serve as the $2M$ linearly independent solutions of the set (5.2.1):

$$\lim_{x\to\infty} F^{\pm}_{\alpha\beta}(k, x)e^{\mp ikx} = \delta_{\alpha\beta} . \tag{5.2.2}$$

This means that for a fixed β and large x the vector solution of (5.2.1) possesses an outgoing or incident wave only in the channel β, while in the remaining channels $F^{\pm}_{\alpha\beta} \xrightarrow[|x|\to\infty]{} 0$. For the M of different β we obtain M vector columns, of which the matrices of the Jost solutions can be composed $F^{+}(k, x) = \| F^{+}_{\alpha\beta}(k, x)\|$; $F^{-}(k, x) = \| F^{-}_{\alpha\beta}(k, x)\|$.

The linear combination of the two matrices $F^{+}(k, x)$, $F^{-}(k, x)$ (i.e., $2M$ vector solutions) is the general solution of (5.2.1). So, in the matrix form the situation recalls the one-channel case: in (5.2.1) and (1.2.1) the fundamental set of solutions includes *two* components, even though they are of different nature. The coefficients in the general solutions for (1.2.1) are scalars, while those for (5.2.1) are matrices.

It is possible, for instance, to choose these coefficients so that from $F^{\pm}(k, x)$ one can construct a matrix solution $\Phi(k, x)$ which is regular at zero [compare with (1.2.4), (1.3.9)]:

$$\Phi(k, 0) = 0; \quad \Phi'(k, 0) = 1 , \tag{5.2.3}$$

$$\text{i.e.,} \ \Phi_{\alpha\beta}(k, 0) = 0 , \quad \Phi'_{\alpha\beta}(k, 0) = \delta_{\alpha\beta} ;$$

$$\Phi(k, x) = [F^{+}(k, x)\tilde{F}^{-}(k) - F^{-}(k, x)\tilde{F}^{+}(k)]/2ik ;$$

$$\Phi_{\alpha\beta}(k, x) = \sum_{\beta'} [F^{+}_{\alpha\beta'}(k, x)F^{-}_{\beta\beta'}(k) - F^{-}_{\alpha\beta'}(k, x)F^{+}_{\beta\beta'}(k)]/2ik . \tag{5.2.4}$$

Here the coefficients $F^{\pm}(k) \equiv \| F^{\pm}_{\alpha\beta}(k)\|$ are matrix Jost functions. In (5.2.4) the order of the multipliers is essential, and the tilde over F indicates transposition. In (5.2.4) scalar coefficients as in (1.3.9) are no longer sufficient, since the M^2 conditions of (5.2.3) must be satisfied by combining $2M$ column solutions from $F^{\pm}(k, x)$.

When no interaction exists in (5.2.1), or when the nondiagonal matrix elements are zero, $\mathring{V}_{\alpha\beta\neq\alpha}(x) = 0$, the respective matrix solutions $\mathring{\Phi}(k, x)$, $\mathring{F}^{\pm}(k, x)$ and the Jost functions $\mathring{F}^{\pm}(k)$ become diagonal:

$$\mathring{\Phi}(k, x) \equiv \| \mathring{\Phi}_{\alpha\alpha}\delta_{\alpha\beta} \| ; \quad \mathring{F}^{\pm}(k, x) \equiv \| \mathring{F}_{\alpha\alpha}\delta_{\alpha\beta} \| ;$$

$$\mathring{F}^{\pm}(k) \equiv \| \mathring{F}^{\pm}_{\alpha\alpha}\delta_{\alpha\beta} \| .$$

As in the case of a single channel, the physical wave function Ψ has a unique (but matrix) boundary condition [the same as that for Φ in (5.2.3)]:

$$\Psi(k, 0) = 0 \quad \text{or} \quad \Psi_{\alpha\beta}(k, 0) = 0 \,. \tag{5.2.3'}$$

Therefore Ψ differs from Φ only by a normalizing factor. Because this is a matrix factor, the separate components of $\Psi_{\alpha\beta}$ and $\Phi_{\alpha\beta}$ are no longer proportional to each other as were ψ and φ in Chap. 1. If Φ in (5.2.4) is multiplied on the right by $k\tilde{F}^{+-1}$, then the matrix Ψ is obtained in the form

$$\Psi(k, x) = \frac{i}{2}\,[F^{-}(k, x) - F^{+}(k, x)S(k)]\,;$$

$$\Psi_{\alpha\beta}(k, x) = \frac{i}{2}\left[F^{-}_{\alpha\beta}(k, x) - \sum_{\beta'}^{M} F^{+}_{\alpha\beta'}(k, x)S_{\beta'\beta}(k)\right], \tag{5.2.5}$$

when in each of the M column solutions an incident wave of unit amplitude is present only in a single channel, the number of which coincides with the number of the solution ($\alpha = \beta$):

$$\Psi_{\alpha\beta}(k, x) \xrightarrow[x \to \infty]{} \frac{i}{2}\,[e^{-ikx}\delta_{\alpha\beta} - e^{ikx}S_{\alpha\beta}(k)]\,, \tag{5.2.6}$$

where the scattering matrix

$$S(k) = \tilde{F}^{-}(k)\tilde{F}^{+-1}(k) \quad \text{or} \quad S_{\alpha\beta}(k) = \sum_{\beta} F^{-}_{\beta'\alpha}(\tilde{F}^{+-1})_{\beta'\beta}\,. \tag{5.2.7}$$

If physical solutions are to be obtained with incident waves in several channels, then (5.2.6) must be multiplied on the right by the vector column of the desired weight amplitudes B_β:

$$\Psi_{\alpha}(k, x) = \sum_{\beta} \Psi_{\alpha\beta}(k, x)B_\beta \xrightarrow[x \to \infty]{} e^{-ikx}B_\alpha - \sum_{\beta} S_{\alpha\beta}(k)B_\beta e^{ikx}\,. \tag{5.2.6'}$$

Unitary transformation of Ψ with the asymptotic behavior of (5.2.6) represents rotation in the M-dimensional vector space of channels. As a result of such transformation, a new set of column solutions is obtained, gathered together in the new Ψ matrix (5.2.6').

Thus, it is sufficient to calculate $S_{\alpha\beta}(k)$ only once, when solving (5.2.1), in order to express the amplitudes of any scattering processes through $S_{\alpha\beta}(k)$.

One can also construct solutions with asymptotic behavior in the form of free standing waves (1.4.36):

$$\Psi_{\alpha\beta}(k, x) \xrightarrow[x \to \infty]{} \delta_{\alpha\beta}\sin kx + K_{\alpha\beta}\cos kx\,, \tag{5.2.8}$$

where the matrix K is related to S:

$$S = (1 - iK/2)(1 + iK/2)^{-1}\,. \tag{5.2.9}$$

Conservation of total flux. To derive this law we multiply (5.2.1) written in matrix form

$$-\frac{\hbar^2}{2m}\Psi''(k, x) + V(x)\Psi(k, x) = E\Psi(k, x) \tag{5.2.10}$$

on the left by the hermitian conjugate matrix Ψ^+ (where the superscript cross indicates transposition and complex conjugation) and the equation for Ψ^+

$$-\frac{\hbar^2}{2m}\Psi^{+''}(k,x) + \Psi^+(k,x)V^+(x) = E\Psi^+(k,x) \tag{5.2.10'}$$

on the right by Ψ. Equation (5.2.10′) is simply the hermitian conjugate of (5.2.10). The order of the multipliers Ψ^+ and V^+ in (5.2.10′) is due to their permutation when the product of the matrices is transposed. Next we subtract the resulting equations from each other. The terms containing E cancel out; the same occurs with the interaction terms under the condition that $V^+ = V$. Integration over x yields for the total flux

$$J(x) = \frac{\hbar^2}{2m}(\Psi^+\Psi' - \Psi^{+\prime}\Psi) = \text{const};$$

$$J_{\beta\beta'}(x) = \frac{\hbar^2}{2m}\sum_\alpha(\Psi^*_{\alpha\beta}\Psi'_{\alpha\beta'} - \Psi^{*\prime}_{\alpha\beta}\Psi_{\alpha\beta'}) = \text{const}. \tag{5.2.11}$$

The elements of (5.2.11) diagonal in the indices of the boundary conditions ($\beta = \beta'$) express conservation of the total flux. Generally, the Wronskian of any two solutions is independent of x:

$$\frac{d}{dx}W[A(k,x),\quad D(k,x)] = 0. \tag{5.2.11'}$$

If Ψ satisfies the boundary condition (5.2.3′), then the coupling constant on the right in (5.2.11) is zero. This corresponds to the equality of the particle flux incident on the interaction region to the flux outgoing in the opposite direction.

When $V_{\alpha\beta\neq\alpha} = 0$ (uncoupled channels), the flux is conserved for each channel. On the other hand, in the general case the channels *exchange fluxes* owing to interaction and only the sum of fluxes is conserved in all the bound channels.

By substitution of the asymptotic expression (5.2.6) into (5.2.11) with the right-hand side set to zero, we obtain the unitary relation for the scattering matrix (as a consequence of conservation of the flux):

$$S^+S = 1 \quad \text{or} \quad \sum_{\beta'} S^*_{\beta'\alpha}(k)S_{\beta'\beta}(k) = \delta_{\alpha\beta}. \tag{5.2.12}$$

In the case of an infinite number of channels one must also require $SS^+ = 1$ (Taylor 1972).

We shall already need the property (5.2.12) in the next section when introducing eigenchannel states. Relations such as (5.2.11) and (5.2.11′) can serve as reliable criteria of the precision of numerical calculations in the direct problem: deviation of the Wronskian from a constant value represents a measure of uncertainty of the calculations.

Eigenphase States. The Levinson Theorem

Consider the case when incident waves are present in all the channels and $B = \{B_\alpha\}$ is the vector of their amplitudes. The vector of scattered wave amplitudes will then, in accordance with (5.2.6′), be $\sum_\beta S_{\alpha\beta} B_\beta$. It turns out that for a given S-matrix the amplitudes B_β can be chosen to be such that the relationship between the intensities of the scattered waves remains the same as in the case of the incident waves:

$$\mathbf{S}\mathbf{B}^\mu = e^{i\delta_\mu}\mathbf{B}^\mu \quad \text{or} \quad \sum_\beta S_{\alpha\beta} B^\mu_\beta = e^{i\delta_\mu} B^\mu_\alpha \ , \tag{5.2.13}$$

i.e., $\mathbf{B}^\mu$ is an eigenvector; $e^{i\delta_\mu}$ are eigenvalues of S.

It is well-known that a unitary $M \times M$ matrix has M eigenvectors ($\mu = 1, 2, \ldots, M$). The corresponding wave functions $\Psi = \| \Psi^\mu_\alpha \|$ of the form (5.2.6′) are called eigenchannel states, with which it is possible to form a matrix $\| \Psi_{\alpha\mu} \|$ where μ is now the index of the boundary conditions, and δ_μ are called the eigenphase shifts.

It is precisely for these eigenphase shifts that the Levinson theorem is formulated in the multi-channel case:

$$\sum_\mu^M [\delta_\mu(0) - \delta_\mu(\infty)] = n\pi \ . \tag{5.2.14}$$

The summation is performed over all the eigenphases.

Relation (5.2.14) requires modification, when in the right-hand sides of the equations for the channel functions (5.2.1) or (5.2.10) there occur different energies (E depends on α). If the number of channels is unlimited, the sum over the eigenphase shifts may diverge. Examples of such problems will be discussed in the following chapters.

In the limit of the uncoupled equations (5.2.1,10) ($V_{\alpha\beta \neq \alpha} = 0$) the eigenphase shifts transform into the usual phase shifts for independent channels, while the equality (5.2.14) transforms into a sum of the one-channel Levinson relations.

5.3 Finite-difference Equations of Motion

The simplest difference analogue of the systems (5.2.1,10) for bound channels is

$$-\frac{\hbar^2}{2m\Delta^2}\{\Psi_{\alpha\beta}(E, n+1) - 2\Psi_{\alpha\beta}(E, n) + \Psi_{\alpha\beta}(E, n-1)\}$$

$$+ \sum_{\alpha'}^M V_{\alpha\alpha'}(n)\Psi_{\alpha'\beta}(E, n) = E\Psi_{\alpha\beta}(E, n) \ . \tag{5.3.1}$$

Let us consider the solutions of this system that satisfy homogeneous boundary conditions such as (1.2.2a):

$$U_\alpha(E, 0) = U_\alpha(E, N+1) = 0 \ . \tag{5.3.2}$$

Together with (5.3.1) they form a set of $N \times M$ homogeneous algebraic equations for $U_\alpha(E, n)$ ($\alpha = 1, 2, \ldots, M$; $n = 1, 2, \ldots, N$). This system has solutions forming an orthonormalized complete set of *vectors* $\mathbf{U}^\lambda(n)$ at $N \times M$ energies $E = E_\lambda(\lambda = 1, 2, \ldots, N \times M)$:

$$\sum_{\alpha=1}^{M}\sum_{n=1}^{N} \Delta U_\alpha^\lambda(n) U_\alpha^{\lambda'}(n) = \delta_{\lambda\lambda'} ; \tag{5.3.3}$$

$$\sum_{\lambda=1}^{N\times M} U_\alpha^\lambda(n) U_\beta^\lambda(m) = \delta_{mn}\delta_{\alpha\beta}/\Delta . \tag{5.3.4}$$

The derivation of the orthonormalization condition (5.3.3) differs from the one-channel case only in additional summation over the indices of the channels α from $\alpha = 1$ to $\alpha = M$ and in taking into account the symmetry $V_{\alpha\alpha'}(n) = V_{\alpha'\alpha}(n)$. As a result, we obtain instead of (1.2.20)

$$(E_\lambda - E_{\lambda'}) \sum_{\alpha=1}^{M}\sum_{n=1}^{N} \Delta U_\alpha^\lambda(n) U_\alpha^{\lambda'}(n) = 0 , \tag{5.3.5}$$

whence follows (5.3.3), since the vectors U_α^λ can be normalized to unity when $\lambda = \lambda'$. For deriving (5.3.4) we multiply (5.3.3) by $U_{\alpha'}^{\lambda'}(m)$, then sum λ' and change the order of summation over α, n, λ':

$$\sum_{\alpha=1}^{M}\sum_{n=1}^{N} \left\{ \Delta \sum_{\lambda'=1}^{N\times M} U_{\alpha'}^{\lambda'}(m) U_\alpha^{\lambda'}(n) \right\} U_\alpha^\lambda(n) = U_{\alpha'}^\lambda(m) . \tag{5.3.6}$$

The expression within the braces acts like $\delta_{mn}\delta_{\alpha\alpha'}$, i.e., we have (5.3.4).

The conditions of orthogonality (1.2.21), (5.3.3) and of completeness (1.2.23), (5.3.4), can be given a more symmetric form if the solutions $\mathbf{U}^\lambda$ at differing energies E_λ are combined in the matrices U($N \times N$ in the one-channel case and $NM \times NM$ in the M-channel case):

$$U^+U = 1; \quad UU^+ = 1 , \tag{5.3.7}$$

where 1 is the unit matrix. In the case of a single channel, λ and n serve as the indices of the matrix elements in (5.3.7), while in the multi-channel case, λ and the double index αn are these indices. The superscript cross in U^+ indicates transposition.

With the aid of (5.3.4) we obtain a set of recursion relations similar to (1.2.8–11):

$$U_\alpha^\lambda(n) = \frac{2m}{\hbar^2}\Delta^2\left[\sum_{\alpha'=1}^{M} V_{\alpha\alpha'}(n-1) U_{\alpha'}^\lambda(n-1) - E_\lambda U_\alpha^\lambda(n-1)\right] + 2U_\alpha^\lambda(n-1) - U_\alpha^\lambda(n-2) ;$$

$$V_{\alpha\alpha'}(n) = \Delta \sum_{\lambda=1}^{N\times M} E_\lambda U_\alpha^\lambda(n) U_{\alpha'}^\lambda(n) - \frac{\hbar^2}{m\Delta^2}\delta_{\alpha\alpha'} ;$$

$$U_\alpha^\lambda(n) = \frac{2m}{\hbar^2}\Delta^2\left[\sum_{\alpha'=1}^{M} V_{\alpha\alpha'}(n+1) U_{\alpha'}^\lambda(n+1) - E_\lambda U_\alpha^\lambda(n+1)\right] + 2U_\alpha^\lambda(n+1) - U_\alpha^\lambda(n+2) . \tag{5.3.8}$$

These relations permit reconstruction of the interaction matrix $V_{\alpha\alpha'}(n)$ and of $U_\alpha^\lambda(n)$ from the set of spectral parameters of the potential well $\{E_\lambda, C_\alpha^\lambda = U_\alpha^\lambda(1)\}$ with infinite walls at the points $r = a = (N+1)\Delta$ and $r = 0$ [the first two equations in (5.3.8)] or from the set of R-matrix resonance parameters $\{E_\lambda, \gamma_\alpha^\lambda = U_\alpha^\lambda(N)\hbar/\sqrt{2ma}\}$ [the last two equations of (5.3.8)], determined from $R_{\alpha\alpha'}(E)$ (or S-matrix):

$$R_{\alpha\alpha'}(E) = \sum_\lambda^{N\times M} \gamma_{\alpha\lambda}\gamma_{\alpha'\lambda}/(E_\lambda - E) . \tag{5.3.9}$$

An algebraic analogue of the Gelfand–Levitan matrix equations. Let us define for the system (5.3.1) the matrix of solutions satisfying the boundary conditions

$$\Phi_{\alpha\beta}(k, 0) = 0; \quad \Phi_{\alpha\beta}(k, 1) = \delta_{\alpha\beta}\Delta . \tag{5.3.10}$$

Like in (1.2.5), the eigen*vector* functions $\mathbf{U}^\lambda(n)$ can be obtained by multiplying the matrix Φ by the normalizing vector $\mathbf{C}^\lambda = \mathbf{U}^\lambda(1)$:

$$U_\beta^\lambda(n) = \sum_\alpha^M \Phi_{\beta\alpha}(E_\lambda, n)C_\alpha^\lambda . \tag{5.3.11}$$

Substituting (5.3.11) into the completeness condition for the eigenfunctions (5.3.4) we obtain

$$\sum_\lambda^{N\times M} \sum_{\delta\eta}^M \Phi_{\alpha\delta}(E_\lambda, n)C_\delta^\lambda C_\eta^\lambda \Phi_{\beta\eta}(E_\lambda, m) = \delta_{mn}\delta_{\alpha\beta}/\Delta . \tag{5.3.12}$$

Note that (5.3.4) signifies mutual orthogonality of the individual components $U_\alpha^\lambda(n)$ as functions of the energy parameter λ for differing fixed values of α and n, while (5.3.12) represents the orthonormalization of the whole vectors

$$\Phi_\alpha(E_\lambda, n) \equiv \{\Phi_{\alpha 1}(E_\lambda, n), \Phi_{\alpha 2}(E_\lambda, n), \ldots, \Phi_{\alpha M}(E_\lambda, n)\}$$

with the weight of the spectral function that is the matrix

$$\rho_{\alpha\beta}(E) = \sum_\lambda \theta(E - E_\lambda)C_\alpha^\lambda C_\beta^\lambda ,$$

where θ is the Heaviside function.

The polynomial dependence of the solutions $\Phi_{\alpha\beta}$ on the energy can be verified just like in the case of the one-channel problem (Sect. 1.3), if the system (5.3.1) is written in matrix form without the indices of the channels. The matrices of the solutions $\Phi(E, n)$ and $\mathring{\Phi}(E, n)$ are polynomials in E of the same degree $n-1$, but with different matrix coefficients. Therefore, by analogy to (1.3.2,3.2a), we construct Φ in the form of linear combinations of $\mathring{\Phi}$:

$$\Phi(E, n) = \mathring{\Phi}(E, n) + \sum_{n'=1}^{n-1} \Delta K(n, n')\mathring{\Phi}(E, n') , \tag{5.3.13}$$

or

$$\Phi_{\alpha\beta}(E, n) = \mathring{\Phi}_{\alpha\beta}(E, n) + \sum_\eta^M \sum_{n'=1}^{n-1} \Delta K_{\alpha\eta}(n, n')\mathring{\Phi}_{\eta\beta}(E, n') , \tag{5.3.13'}$$

which represents the matrix finite-difference analogue of the Volterra integral equation (1.3.2).

The double summation occurring in (5.3.13′) might, however, seem to be not quite justified. Indeed, to form each separate element $\Phi_{\alpha\beta}(E, n)$ representing a polynomial of degree $n - 1$, a linear combination of only n polynomials, $\mathring{\Phi}_{\alpha\beta}(E, m)$, $m = 1, 2, \ldots, n$ is sufficient. Unlike (5.3.13′), however, in this case the coefficients K will depend on the index β and this hinders application of the completeness relations (5.3.12) in deriving the equations of the inverse problem. (Since in (5.3.12) summation is performed over the indices of the channels, it would also involve the K in the course of orthonormalization of $\Phi(E, n)$ with respect to $\mathring{\Phi}(E, m)$, $m \neq n$). In contrast, in (5.3.13,13′) the *same* coefficients K serve for all values of β.

Following the derivation of equations for the K from the orthonormality conditions of the functions $\Phi(E, n)$ in the one-channel case (1.3.2–6) we obtain the matrix algebraic analogue of the Gelfand–Levitan equations:

$$K_{\alpha\beta}(n, n') + Q_{\alpha\beta}(n, n') + \sum_{p=1}^{n-1} \Delta \sum_{\eta}^{M} K_{\alpha\eta}(n, p) Q_{\eta\beta}(p, n') = 0 \,, \tag{5.3.14}$$

where

$$Q_{\alpha\beta}(n, n') = \sum_{\lambda}^{NM} \sum_{\alpha'\beta'}^{M} \mathring{\Phi}_{\alpha\alpha'}(E_\lambda, n) C^\lambda_{\alpha'} C^\lambda_{\beta'} \mathring{\Phi}_{\beta'\beta}(E_\lambda, n) - \frac{\delta_{nn'}\delta_{\alpha\beta}}{\Delta} \,. \tag{5.3.15}$$

The relationship between the desired interaction matrix and K is derived just like in the one-channel case:

$$V_{\alpha\beta}(n) = \frac{\hbar^2}{2m\Delta}\{K_{\alpha\beta}(n + 1, n) - K_{\alpha\beta}(n, n - 1)\} + \mathring{V}_{\alpha\beta}(n) \,. \tag{5.3.16}$$

In the limit $\Delta \to 0$ (5.3.13–16) transform into matrix generalizations of the equations in the Gelfand–Levitan approach:

$$V_{\alpha\beta}(r) = \mathring{V}_{\alpha\beta}(r) + \frac{\hbar^2}{m}\frac{d}{dr} K_{\alpha\beta}(r, r) \,, \tag{5.3.16′}$$

$$\Phi_{\alpha\beta}(E, r) = \mathring{\Phi}_{\alpha\beta}(E, r) + \sum_{\eta}^{M} \int_{0}^{r} K_{\alpha\eta}(r, r') \mathring{\Phi}_{\eta\beta}(r')\, dr' \,; \tag{5.3.17}$$

$$K_{\alpha\beta}(r, r') + Q_{\alpha\beta}(r, r') + \sum_{\eta}^{M} \int_{0}^{r} K_{\alpha\eta}(r, r'') Q_{\eta\beta}(r'', r')\, dr'' = 0 \,; \tag{5.3.18}$$

where

$$\begin{aligned} Q_{\alpha\beta}(r, r') = {} & \sum_{\lambda} \sum_{\beta'\alpha'} \mathring{\Phi}_{\alpha\beta'}(E_\lambda, r) C^\lambda_{\beta'} C^\lambda_{\alpha'} \mathring{\Phi}_{\beta\alpha'}(E_\lambda, r') \\ & - \sum_{\lambda} \sum_{\beta'\alpha'} \mathring{\Phi}_{\alpha\beta'}(\mathring{E}_\lambda, r) \mathring{C}^\lambda_{\beta'} \mathring{C}^\lambda_{\alpha'} \mathring{\Phi}_{\beta\alpha'}(\mathring{E}_\lambda, r') \,. \end{aligned} \tag{5.3.19}$$

Continuing as in Sect. 1.3, we can obtain the multi-channel formalism of the R-matrix theory and equations such as (5.3.13–19) for scattering on the semiaxis in the *Marchenko* (1955) approach.

5.4 Exactly Solvable Models

The one-channel theory presented in Chap. 2 has a matrix generalization. The value of the multi-channel Bargmann potentials is enhanced by the fact that without them the class of problems permitting solution in a closed analytical form is significantly more narrow for systems of coupled equations, than in the one-channel case. At the same time the necessity of simple models for investigating complex objects involving strong coupling between channels is even more compelling. We shall note the nontrivial points of the indicated generalization.

Total Factorization of the Kernel Q^{GL}

The matrix analogue of the simplest factorized one-channel kernel Q^{GL} (2.2.1) is

$$Q^{\mathrm{GL}}_{\alpha\alpha'}(r, r') = \sum_{\beta\beta'} \mathring{\Phi}_{\alpha\beta}(\mathrm{i}\kappa, r) O_{\beta\beta'} \mathring{\Phi}_{\alpha'\beta'}(\mathrm{i}\kappa, r')\,, \quad Q^{\mathrm{GL}}(r, r') = \mathring{\Phi}(r) O \tilde{\mathring{\Phi}}(r') \qquad (5.4.1)$$

where $\mathring{\Phi}$ are the matrices of the regular solutions for the reference interaction matrix $\| \mathring{V}_{\alpha\beta} \|$ permitting exact solution of the systems (5.2.1,10); $E = -K^2\hbar^2/2m$ is the energy of the bound state (assumed to be the only contributing to Q, and O is substituted for the normalizing constant c^2 in (2.2.1). In the right-hand side of (5.4.1) the order of the factors corresponds to the correct arrangement of the matrices. Owing to the double summation over the indices of the channels, the kernel Q is generally *multi-term*, even though it is constructed like the one-term one-channel kernel (2.2.1). For this reason it was at first not understood that there also exist one-term matrix kernels Q. They are obtained when the normalizing matrix is factorized. Thus, for example, in the case of a nondegenerate bound state

$$O_{\beta\beta'} = C_\beta C_{\beta'}\,. \qquad (5.4.2)$$

This may be explained as follows. The wave function of a multi-channel bound state $\Psi(E, r) \equiv \{\Psi_\alpha(E, r)\}$ is a *vector* (but not a matrix) with respect to the channel indices α. Indeed, of the M column solutions of the matrix $\Phi(\mathrm{i}\kappa, r)$, satisfying $\Phi(\mathrm{i}\kappa, 0) = 0$, such a linear combination is constructed, in which the functions increasing with r in each of the M components are absent. The coefficients of the combination form a vector $\mathbf{C} \equiv \{C_\alpha\}$:

$$\Psi = \Phi \mathbf{C}\,. \qquad (5.4.3)$$

Here the order of the factors is also important: the product of the matrix Φ and the vector $\mathbf{C}$ yields the vector Ψ.

With O from (5.4.2) the kernel Q^{GL} is expressed in a totally factorized form:

$$Q^{\mathrm{GL}}_{\alpha\alpha'}(r, r') = \sum_{\beta} \mathring{\Phi}_{\alpha\beta}(\mathrm{i}\kappa, r) C_\beta \sum_{\beta'} C_{\beta'} \mathring{\Phi}_{\alpha'\beta'}(\mathrm{i}\kappa, r') = \mathring{\Psi}_\alpha(\mathrm{i}\kappa, r) \mathring{\Psi}_{\alpha'}(\mathrm{i}\kappa, r')\,, \qquad (5.4.4)$$

i.e., the dependences of Q^{GL} are simultaneously factorized both on the continuous space variables and on the discrete indices of the channels. Expression

(5.4.4) for Q^{GL} also causes factorization of K:

$$K^{GL}_{\alpha\alpha'}(r, r') = -\Psi_\alpha(i\kappa, r)\mathring{\Psi}_{\alpha'}(i\kappa, r'),\ K^{GL}(r, r') = -\Psi(i\kappa, r)\tilde{\mathring{\Psi}}(i\kappa, r') .$$

Substitution of these Q^{GL}, K^{GL} into the matrix Gelfand–Levitan equation reduces it to the algebraic equation for $\Psi_\alpha(i\kappa, r)$. As in the one-channel case, we further obtain [compare (2.2.6–7)]:

$$\Psi_\alpha(i\kappa, r) = \sum_{\alpha'}\Phi_{\alpha\alpha'}(i\kappa, r)C_{\alpha'} = \sum_{\alpha'}\mathring{\Phi}_{\alpha\alpha'}(i\kappa, r)C_{\alpha'}/p(r) = \mathring{\Psi}_\alpha(i\kappa, r)/p(r) ;$$

$$V_{\alpha\alpha'}(r) = \mathring{V}_{\alpha\alpha'}(r) - \frac{\hbar^2}{m}\frac{d}{dr}\sum_\beta \mathring{\Phi}_{\alpha\beta}(i\kappa, r)C_\beta \sum_{\beta'} C_{\beta'}\mathring{\Phi}_{\alpha'\beta'}(i\kappa, r)/p(r)$$

$$= \mathring{V}_{\alpha\alpha'}(r) - \frac{\hbar^2}{m}\frac{d}{dr}[\mathring{\Psi}_\alpha(i\kappa, r)\mathring{\Psi}_{\alpha'}(i\kappa, r)/p(r)] ;$$

$$\Phi_{\alpha\alpha'}(k, r) = \mathring{\Phi}_{\alpha\alpha'}(k, r) - \sum_\beta \mathring{\Phi}_{\alpha\beta}(i\kappa, r)C_\beta \int_0^r \sum_\eta\sum_{\beta'} C_{\beta'}\mathring{\Phi}_{\eta\beta'}(i\kappa, t)\mathring{\Phi}_{\eta\alpha'}(k, t)\,dt/p(r) , \tag{5.4.5}$$

where

$$p(r) = 1 + \sum_{\alpha\beta\beta'}\int_0^r dt\,\mathring{\Phi}_{\alpha\beta}(i\kappa, t)C_\beta C_{\beta'}\mathring{\Phi}_{\beta'\alpha}(i\kappa, t) .$$

It is not difficult to separate in the asymptotic expression for the solution $\Phi(k, r)$ the incident and outgoing waves in accordance with (5.2.4). The amplitudes of these waves form Jost matrices. Multiplying $\Phi(k, r)$ on the right by $\tilde{F}^{-1}(k)k$ we obtain the scattering wave function $\Psi(k, r)$. For $\mathring{V}(r) = 0$ the formulas are simplified since $\mathring{\Phi}(i\kappa, r)$ is then diagonal

$$\mathring{\Phi}_{\alpha\beta}(i\kappa, r) = \delta_{\alpha\beta}\mathring{\Phi}_{\alpha\alpha}(i\kappa, r) .$$

As for (5.4.5), the simplest solutions are obtained within the Marchenko approach (on the semiaxis and on the entire axis) and within the framework of the R-matrix theory.

Degenerate Levels

When several bound states possess the same energy E, the elements of the normalizing matrix 0 can be represented in the form of products of the orthonormalized eigenvectors $\mathbf{C}^s$:

$$O_{\alpha\beta} = \sum_{s=1}^{p} \mu_s C^s_\alpha C^s_\beta , \tag{5.4.6}$$

where

$$\sum_\beta^M O_{\alpha\beta}C^s_\beta = \mu_s C^s_{\alpha'} \quad p \leqslant M .$$

In such cases Q^{GL} can be represented in the form of the sum of p factorized terms (as previously, we shall assume that only one energy level contributes to Q^{GL}):

$$Q_{\alpha\alpha'}(r,r') = \sum_{s=1}^{p} \mu_s \left\{ \sum_{\beta}^{M} \mathring{\Phi}_{\alpha\beta}(i\kappa,r) C_\beta^s \sum_{\beta'}^{M} C_{\beta'}^s \mathring{\Phi}_{\beta'\alpha'}(i\kappa,r') \right\}$$

$$\equiv \sum_{s=1}^{p} \mathring{\Psi}_\alpha^s(i\kappa,r) \mathring{\Psi}_{\alpha'}^s(i\kappa,r') .$$

We write the kernel K in a similar manner:

$$K_{\alpha\alpha'}(r,r') = -\sum_{s=1}^{p} \Psi_\alpha^s(i\kappa,r) \mathring{\Psi}_{\alpha'}^s(i\kappa,r') , \tag{5.4.7}$$

where

$$\mathring{\Psi}_\alpha^s(i\kappa,r) = \sum_{\beta}^{M} \mathring{\Phi}_{\alpha\beta}(i\kappa,r) C_\beta^s \sqrt{\mu_s}; \quad \Psi_\alpha^s(i\kappa,r) = \sum_{\beta}^{M} \Phi_{\alpha\beta}(i\kappa,r) C_\beta^s \sqrt{\mu_s} .$$

The equations (5.3.18) with such kernels Q and K reduce to a set of p algebraic equations for the *vector* functions $\Psi^s(i\kappa,r)$, which, instead of (5.4.5), yield

$$\Psi_\alpha^s(\kappa,r) = \sum_{s'}^{p} \mathring{\Psi}_\alpha^{s'}(i\kappa,r)(P^{-1}(r))_{s's} ;$$

$$P_{ss'}(r) = \delta_{ss'} + \sum_{\alpha}^{M} \int_0^r \mathring{\Psi}_\alpha^s(i\kappa,t) \mathring{\Psi}_\alpha^{s'}(i\kappa,t)\,dt ;$$

$$\Phi_{\alpha\alpha'}(k,r) = \mathring{\Phi}_{\alpha\alpha'}(k,r) - \sum_{\beta}^{M} \sum_{ss'}^{p} \mathring{\Psi}_\alpha^{s'}(i\kappa,r)(P^{-1}(r))_{s's}$$

$$\times \int_0^r \mathring{\Psi}_\beta^s(i\kappa,t) \mathring{\Phi}_{\beta\alpha'}(k,t)\,dt ; \tag{5.4.8}$$

$$V_{\alpha\alpha'}(r) = \mathring{V}_{\alpha\alpha'}(r) + \Delta V_{\alpha\alpha'}(r) = \mathring{V}_{\alpha\alpha'}(r) - \frac{\hbar^2}{m}\frac{d}{dr} \left\{ \sum_{ss'}^{p} \mathring{\Psi}_\alpha^{s'}(i\kappa,r)(P^{-1}(r))_{ss'} \mathring{\Psi}_{\beta'}^s(i\kappa,r) \right\} .$$

In a similar manner formulas can be obtained when several bound states or resonances contribute to Q [see the matrix generalization of (2.2.2–4)]. In this case in the sum over s occurring in (5.4.7), it is necessary to take into account not only degenerate states but also various bound states and resonances.

5.5 Notes on the Literature

For details concerning the approximate solution of systems such as (5.2.1,10) with the aid of computers we refer the reader to the book by *Zhigunov* and *Zakhariev* (1974). Very few works on the reconstruction of interaction matrices exist. *Newton* and *Fulton* (1956, 1957) succeeded in finding tensor and spin-orbital n–p forces using the Bargmann matrices $V_{\alpha\beta}$ for a system consisting of two bound Schrödinger equations with differing l. A similar result was obtained by *Wiesner* et al (1979) by multiple solution of the direct problem. But until now no model calculations have been performed with an a priori known interaction to verify the convergence of approximations, as it was done in the one-channel case (Chap. 3).

The multi-channel problem on the whole axis was considered by *Wadati* and *Kamija* (1974). There the relationship between the amplitudes of the passing and reflected waves for two processes was established, when incident waves arrive from one of the two directions. The inverse problem for the difference matrix Schrödinger operator has been examined by *Berezansky* (1965).

The completeness of multi-channel solutions was proved by *Cox* (1964, 1966), see also *Cox* and *Garcia* (1975). By the way, the formulas therein are written for the case when the reduced mass m may be dependent on the channel indices. A matrix generalization of the Crum–Krein transformations for identical thresholds in all channels is given in the book by *Agranovich* and *Marchenko* (1963) and for different thresholds by *Humi* (1985, 1986) and *Suzko* (1986).

There seems to be no matrix equivalent to the theorem of two spectra. In the presence of degeneracy there can be no intermittent eigenvalues corresponding to different boundary conditions. The possible existence of bound states in the continuous spectrum in the two-channel problem with coupling matrix elements $V_{\alpha\beta}$ of rectangular form has been considered by *Newton* (1982), and corresponding exactly solvable multi-channel Bargmann models are given by *Pivovarchik* et al (1986).

The separation of the motion of waves from channel to channel or over the periods of the external field (along the discrete variable numbering the channels or periods), leads to new insight into multi-channel processes and the notions of quasimoment and quasienergy for a particle in the periodic potential. This was proposed by *Zakhariev* and *Zastavenko* (1987) and is discussed in Chap. 6.

5.6 Exercises

1. The levels of bound states of an M-channel system may be N times degenerate. Verify the above for the simplest example of a system with a diagonal $N \cdot M \times N \cdot M$ interaction matrix (split channels).
2. Obtain the 2×2 solution matrix for a two-channel system with $V_{\alpha\beta}(r) = \text{const} + \delta(r - r_0)$.
3. Find the simplest Bargmann solution for a set of equations with a finite number of channels coupled to the continuum of channels (let the channel indices run through discrete values, $\alpha = 1, 2, \ldots$ as well as continuous ones, $\alpha_{\min} < \alpha < \alpha_{\max}$).
4. How many arbitrary constants are there in the general solution of the N-channel problem? Determine them by writing down the boundary conditions: a) for a bound state, b) for scattering when the incident wave is present in the first and/or the second channels. How must the relative weights of the incident waves be chosen to provide for the outgoing waves having the same absolute values of the amplitudes (eigenchannel states)?

Chapter 6

Multi-dimensional Problems

6.1 General Remarks

The special case of three-dimensional quantum systems with a spherically symmetric interaction $V(\mathbf{r}) = V(r)$ permitting separation of the variables in the Schrödinger equation when it reduces to separate equations for the partial waves was already considered in Part I. There all attention was concentrated on the aspects of radial motion in a single independent channel.

An essentially new aspect in the case of several dimensions is that the numbers of variables on which the local potential and the scattering function depend differ from each other. In three-dimensional space there are three and five variables, respectively:

$$V(\mathbf{r} \equiv \{r, \theta, \varphi\}) \quad \text{and} \quad s(\mathbf{k} \equiv \{k, \theta, \varphi\}\ ; \quad \mathbf{k}' \equiv \{k, \theta', \varphi'\})\ ,$$

where the angles θ, φ; θ', φ' determine the direction of the incident and outgoing waves. In one-dimensional space there is a single variable r for V and k for s.

$$V(r) \leftrightarrow s(k)\ .$$

Overdetermination of the initial data prevents direct generalization of the Gelfand–Levitan–Marchenko formalism to the multi-dimensional case.

It is remarkable that, owing to the redundancy of the scattering data set, in the multi-dimensional case the parameters of the bound states are not necessary for reconstruction of the potential: the information corresponding to $E > 0$ is sufficient.

In the one-channel finite-difference version of the R-matrix theory the situation was already encountered when N values of $V(n)$ $(n = 1, 2, \ldots, N)$ determined the $2N$ spectral parameters $\{E_\lambda, \gamma_\lambda\}$ $(\lambda = 1, 2, \ldots, N)$ used for reconstruction of $V(n)$. Because the potential is local severe constraints on $\{E_\lambda, \gamma_\lambda\}$ are imposed, while no local $V(n)$ can be made to correspond to arbitrary $\{E_\lambda, \gamma_\lambda\}$. A complete correspondence between the scattering and interaction parameters is established for a quasilocal $V(\hat{p}^2, n)$ depending on the velocity.

One can proceed with violation of locality in three-dimensional problems as did *Kay* and *Moses* (1961) who proposed a direct analogue of the Gelfand–Levitan theory for potentials nonlocal with respect to part of the variables [namely, the angles $V(r, \theta, \varphi, \theta', \varphi')$]. However, the applicability of this approach to the special case of local forces seems dubious.

Another way of achieving correspondence between the numbers of variables of V and S is to reduce the three-dimensional direct and inverse problems to the multi-channel problems considered in Chap. 5. We expand $\psi(r)$ in the complete set of spherical harmonics $Y_{lm}(\theta, \varphi)$. Then, for the partial components of the wave function $\Psi_{lm}(r)$ we obtain a set of equations such as (5.2.1):

$$-\frac{\hbar^2}{2m}\Psi''_{lm}(r) + \frac{l(l+1)}{r^2}\Psi_{lm}(r) + \sum_{l'm'} V_{lml'm'}(r)\Psi_{l'm'}(r) = E\Psi_{lm}(r) \qquad (6.1.1)$$

with the interaction matrix

$$V_{lml'm'}(r) = \int Y^*_{lm}(\boldsymbol{\Omega})V(r, \boldsymbol{\Omega})Y_{l'm'}(\boldsymbol{\Omega})\, d\boldsymbol{\Omega}, \boldsymbol{\Omega} \equiv \{\theta, \varphi\}\ , \qquad (6.1.2)$$

which, unlike the initial potential $V(\boldsymbol{r})$, is already a function of five variables: four discrete variables, l, m, l', m', and one continuous variable, r. Such, also, is the scattering matrix $S_{lml'm'}(k)$. Expression (6.1.2) can be understood to be a function "nonlocal with respect to the channel indices". As will be discussed below, in the general case a potential nonlocal in the angles (6.3.26) corresponds to the matrices $S_{lml'm'}(k)$ and $V_{lml'm'}(k)$. This also concerns the diagonal

$$V_{lml'm'}(r) = V_{lm}(r)\delta_{ll'}\delta_{mm'}\ ; \quad S_{lml'm'}(k) = S_{lm}(k)\delta_{ll'}\delta_{mm'}$$

when the equations for the partial waves split up, and the form of $V_{lm}(r)$ depends on lm. Reconstruction of the potential from the interaction matrix is performed by (6.3.26,27).

Using the procedure developed for nonlocal V is most likely not the optimum method of searching for local potentials. Hence, the investigations of the reconstruction of local multi-dimensional $V(\boldsymbol{r})$ by *Faddeev* (1976), *Newton* (1982, 1983, 1985), *Nachman* and *Ablowitz* (1984), *Novikov* and *Henkin* (1987), and others are of great significance. However, no numerical solutions have yet been obtained using the algorithms proposed by these authors. This theory will not be discussed here in detail; in Sect. 6.4 we shall only present its main results, following the review by *Novikov* and *Henkin* (1987).

It is interesting that within the finite-difference approach it turns out to be possible to construct a multi-dimensional analogue of the Gelfand–Levitan formalism without any particular difficulties that are inherent in the theory with continuous space variables. This simple discrete model is considered in Sect. 6.2.

The nature of multi-dimensional problems is best understood by considering the motion of a particle in the field of a noncentral potential, which permits separation of variables in spheroidal coordinates (Sect. 6.5). Such a model is of special interest, since it bridges the thoroughly investigated case of an external field with central symmetry and the case of systems in which the variables cannot be separated.

6.2 The Finite-difference Formalism

To achieve an understanding of the essence of multi-dimensional problems, it is sufficient to limit the consideration to two dimensions. Let us compare the partial differential Schrödinger equation

$$-\frac{\hbar^2}{2m}\left[\frac{\partial^2}{\partial x^2}\psi(E,x,y)+\frac{\partial^2}{\partial y^2}\psi(E,x,y)\right]+V(x,y)\psi(E,x,y)=E\psi(E,x,y)\,. \tag{6.2.1}$$

with its algebraic analogue involving partial differences [compare (1.2.1,1a)]:

$$-\frac{\hbar^2}{2m\Delta^2}[\psi(E,n,m+1)+\psi(E,n+1,m)-4\psi(E,n,m)+\psi(E,n,m-1)$$
$$+\psi(E,n-1,m)]+V(n,m)\psi(E,n,m)=E\psi(E,n,m)\,. \tag{6.2.2}$$

This equation exhibits certain features in common with the set of one-dimensional multi-channel difference equations (5.3.1), if one of the space arguments, for instance, m, is considered to be the channel index, while the other index n is the single coordinate. But, unlike (5.3.1), the coupling of "channels" in (6.2.2) is not realized by means of the interaction potential, but by part of the operator of kinetic energy.

As for the one-dimensional problem, we shall first consider the case of a restricted space region $G(0 \leqslant n < N+1;\ 0 \leqslant m < M = 1)$ and of a purely discrete spectrum. We shall define homogeneous boundary conditions for the eigenfunctions $u_\lambda(n,m)$ (bound states or R-resonance states):

$$u(E,n,M+1)=u(E,N+1,m)=u(E,0,m)=u(E,n,0)=0\,;$$
$$n=0,1,\ldots,N\,;\quad m=0,1,\ldots,M\,. \tag{6.2.3}$$

The conditions (6.2.3) single out from the infinite set of homogeneous algebraic equations (6.2.2) an $N\times M$ set for the unknown $u(E,n,m)$. For $E=E_\lambda$ $(\lambda=1,2,\ldots,N\times M)$ there exist $N\times M$ solutions $u_\lambda(n,m) = u(E_\lambda,n,m)$ that form an orthonormalized complete set [the derivation is similar to (5.3.1–6)]:

$$\sum_{n=1}^{N}\sum_{m=1}^{M}u_\lambda(n,m)u_{\lambda'}(n,m)\Delta^2=\delta_{\lambda\lambda'}\,; \tag{6.2.4}$$

$$\sum_{\lambda=1}^{NM}u_\lambda(n,m)u_\lambda(n',m')=\delta_{nn'}\delta_{mm'}/\Delta^2\,. \tag{6.2.5}$$

Reconstruction of $V(n,m)$ by means of recursion relations. For solution of the inverse problem it is necessary to know all the E_λ and of all the values of the eigenfunctions u_λ along one of the sides of the boundary of the region G, for example, $u_\lambda(1,m)$ $(m=1,2,\ldots,M)$, but not at a single point, as required in the one-dimensional case. Due to the condition $u_\lambda(0,m)=0$ $(m=1,2,\ldots,M)$ we have the known values of u_λ on the two adjacent starting lines $n = 0; 1$ at one edge of the interaction region (similarly at two points, but for all the channels in the multi-channel difference case).

With the aid of the Schrödinger equation (6.2.2) and the completeness condition (6.2.5) we obtain, like (1.2.7–11), the recursion relations (6.2.6,7) permitting solution of the inverse problem.

For any arbitrary fixed n and m, (6.2.2) relates $u_\lambda(n, m)$ to the values of u_λ at four adjacent points $(n \pm 1; m \pm 1)$ of the difference net. Multiplying (6.2.2) by $u_\lambda(n, m)$, summing over λ, and taking into account the orthogonality (6.2.5) of the functions u_λ at difference nodes of the net, we obtain the expression for $V(n, m)$ through u_λ at a single point (n, m):

$$V(n, m) = \Delta^2 \sum_{\lambda=1}^{NM} E_\lambda u_\lambda^2(n, m) - 2\hbar^2/m\Delta^2 , \tag{6.2.6}$$

while from (6.2.2,3) we find the values $u_\lambda(n + 1, m)$ on the line $n + 1$ through the values u_λ on the two adjacent lines $n, n - 1$ where the u_λ are already assumed to be known:

$$u_\lambda(n + 1, m) = \left\{\frac{2m}{\hbar^2}\Delta^2[v(n, m) - E_\lambda] + 4\right\}u_\lambda(n, m) - u_\lambda(n, m + 1)$$
$$- u_\lambda(n, m - 1) - u_\lambda(n - 1, m - 1) . \tag{6.2.7}$$

Passing from line to line in succession and applying (6.2.6,7) in alternation we determine V and u_λ throughout the entire region G, using as the initial spectral data

$$\{E_\lambda, u_\lambda(1, m)\} \quad (\lambda = 1, 2, \ldots, NM; \quad m = 1, 2, \ldots, M) .$$

An alternative approach: the Gelfand–Levitan equations. The same problem may also be solved with the aid of the two-dimensional difference analogue of the Gelfand–Levitan formalism. To this end we shall consider the auxiliary solutions φ_s and $\mathring{\varphi}_s$ [the latter satisfy the same equation (6.2.2), except with $\mathring{V}(n, m) \equiv 0$ or with a known $\mathring{V}(n, m)$] given by the boundary conditions:

$$\varphi_s(E, 0, m) = \mathring{\varphi}_s(E, 0, m) = 0 ;$$
$$\varphi_s(E, 1, m) = \mathring{\varphi}_s(E, 1, m) = \Delta\delta_{ms} . \tag{6.2.8}$$

Here s is a point on the m axis. We shall assume that the lines $n = 0; 1$ where the conditions (6.2.8) are given pass close to the region G where the potential $V(n, m)$ differs from zero [$V(0, m) \equiv 0$, $V(1, m) \equiv 0$, Fig. 6.1]. As in the one-dimensional case, for derivation of the equations of the inverse problem it is essential that φ_s and $\mathring{\varphi}_s$ be polynomials in the variable E.

In accordance with the conditions of (6.2.8) functions φ_s and $\mathring{\varphi}_s$ equal zero on the lines $n = 0; 1$, except at the point $n = 1$, $m = s$ where $\varphi_s(E, 1, s) = \mathring{\varphi}_s(E, 1, s) = 1$. On the basis of these values we find from the Schrödinger equation (6.2.2) the functions φ_s and $\mathring{\varphi}_s$ at all the other points of the coordinate net. The functions φ_s and $\mathring{\varphi}_s$ differ from zero at points inside the cone (Fig. 6.1a) with its

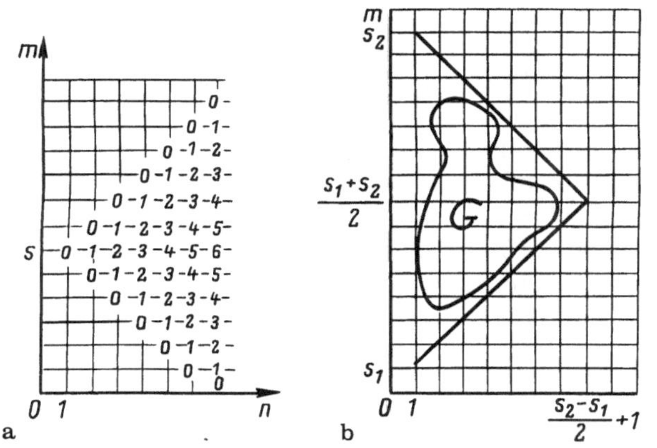

Fig. 6.1. The regions of essential values of s, m, n of the function $\varphi_s(n, m)$: **a** the region where $\varphi_s(n, m)$ polynomials in E differ from zero; the numbers at the nodes of the net indicate the degrees of these polynomials; **b** functions φ_s differing from zero in the interaction region G correspond to the values of s from the interval $[s_1, s_2]$

apex at the point $(1, s)$. At each fixed point (n, m) the functions $\varphi_s(E, n, m)$ and $\mathring{\varphi}_s(E, n, m)$ represent polynomials in E of the degree $n - (s - m)$.

For clarity we shall now point out the common features of the multi-dimensional and the multi-channel problems. This, in particular, will clarify how φ is constructed from $\mathring{\varphi}$. We shall compare the variable m in the function $\varphi_s(E, n, m)$ to the channel index α occurring in the solution $\varphi_{\alpha\beta}(E, n)$ of the system (5.3.1) and s to the index β. Coupling of the equations (6.2.2) with respect to m occurs only between immediate neighbors (the difference operator of kinetic energy relates the functions at m; $m + 1$). On the other hand, the interaction matrix in (5.3.1) is "nonlocal" with respect to the channel indices α: it relates all the M channels in the general case. Thus, (6.2.2) can be considered to be a special case of (5.3.1).

Analogously as in Chap. 5 it is easy to formulate the problem of reconstruction of the potential $V(n, m, m')$ which is nonlocal with respect to one of the variables. In the case of continuous coordinates the inverse problem for a potential, nonlocal relative to angles, was considered by *Kay* and *Moses* (1961).

Now let us construct $\varphi_s(E, n, m)$ in the form of a linear combination of $\mathring{\varphi}_s(E, n', m')$ by analogy with (5.3.13′), and provide for the coefficients K to be independent of s and to be common for all the expressions of φ in terms of $\mathring{\varphi}$ involving any values of s. We shall only take into account that because of the different couplings in (6.2.2) and (5.3.1), instead of summation over all the M channels in (5.3.13′), summation over m' must be performed from 0 to $m \pm (n - n')$. Indeed, for a fixed point (n, m) the vector of solutions φ has components in s differing from zero within the interval $s_1 \leqslant s \leqslant s_2$ (Fig. 6.1b). Naturally, it must be composed of the vectors $\mathring{\varphi}$ which possess no superfluous components in s. These include $\mathring{\varphi}(E, n', m')$ at the nodes (n', m') situated inside the triangle $(1, s_1)$, $(1, s_2)$, (n, m). As a result, we have:

$$\varphi_s(E, n, m) = \sum_{n'=1}^{n} \sum_{m'=m-n+n'}^{m+n-n'} \Delta^2 K(n, m; n', m') \mathring{\varphi}_s(E, n', m') . \tag{6.2.9}$$

When $s = m$, the coefficients of the higher powers of E in $\varphi_m(E, n, m)$ are independent of V and, consequently, they coincide in $\varphi_m(E, n, m)$ and $\mathring{\varphi}_m(E, n, m)$, therefore, $K(n, m; n, m) = 1/\Delta^2$. In a similar manner, in the one-dimensional case we have $K(n, n) = 1/\Delta$.

For the determination of the remaining coefficients K we shall take advantage of the completeness relation (6.2.5) written for the solutions $\boldsymbol{\varphi}$. To do this we shall express $u_\lambda(n, m)$ through $\varphi_s(E_\lambda, n, m)$, like in (5.3.11). We shall define homogeneous boundary conditions $u_\lambda(n, m) = 0$ at the nodes of the net, situated on the sides of the isosceles triangle with vertices $(0, s_1)$; $(0, s_2)$; $\left(\frac{s_2 - s_1}{2}, \frac{s_2 + s_1}{2}\right)$ shown in Fig. 6.1b. The values of s_1 and s_2 must be such that the interaction region G [where $V(G) \neq 0$] lies entirely within the indicated triangle. Then

$$u_\lambda(n, m) = \sum_{s=s_1+1}^{s_2-1} \Delta \varphi_s(E_\lambda, n, m) c_s^\lambda , \tag{6.2.10}$$

where

$$c_s^\lambda = u_\lambda(1, s) . \tag{6.2.11}$$

The following arguments corroborate the validity of (6.2.10). The solutions of (6.2.2) in the triangle are uniquely determined by their values on the lines $n = 0$; 1 at points lying on the triangle. In addition the sum in the right-hand side of (6.2.10) coincides with u_λ at these points in accordance with (6.2.8,11), and it satisfies (6.2.2) since it is made up of the solutions φ. Hence, both parts of (6.2.10) coincide not only on the lines $n = 0$; 1, but at all of the remaining nodes of the triangle.

Equations of the inverse problem. By substituting (6.2.10) into (6.2.5) we obtain the condition of orthonormalization of the vectors $\boldsymbol{\varphi}$ with respect to the energy variable (Zakhariev et al 1977):

$$\sum_{\lambda=1}^{NM} \sum_{s=m-n+1}^{m+n-1} \sum_{s'=m'-n'+1}^{m'+n'-1} \Delta^2 \varphi_s(E_\lambda, n, m) \times c_s^\lambda c_{s'}^\lambda \varphi_{s'}(E_\lambda, n', m') = \frac{\delta_{nn'} \delta_{mm'}}{\Delta^2} . \tag{6.2.12}$$

From (6.2.12) and from the inverse transformation relative to (6.2.9) it follows that for $n' < n$; $m - n + n' \leqslant m' \leqslant m + n - n'$ the vectors $\boldsymbol{\varphi}$ are orthogonal to $\mathring{\boldsymbol{\varphi}}$ with the weight of the matrix $\| c_s^\lambda c_{s'}^\lambda \|$:

$$\sum_{\lambda=1}^{NM} \sum_{s=m-n+1}^{m+n-1} \sum_{s'=m'-n'+1}^{m'+n'-1} \Delta^2 \varphi_s(E_\lambda, n, m) \times c_s^\lambda c_{s'}^\lambda \mathring{\varphi}_{s'}(E_\lambda, n', m') = 0 . \tag{6.2.13}$$

Substituting in φ in the form of (6.2.9) we obtain a multi-dimensional analogue of the Gelfand–Levitan equations [compare with (5.3.14)]:

$$K(n, m; n', m') + Q(n, m; n', m') + \sum_{n''=1}^{n-1} \Delta \sum_{m''=m-n+n''}^{m+n-n''} \Delta \\ \times K(n, m; n'', m'')Q(n'', m''; n', m') = 0 , \tag{6.2.14}$$

where

$$n' < n \,; \quad m - n + n' \leqslant m' \leqslant m + n - n'$$

while

$$Q(n, m; n', m') = \sum_{\lambda} \sum_{ss'} \Delta^2 \mathring{\varphi}_s(E_\lambda, n, m) c_s^\lambda c_{s'}^\lambda \mathring{\varphi}_{s'}(E_\lambda, n', m') - \frac{\delta_{nn'}\delta_{mm'}}{\Delta^2} . \tag{6.2.15}$$

We find the relationship between $V(n, m)$ and K following the derivation of (1.3.7) and (5.3.16):

$$V(n, m) = \frac{\hbar^2}{2m\Delta} \{K(n, m; n-1, m) - K(n+1, m; n, m)\} . \tag{6.2.16}$$

Unlike the one-dimensional case, here it is impossible to obtain the formalism of the inverse problem with continuous space variables by transition to the limit $\Delta \to 0$. For example, there exists no analogue of solutions such as $\varphi_s(n, m)$ and $\mathring{\varphi}_s(n, m)$. (For elliptic partial differential equations, including the stationary Schrödinger equation, the Cauchy problem is in general ill-posed).

6.3 Reduction of Multi-dimensional Problems to Multi-channel Problems

In the case of short-range potentials it is convenient to expand the wave function $\psi(r)$ in spherical harmonics:

$$\psi(r) = \sum_{lm} \Psi_{lm}(r) Y_{lm}(\Omega) . \tag{6.3.1}$$

(The forces of finite range a mainly perturb only the partial waves with limited values of the angular momentum, $l \leqslant ka$.) Then, for the partial radial waves $\Psi_{lm}(r)$ we obtain the infinite set (6.1.1) with the interaction matrix (6.1.2).

As it was noted in Chap. 5, with the functions $\Psi_{lm}(r)$ corresponding to different boundary conditions it is possible to form the solution matrix. To do this the $\Psi_{lm}(r)$ are assigned additional indices $l'm'$ indicating the channel in which the incident waves are present:

$$\Psi_{lml'm'}(k, r) \underset{r\to\infty}{\approx} (\mathrm{e}^{-\mathrm{i}kr}\delta_{ll'}\delta_{mm'} - S_{lml'm'}(k)\mathrm{e}^{\mathrm{i}kr}\mathrm{e}^{-\mathrm{i}\pi l}) . \tag{6.3.2}$$

With the aid of $\Psi_{lml'm'}(r)$ a wave function that corresponds to arbitrarily given asymptotics is readily constructed. For example, when a plane wave is incident with momentum $p = \hbar k$:

$$\psi(\mathbf{k}, \mathbf{r}) \underset{r\to\infty}{\sim} e^{i\mathbf{kr}} + T(\mathbf{k}, \mathbf{k}')\frac{e^{ikr}}{r} ; \tag{6.3.3}$$

$$\psi(\mathbf{k}, \mathbf{r}) = \left(\frac{2mk}{\pi}\right)^{1/2} (kr)^{-1} \sum_{lm\, l'm'} \psi_{lml'm'}(k, r) Y^*_{lm}(\Omega_k) Y_{l'm'}(\Omega_r) . \tag{6.3.4}$$

In the case of an infinite number of coupled equations it is necessary to address the problem of approximation: the expansion of the wave function must be limited to N terms and the multi-channel system has to be cut off correspondingly. In this case the cut off of (6.1.1) is facilitated by the presence of centrifugal terms that suppress the coupling of equations as l increases. But the convergence of the approximations with increasing N must be rigorously analyzed just like in the case of the cut off of infinite systems of algebraic equations, as mentioned in Sect. 3.1.

Generalization of the Levinson theorem to the multi-dimensional case is nontrivial (Newton 1977). This is because conventional phase shifts are inapplicable to the case of a nonspherical potential and, even for $V(|\mathbf{r}|)$, summation over l, m of the Levinson relations for the individual partial channels

$$\delta_l(0) - \delta_l(\infty) = n_l \pi$$

leads to divergence in the left-hand side of the equality. It has been demonstrated (Dreyfus 1978; and bibliography therein), that for integrable potentials satisfying the Rolnik condition

$$\int d\mathbf{r} |V(\mathbf{r})| < \infty ; \quad \int |V(\mathbf{r})| |r - r'|^{-2} |V(\mathbf{r}')| d\mathbf{r} d\mathbf{r}' < \infty \tag{6.3.5}$$

the following generalized relation of the Levinson type holds:

$$\ln \det S(0) = 2\pi i n , \tag{6.3.6}$$

where n is the total number of bound states, and $\ln \det S(0)/2i$ is the sum of all the eigenphase shifts at $k = 0$.

It turns out that multi-dimensional problems for the motion of a particle in an external field are sometimes equivalent to multi-particle problems. Therefore, the first may be used as models for learning the methods of solving the latter. We shall consider a number of such examples below.

Different Thresholds of Channel Excitation

We shall proceed with applying the simplest version of the technique of the unified reaction theory for solution of the two-dimensional problem of a particle in the field $V(y) + V(x, y)$ when motion along x is not restricted and motion along y is limited by an infinitely deep potential well $V(y)$.

Now we shall expand the total wave function $\Psi(x, y)$ in the quantum states of transverse oscillations $\chi_\alpha(y)$:

$$\Psi(x, y) = \sum_\alpha \Psi_\alpha(x)\chi_\alpha(y) , \tag{6.3.7}$$

where

$$-\frac{\hbar^2}{2m}\chi_\alpha''(y) + V(y)\chi_\alpha(y) = \varepsilon_\alpha\chi_\alpha(y) . \tag{6.3.8}$$

For the coefficients Ψ_α which can be considered as the wave functions of longitudinal motion for the states α, we obtain a set of multi-channel equations from the partial differential Schrödinger equation (following the Bubnov–Galerkin principle):

$$-\frac{\hbar^2}{2m}\Psi_\alpha''(K, x) + \sum_{\alpha'} V_{\alpha\alpha'}(x)\Psi_{\alpha'}(K, x) = E_\alpha\Psi_\alpha(K, x);\ E_\alpha = E - \varepsilon_\alpha , \tag{6.3.9}$$

where the elements of the interaction matrix are determined by the integral

$$V_{\alpha\alpha'}(x) = \int \chi_\alpha(y)V(x, y)\chi_{\alpha'}(y)dy , \tag{6.3.10}$$

and the argument K is the diagonal matrix

$$K_{\alpha\beta} = k_\alpha\delta_{\alpha\beta};\quad k_\alpha = \sqrt{2mE_\alpha}/\hbar . \tag{6.3.11}$$

It is interesting that such equations may also correspond to a system of two particles bound by the potential $V(x_1 - x_2 \equiv y)$ (quasideuteron) and taking part in one-dimensional motion along x (Chap. 7). Thus, the descriptions of physical objects of differing characteristics are bridged.

What is essentially new in (6.3.9), in comparison with the multi-channel equations (5.2.1), is the dependence of the energy factor E_α in the right-hand part of the equation on the channel index α. The energy in the channel is obtained by subtraction from the total energy E of the energies of transverse oscillations ε_α.

We shall assume that $V(x, y)$ vanishes at large $|x|$: $V(x, y) \xrightarrow[x\to\pm\infty]{} 0$ and, consequently, $V_{\alpha\alpha'}(x) \xrightarrow[x\to\pm\infty]{} 0$, while the equations (6.3.9) are decoupled and transformed into equations of free motion along x with energy E_α. Those channels in which $E_\alpha > 0$ are called *open* channels. On the other hand, those in which $E_\alpha < 0$, i.e., the excitation energy ε_α of transverse oscillations is higher than E and longitudinal motion takes place with negative energy at large $|x|$, are called *closed* channels.

At large x the functions $\Psi_\alpha(x)$ become linear combinations of free solutions: for example, $\exp(\pm ik_\alpha x)$ with real k_α for open and $\exp \pm \kappa_\alpha x$, with real $\kappa_\alpha = ik_\alpha$, $\kappa_\alpha^2 = -2mE_\alpha/\hbar^2 > 0$ for closed channels. As physical boundary conditions it is natural to require that the coefficients of the solutions in closed channels, increasing exponentially at large distances, be zero. Incident waves in open channels should only occur in entrance channels. In the case of motion along the semi-axis $x \geqslant 0$ we require, instead of the conditions at $x \to -\infty$ that $\Psi_\alpha(0) = 0$.

In this case, we assume the wave incident from the right to be present only in the α' state of transverse oscillations. The asymptotic behavior of Ψ_α then has the form:

$$\Psi_{\alpha\alpha'}(K,x) \underset{x\to\infty}{\longrightarrow} \begin{cases} \exp(-ik_\alpha x)\delta_{\alpha\alpha'} - S_{\alpha\alpha'}(K)\exp(ik_\alpha x) \\ \text{when } E_\alpha \geqslant 0; \\ \exp(\kappa_\alpha x)\delta_{\alpha\alpha'} - S_{\alpha\alpha'}(K)\exp(-\kappa_\alpha x) \\ \text{when } E_\alpha \leqslant 0\,, \end{cases} \tag{6.3.12}$$

where the channel function $\Psi_{\alpha\alpha'}$ has been assigned the additional index α', precisely indicating which boundary condition has been chosen, so that $\Psi_{\alpha\alpha'}$ form a matrix of solutions [compare (5.2.6)]. The total matrix $\| \Psi_{\alpha\alpha'}(x) \|$ in (6.3.12) also includes nonphysical solutions without incident waves in open channels, but with components corresponding to $E_{\alpha'} \leqslant 0$ that increase as $x \to \infty$.

Actually, in inelastic scattering processes only those states that correspond to open channels can be excited. But in the vicinity of the region where $V_{\alpha\alpha'}(x) \neq 0$ closed channels also take part (virtually) in the interaction mechanism.

With the equations (6.3.9) it is also possible to describe the motion of charged particles through crystal lattices (canalling).

Adiabatic expansion. We shall again restrict our consideration to the two-dimensional case of a particle in a field to demonstrate one more method for reducing the partial differential Schrödinger equation to a set of ordinary differential equations.

Each essentially new basis of auxiliary functions $\chi_{\alpha'}$ used to construct $\psi(x, y)$ provides not only a more or less effective method of approximate solution of the problem, but also the possibility of a new insight into the complicated mechanism of quantum motion.

If we add the potential $V(x, y)$ to (6.3.8) then χ_α and ε_α will depend on x as a parameter:

$$-\frac{\hbar^2}{2m}\frac{d^2}{dy^2}\chi_\alpha(x,y) + [V(y) + V(x,y)]\chi_\alpha(x,y) = \varepsilon_\alpha(x)\chi_\alpha(x,y)\,, \tag{6.3.13}$$

i.e., $\chi_\alpha(x, y)$ describes motion parallel to the y axis in the field $V(x) + V(x, y)$ at a fixed value of x. Now we expand $\psi(x, y)$ at each given x in a set of states $\chi_\alpha(x, y)$:

$$\psi(x,y) = \sum_\alpha \Psi_\alpha(x)\chi_\alpha(x,y)\,, \tag{6.3.14}$$

Substituting $\psi(x, y)$ in this form into the Schrödinger equation (6.2.1) with

$$V(x,y) \to V(y) + V(x,y)\,,$$

multiplying the obtained equation on the left by χ_α, integrating both its parts over y, and taking into account that the states χ_α are orthonormalized

($\int \chi_\alpha(x, y)\chi_{\alpha'}(x, y)\,dy = \delta_{\alpha\alpha'}$) we find the following:

$$-\frac{\hbar^2}{2m}\frac{d^2}{dx^2}\Psi_\alpha(x) + \sum_{\alpha'} \hat{\Lambda}_{\alpha\alpha'}(x)\Psi_{\alpha'}(x) + \varepsilon_\alpha(x)\Psi_\alpha(x) = E\Psi(x)\,, \tag{6.3.15}$$

where $\hat{\Lambda}_{\alpha\alpha'}(x)$ is a differential operator of the first order,

$$-\frac{2m}{\hbar^2}\hat{\Lambda}_{\alpha\alpha'}(x) = \int \chi_\alpha(x, y)\left\{\frac{d^2}{dx^2}\chi_{\alpha'}(x, y)\right\}dy + 2\int dy\,\chi_\alpha(x, y) \times \left\{\frac{d}{dx}\chi_{\alpha'}(x, y)\right\}\frac{d}{dx}\,. \tag{6.3.16}$$

The set (6.3.15) differs significantly from (6.3.9) in that the coupling of channels occurs not through the matrix elements of the potentials, but of the operator of kinetic energy $-\frac{\hbar^2}{2m}\frac{d^2}{dx^2}$. The part of the potentials is played by the terms $\varepsilon_\alpha(x)$.

Restricting the expansion (6.3.14) to a single term corresponds to the genuine adiabatic approximation. Then the system (6.3.15) reduces to a single equation. Such an approximation is justified when the potential and, consequently, the functions χ_α depend smoothly on x (weak coupling $\hat{\Lambda}_{\alpha\alpha'}$).

The inverse problem for the set of multi-channel equations (6.3.15) was formulated by *Vinitsky* and *Suzko* (1990), *Dubovik* et al, and will be considered in Chap. 7 with some additional interaction terms.

Basis sets $\{\phi_\alpha\}$ with discrete and continuous values of α. When the motion of a particle along y is limited by infinite potential walls, it is natural to use as a basis the purely discrete states of transverse oscillations $\chi_\alpha(y)$. If these potential walls were of a *finite* height, a continuous part in the spectrum of states of transverse motion would appear. Then, besides the sum over bound states with $\varepsilon_\alpha < 0$, the expansion of $\psi(x, y)$ in $\chi_\alpha(y)$, unlike (6.3.7), would involve integration over the continuum of α values:

$$\psi(x, y) = \left\{\sum_\alpha + \int d\alpha\right\}\Psi_\alpha(x)\chi_\alpha(y)\,. \tag{6.3.17}$$

[One must not confuse states $\chi_\alpha(y)$ of the subsystem describing the motion solely along y, and $\psi(x, y)$ of the entire system for a particle in the two-dimensional field $V(x, y)$.]

One obtains a continuum of differential equations for $\Psi_\alpha(x)$ from (6.3.17):

$$-\frac{\hbar^2}{2m}\Psi''_\alpha(x) + \left\{\sum_{\alpha'}\int d\alpha'\right\}V_{\alpha\alpha'}(x)\Psi_{\alpha'}(x) = (E - \varepsilon_\alpha)\Psi_\alpha(x)\,. \tag{6.3.18}$$

The solution of these equations is significantly more difficult than of (6.3.9), even if integration over α in (6.3.17,18) is approximately restricted to a finite interval up to α_{max}.

Such a situation is quite typical when for calculations a discrete set $\{\chi_\alpha\}$ is desirable, but the direct choice of the eigenstates of the asymptotic Hamiltonian of the subsystem as a basis leads to χ_α with a continuous spectrum of α values. We shall point out some methods of solving such problems. In the special case of purely elastic scattering the following approach was proposed.

The Sturm functions as a basis in scattering problems. If in (6.3.17) consideration is limited only to the sum, then, instead of (6.3.18), the previous system (6.3.9) would be obtained. However, the integral neglected in (6.3.17) may turn out to be too rough an approximation, especially when there exists a single weakly bound state χ_1. Is it possible to satisfy such seemingly contradictory requirements as follows: first, the set χ_n should be countable (without a continuous part) and, second, it should include the ground bound state of transverse motion, which usually belongs to the set of discrete *plus* continuous physical states and which enters into the asymptotic form of ψ:

$$\psi(x, y) \xrightarrow[x \to \pm\infty]{} \Psi_1(x)\chi_1(y) ,$$

when the sole elastic scattering channel is open. It turns out that this may be achieved with the aid of the Sturm functions satisfying the equation

$$-\frac{\hbar^2}{2m} s_n''(y) + \lambda_n V(y) s_n(y) = \varepsilon_1 s_n(y) , \tag{6.3.19}$$

where ε_1 is the energy of the ground state of transverse motion (along y) as $x \to \pm\infty$. This ε_1 *remains unchanged* for all the solutions s_n involving different n.

We shall search for the solutions of (6.3.19) turning to zero at large y. When $\lambda = 1 \equiv \lambda_1$, the function s_1 coincides with the function of the ground state $\chi_1(y)$, as was required. This is the first Sturm function.

Multiplication of $V(y)$ by $\lambda > 1$ leads to the potential well $\lambda V(y)$ becoming deeper than $V(y)$. As a result, the levels in $\lambda V(y)$ are lowered with respect to the corresponding levels in $V(y)$. As α is increased continuously, the levels in $\lambda V(y)$ are lowered more and more, and new discrete states are detached from the continuum. Therefore, it is possible to pick out the discrete values of λ_n for which the energies of the levels with numbers coming next in order, in the wells $\lambda_n V(y)$, are equal to ε_1. The solutions $s_n(y)$ corresponding to these λ_n form an infinite *countable set* of Sturm functions (Fig. 6.2). They compose a complete system. The condition of orthonormalization for the functions s_n differs from the conventional one in that it is written with the weight of the potential

$$-\int s_n(y) V(y) s_{n'}(y) dy = \delta_{nn'} , \tag{6.3.20}$$

where the sign "minus" is introduced to make the weight of $-V(y)$ a positive quantity ($V(y) \leqslant 0$). (It is assumed that at all y the potential $V(y) \leqslant 0$. Then the system $\{s_n(y)\}$ is complete in the space of functions with the modulus $\|\chi\| = -\int \chi^2(y) V(y) dy$. Note that the appearance in (1.4.30) of the weight factors depending on r can be related to the Sturm set, while λ_n is replaced by $l(l+1)$ in the centrifugal potential.)

Only one of the Sturm functions coincides with the physical wave function of transverse motion. The other functions correspond to artificial potentials

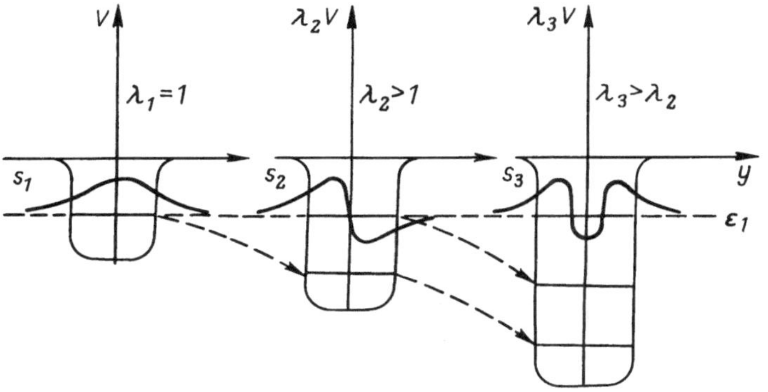

Fig. 6.2. The Sturm functions $s_1, s_2, s_3, \ldots, s_n \ldots$ represent the wavefunctions of bound states of the *same* energy ε_1 in potential wells of varying depth $V, \lambda_2 V, \lambda_3 V \ldots \lambda_n V \ldots$.

$\lambda_n V(y)$. But this does not prevent the description of elastic scattering, since below the threshold of inelastic scattering only the function s_1, which has physical meaning, contributes to the asymptotic behavior of ψ.

For the coefficients of the expansion of $\psi(x, y)$ in $s_n(y)$ equations similar to (5.2.1) are obtained, except that they *do not decouple at large* x, but transform into a set of differential equations with constant coefficients. The analytic forms of the solutions of the latter are known, so no real difficulties are encountered in fixing asymptotic conditions. For details the reader is referred to *Zhigunov* and *Zakhariev* (1974).

The reason that the channels are coupled asymptotically is that with the exception of $s_1(y)$, the set $\{s_n(y)\}$ contains no eigenfunctions of that part of the asymptotic Hamiltonian which corresponds to transverse motion.

The Bubnov–Galerkin (BG) *method.* Consider the general case of the solution of a linear differential equation:

$$L\psi = J\ . \tag{6.3.21}$$

The source function J, the inhomogeneous term, is often absent in real problems, e.g., the Schrödinger equation. But we can go from a scattering boundary condition for ψ with an inhomogeneous term corresponding to the incident wave $\mathring{\psi}$, to a homogeneous boundary condition for $\psi - \mathring{\psi}$. Then a source term $L\mathring{\psi} \equiv (H - E)\mathring{\psi}$ appears in equation of motion for $(\psi - \mathring{\psi})$.

We need to find the function ψ satisfying certain conditions on the boundary S of the region D:

$$\mathscr{L}\psi|_S = g|_S \tag{6.3.21'}$$

($\mathscr{L}$ may be a differential operator, like L.) If the boundary conditions (6.3.21′) are sufficiently simple and homogeneous ($g|_S = 0$) so that a complete set of functions χ_α defined within the region D and satisfying (6.3.21′) with $g|_S = 0$

$$\mathscr{L}\chi_\alpha|_S = 0 \tag{6.3.22}$$

can be found, then it is convenient to search for the approximate solution ψ^N of the equation (6.3.21) in the form of an expansion

$$\psi^N = \sum_\alpha^N \Psi_\alpha^N \chi_\alpha . \tag{6.3.23}$$

The functions Ψ_α^N can be found from the requirement of orthogonality of the basis functions χ_α to the discrepancy of the equation (6.3.21) when ψ^N is substituted into (6.3.21):

$$\left\langle \chi_\alpha | L \sum_{\alpha'}^N \Psi_{\alpha'}^N \chi_{\alpha'} - J \right\rangle_D = 0 . \tag{6.3.24}$$

This is the BG equation for Ψ_α^N which satisfies exactly the boundary conditions of (6.3.21′), in accordance with (6.3.22,23), and only in the *generalized* sense of (6.3.24) satisfies (6.3.21).

Several other methods follow the same logic as the BG method: the Trefz method, the generalized BG method (GBG) and the Petrov–Bubnov–Galerkin method (PBG) (Mikhlin 1966, 1970; Krasnoselsky et al 1969; Zhigunov, Zakhariev 1974). The Trefz method is more suitable when there are some known solutions χ_α of (6.3.21) with $J = 0$ which form a complete set on the boundary S. The function ψ^N is then determined in the form of a linear combination of such solutions χ_α from the condition of orthogonality of the discrepancy of the boundary conditions (6.3.21′) for ψ^N to these solutions χ_α. In the GBG method the complete set within the region D and on its boundary S is chosen as a basis, while the expansion coefficients are found from the requirement that the projections onto the first basis functions of the combinations of discrepancies of the (6.3.21,21′) become zero. In the PBG method orthogonality is required between the discrepancy of (6.3.21) and the functions of a basis differing from the one in which the desired solution was expanded. Another version of the PBG method is the method of moments, in which (6.3.21) is projected onto functions derived from the basis functions by the action of a certain operator L'. In the case that L' is chosen to be L, then this is known as the least-squares method.

A special case of the BG method is the method of finite elements (Strang, Fix 1973) mentioned in Chap. 3 and which is close to the "collocation" method (Mikhlin 1970) requiring the discrepancy to become zero only at certain points, the collocation nodes.

Reconstruction of the initial potential V from $V_{\alpha\beta}(x)$. In the general case, it is possible to construct a potential nonlocal in the variables ξ of the basis functions $\chi_\alpha(\xi)$ from the interaction matrix $\| V_{\alpha\beta}(x) \|$. Let

$$V_{\alpha\beta}(x) = \int\int \chi_\alpha^*(\eta) V(x, \eta, \eta') \chi_\beta(\eta') d\eta \, d\eta' . \tag{6.3.25}$$

Now we multiply (6.3.25) by $\chi_\alpha(\xi)$ and $\chi_\beta^*(\xi')$ and perform summation over

α and β, taking into account the property of completeness of the set χ_α: $\sum_\alpha \chi_\alpha(\xi)\chi_\alpha^*(\eta) = \delta(\xi - \eta')$.

As a result, we obtain

$$V(x, \xi, \xi') = \sum_{\alpha\beta} \chi_\alpha(\xi) V_{\alpha\beta}(x) \chi_\beta^*(\xi') . \tag{6.3.26}$$

If V is local, that is, $V = V(x, \xi)\,\delta(\xi - \xi')$, then, by integrating (6.3.25) over one of the variables we find

$$V(x, \xi) = \sum_{\alpha\beta} \chi_\alpha(\xi) V_{\alpha\beta}(x) \int \chi_\beta(\xi') d\xi' . \tag{6.3.27}$$

In the special case of $\chi_\alpha = Y_{lm}(\boldsymbol{\Omega})$ we have $V(x, \boldsymbol{\Omega}) = \sum_{lm} \chi_{lm}(\boldsymbol{\Omega}) Y_{lmoo}/4\pi$.

6.4 The Multi-dimensional Inverse Problem

We base this discussion on the review by *Novikov* and *Henkin* (1987). *Faddeev* (1976, and references therein) was the first to introduce the multi-dimensional *Volterra* operators (unity + triangle) of generalized shift which play a key role in the inverse problem.

Faddeev also obtained the generalized Gelfand–Levitan equation into which the scattering data enter through the function $h(k, k')$ which is determined from the scattering amplitude $f(k, k')$ by solving the integral equation. Faddeev also formulated the conditions to be imposed on the scattering data to make them correspond to the local potential (actually, only in the absence of bound states).

Part of Faddeev's work was obtained independently by *Newton* (1974; 1982) who then proposed a more simple approach (1983, 1985) which requires solution of equations that are not less complicated, however.

Nachman and *Ablowitz* (1984) simplified the Faddeev procedure by taking advantage of the "$\bar{\delta}$" (d-bar) method. Newton and, independently, *Novikov* and *Henkin* (1987) gave a still more simple formula relating the Fourier transform of the potential $V(p)$ and the auxiliary function h.

A complete description of the scattering data for any smooth, rapidly decreasing potentials (without restrictions imposed on the number of bound states) was given by *Novikov* and *Henkin* (1987) who also indicated the minimal class of such data.

Moses (1956) and *Prosser* (1980) proposed a method for reconstruction of a potential from nonoverdetermined scattering data by using the backward scattering amplitude ("the reflection coefficient") $R(\mathbf{p})$ which is a function of *three* variables, just like the potential $V(\mathbf{r})$. This method requires solution of a *nonlinear* integral equation for auxiliary amplitudes.

Although these methods permit solution of the multi-dimensional inverse problem, they are not directly applicable to the solution of multi-dimensional nonlinear equations as in the one-dimensional case.

Dubrovin et al (1976), *Manakov* (1985) and others [see references in the review by *Novikov* and *Henkin* (1987)] have solved the two-dimensional inverse (not overdetermined) problem for a fixed energy. A generalization to the case of three dimensions was given by *Novikov* and *Henkin* (1987) for sufficiently small and rapidly decreasing potentials.

Grinevitch (1986) obtained examples of exactly solvable two-dimensional problems with potentials that are transparent at a fixed energy and represent soliton solutions of 2 + 1-dimensional nonlinear Veselov–Novikov equations. These potentials have the form of simple rational expressions.

6.5 Separation of Variables in a Noncentral Field

These problems have been discussed by *Pivovarchik* et al (1986), *Funke* and *Zakhariev* (1985; 1987), and *Suzko* (1986).

In the spheroidal coordinates ξ, η, φ (Fig. 6.3) or $\rho = a(\xi^2 - 1)$, $\Theta = \arccos \eta, \varphi$ that are analogous to the conventional spherical coordinates r, θ, φ (Funke, Zakhariev 1985–87), potentials of the form

$$V(\rho, \Theta) = u(\rho)/(\rho^2 + a^2 \sin^2 \Theta) \tag{6.5.1}$$

with the arbitrary function $u(\rho)$ allow reduction of the Schrödinger equation for $\Psi(\rho, \Theta, \varphi) = R(\rho)S(\Theta)\exp(\mathrm{im}\, \varphi)$, where S is a spheroidal function, to the one-dimensional equation

$$R''(\rho) + \left[K^2 - \frac{\lambda_{lm} + u(\rho)}{\rho^2 + a^2} - \frac{a^2}{\rho^2(\rho^2 + a^2)} \left(m^2 + \frac{2\rho^2 - a^2}{4(\rho^2 + a^2)} \right) \right] R(\rho) = 0 \, . \tag{6.5.2}$$

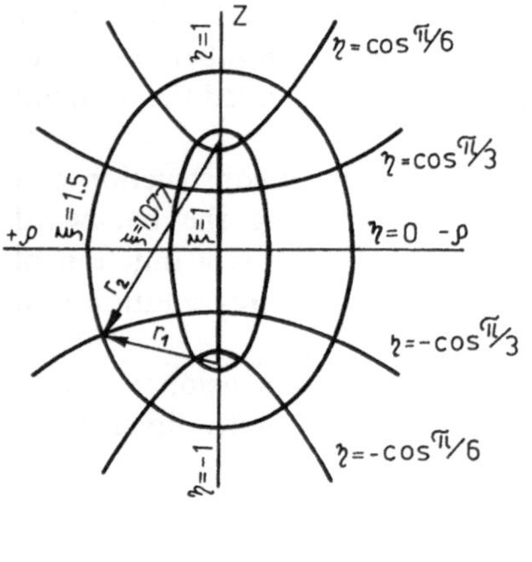

Fig. 6.3. Spheroidal coordinates

This equation resembles the radial equation except for the occurrence of two centrifugal barriers depending on λ_{lm} and m, and the fact that the parameter $\lambda_{lm}(k)$ which replaces $l(l+1)$ depends on the energy.

For states belonging to the continuous spectrum the asymptotic behavior of $R(\rho)$ at large ρ is determined by the phase shift:

$$R(\rho) \xrightarrow[\rho \to \infty]{} \sin(k\rho - \tfrac{1}{2}l\pi + \delta_{lm}) ,$$

where $1 - |m|$ is the number of nodes in the angular behavior of the wave function $S(\Theta)$.

At a fixed energy it is possible to obtain the modified Gelfand–Levitan (Regge–Newton–Sabatier equation corresponding to the "radial" equation (6.5.2)):

$$K(\rho, \rho') + Q(\rho, \rho') + \int_0^\rho K(\rho, \rho'')Q(\rho'', \rho') \frac{d\rho''}{(\rho''^2 + a^2)} = 0 , \tag{6.5.3}$$

where K is the kernel of the operator of generalized shift,

$$R(\rho) = \mathring{R}(\rho) + \int_0^\rho K(\rho, \rho')\mathring{R}(\rho') \frac{d\rho'}{(\rho'^2 + a^2)}$$

and Q is determined from the scattering data (Newton 1983):

$$Q(\rho, \rho') = \sum_{lm} C_{lm} \mathring{R}_{lm}(\rho)\mathring{R}_{lm}(\rho') ,$$

with the coefficients C given by the equations

$$\sin(K\rho - \pi l/2 + \delta_{lm}) = \sin(K\rho - \pi l/2)$$
$$- \sum_{l'm'} C_{l'm'} \sin(K\rho - \pi l/2 + \delta_{lm}) L_{ll'mm'}(\rho)$$

$$L_{ll'mm'} = \int_0^\rho \mathring{R}_{l'm'}(\rho')\mathring{R}_{lm}(\rho')d\rho'(\rho'^2 + a^2)r' .$$

The potential is determined by a formula similar to (1.4.34):

$$u(\rho) = -2\sqrt{\rho^2 + a^2} \frac{d}{d\rho}[K(\rho, \rho)/\sqrt{\rho^2 + a^2}] . \tag{6.5.4}$$

The appearance of the new factors $(\rho^2 + a^2)^{-1}$, $(\sqrt{\rho^2 + a^2})^{\pm 1}$ in (6.5.3,4) is related to the Sturm functions being orthonormalized with the weight of the expression which occurs in front of the eigenvalue [see (6.3.19,20) where $V(y)$ served as this factor]. The peculiarity of (1.4.34) and (2.5.10), these factors, is explained in the same way, as is the factor $1 + \alpha r^2$ (Rudyak et al 1984; 1987; Suzko 1985; 1986) within the approach to the inverse problem with $(E - E')/(\lambda^2 - \lambda^{2\prime}) = \text{const}$ and the corresponding $1/r$ factors in the approaches using variable charge (Poplavsky 1986; Popushoy 1985; 1986).

The dependence of $\lambda(k)$ on energy seems to prevent the solution of the inverse problem with a fixed λ and variable energy just because there simply

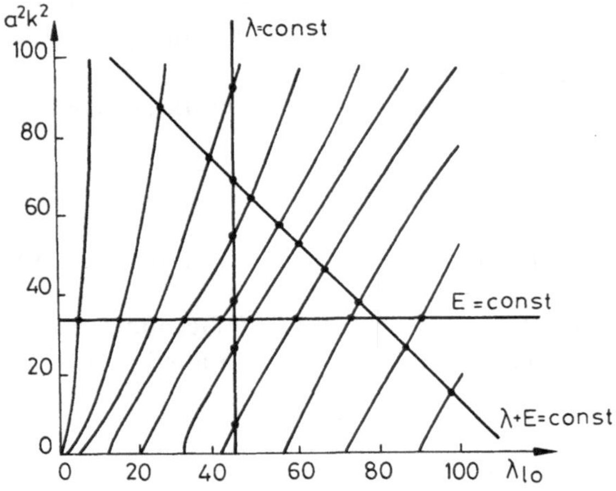

Fig. 6.4. Dependence of $\lambda(k)$ on energy

exist no corresponding scattering data. However, in approximate methods one could find the scattering phases by interpolation from the lines $\lambda(k)$ on which they were measured (Fig. 6.4) to $\lambda = \text{const}$.

6.6 Exactly Solvable Models

The multi-channel technique can be applied for construction of Bargmann potentials in the case of several dimensions (Plekhanov et al 1982). Thus, constructing the interaction matrix (6.1.2) in accordance with (5.4.5)

$$V_{lml'm'}(r) = -2\frac{d}{dr}\left\{j_l(\mathrm{i}\kappa_\lambda, r)\gamma^\lambda_{lm}\gamma^\lambda_{l'm'}j_{l'}(\mathrm{i}\kappa_\lambda, r)(\mathrm{i}\kappa_\lambda)^{-2l-2}\right.$$
$$\left.\times\left[1+\sum_{LM}\kappa_\lambda^{-2l-2}\int_0^r|\gamma_{LM}j_L(\mathrm{i}\kappa_\lambda, r')|^2dr'\right]^{-1}\right\}, \tag{6.6.1}$$

where j_l are Bessel functions and γ^λ_{lm} are constants, we obtain following (6.3.26), a potential nonlocal in angles

$$v(r, \vartheta, \varphi, \vartheta', \varphi') = \sum_{lml'm'} V_{lml'm'}(r)Y_{lm}(\vartheta, \varphi)Y^*_{l'm'}(\vartheta', \varphi') \tag{6.6.2}$$

which permits exact solution of the multi-dimensional Schrödinger equation.

Matrices of a more general Bargmann type yield a wide family of solutions in explicit analytic forms. Thus, by this method it is possible to also construct transparent, spherically nonsymmetric nonlocal potentials. These were found earlier by another method by *Moses* (1979).

Besides phase- and spectral-equivalent potentials, it is possible to introduce different potentials to which the same wave function Ψ corresponds, throughout the entire space and not only in the asymptotic region, as in the case of phase-equivalent potentials.

Bargmann potentials corresponding to noncentral potentials permitting separation of variables in spheroidal coordinates were constructed by *Pivovarchik* et al (1986), *Suzko* (1986) and *Funke* and *Zakhariev* (1987) and in the adiabatic representation by *Vinitsky* and *Suzko* (1990) in the general case.

Nonstationary Solutions

The generalized analogues of Bargmann potentials were given for the time-dependent Schrödinger equation in the two-dimensional case (Dubrovin et al 1988; Krichever 1989; Novikov 1988).

For example, as the generalization of the system of algebraic equations (2.3.11) for the stationary problem we get [see equations (44), (36), (37) in the review by Dubrovin et al 1988]:

$$\sum_{j=1}^{N}\left(C_{ij}+\frac{\mathrm{e}^{\mathrm{i}(\bar{\chi}_i x+\bar{\chi}_i^2 t-\chi_j x+\chi_j^2 t)}}{\bar{\chi}_i-\chi_j}\right)\Psi_j = -\mathrm{e}^{\mathrm{i}\bar{\chi}_i x+\bar{\chi}_i^2 t}\,. \tag{6.6.3}$$

The corresponding time-dependent potential has a form similar to (2.3.14):

$$V(x,t)=2d^2/dx^2\ln\det P(x)\,, \tag{6.6.4}$$

where $P(x)$ is the matrix of coefficients in (6.6.3). The wave function can also be expressed through P:

$$\Psi(x,t,k)=\frac{\det\hat{P}(x,t,k)}{\det P(x,t)}\,, \tag{6.6.5}$$

where $\hat{P}_{ij}=P_{ij}$ for $i,\ j=1,\ldots N$; $\hat{P}_{00}=1$; $\hat{P}_{i0}=\exp(\mathrm{i}\chi_i x+\mathrm{i}\chi_i^2 t)$; $\hat{P}_{0i}=(k-\chi_i x-\chi^2 t)\exp(-\mathrm{i}\chi_i x-\mathrm{i}\chi_i^2 t)$ for $i=1,\ldots,N$. Expression (6.6.5) is similar to the stationary one-channel formula (4.3.19a) in the book by *Chadan* and *Sabatier* (1989) and even more so to the multi-channel Bargmann-type wave function (Amirkhanov et al 1989). This similarity exists in spite of the difference in the number of variables.

Bound states in the continuum. A broad class of multi-channel Bargmann models with bound states E_b embedded in the continuous spectrum (Pivovarchik et al 1986) are obtained from (5.4.5), in which it is no longer necessary to assume the thresholds to be identical in all the channels. If E_b is chosen to be higher than the elastic scattering threshold $E_1 \equiv E-\varepsilon_1=0$, but such that at least one channel remains closed, then in the expression for $p(r)$ there will appear exponentially growing terms which make the matrix $V_{\alpha\beta}$ in (5.4.5) *short-range* and induce the wave function (vector) Ψ of the bound state in the continuum to decay rapidly with distance. If, on the contrary, E_b is chosen to be higher than the uppermost threshold, then the integral in $p(r)$ over the bilinear combinations of the oscillating functions $\mathring{\Phi}_{\alpha\beta}$ will increase as $\sim r$, while in V and Ψ there will appear oscillating tails falling off as $\sim\sin(\kappa r)/r$, as in the one-channel case (2.2.29,30), see Fig. 1.6.

It is shown in Fig. 6.5 how the 2 × 2 potential matrix $V_{\alpha\beta}(r)$ is modified with raising of a bound state through the channel thresholds $E_\alpha = E - \varepsilon_\alpha = 0$ (Zakhariev et al 1989). Unlike the one-channel case, the bound state wave can be confined in the continuum by the short-range interaction for energy values below the upper threshold. Figure 6.5a shows the potential matrix $V_{\alpha\beta}(r)$ for the two-channel system with a bound state below both thresholds $E_1, E_2 = 0$. For a bound state between the thresholds the matrix $V_{\alpha\beta}(r)$ does not change very much, as shown in Fig. 6.5b,c. The interaction remains short-range and can

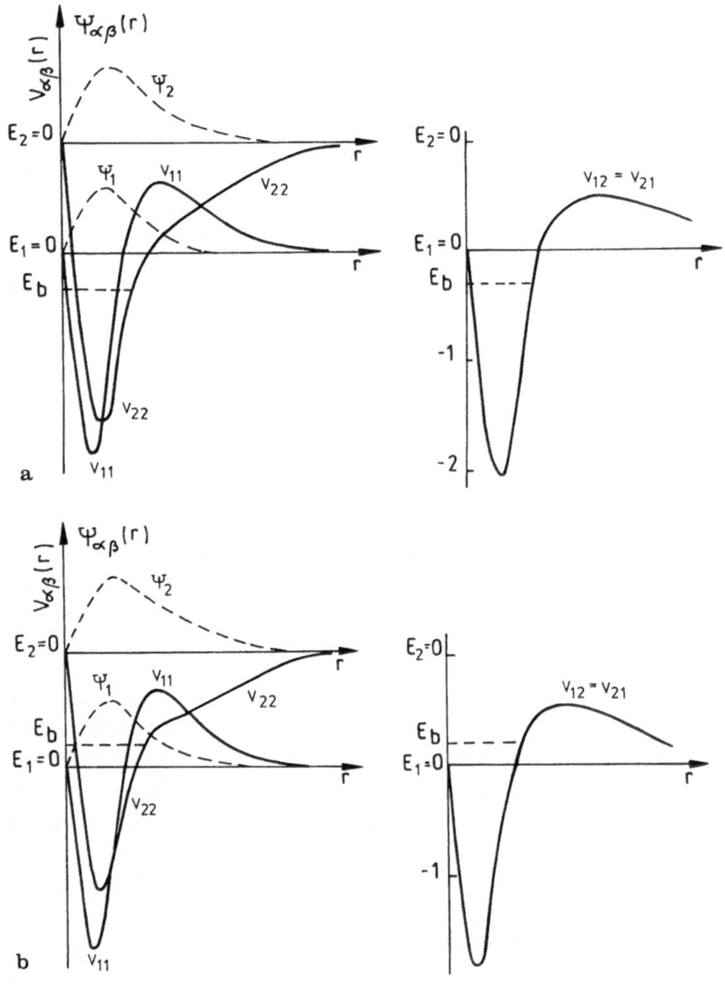

Fig. 6.5 (*continued*)

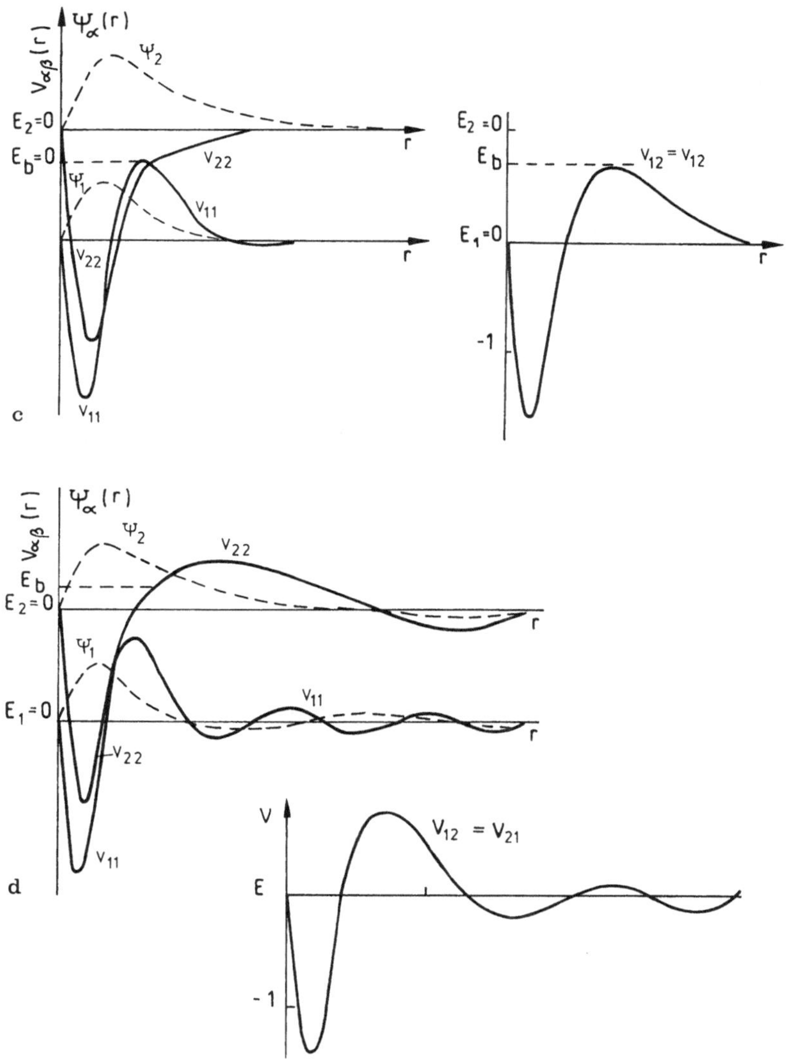

Fig. 6.5a–d. Deformations of the interaction matrix $V_{\alpha\beta}(r)$ of the two-channel system caused by shifting the bound state energy E_b into the continuum. **a** E_b is below the continuum spectrum. **b,c** E_b is above the threshold $E_1 = 0$ where the first channel becomes open. **d** E_b is above the highest threshold. Compare with Fig. 1.6

confine the wave function near the origin although there are scattering states below, above and at the same energy E_b. Only when E_b is shifted above the upper threshold $E_2 = 0$ oscillations in $V_{\alpha\beta}(r)$ appear, to prevent the decay of the bound state.

Particle in an external periodical potential. The separation of the motion of waves inside a single period from the motion over the periods of an external field $V(x)$ (along the discrete α-variable numbering the periods) allows a new formulation of the quasi-momentum and quasi-energy, which are fundamental concepts in solid state physics (Zakhariev, Zastavenko 1989).

It may seem curious that the one-dimensional Schrödinger equation for this problem can be written in a form equivalent to the two-dimensional differential-difference equation in (α, x)-space.

The periodical intervals of the potential $V(x \pm X) = V(x)$ can be numbered by $\alpha = 0, \pm 1, \pm 2 \ldots$. Introducing the index α in the wave function, we can consider $\Psi_\alpha(x)$ as a function in two-dimensional (α, x)-space, Fig. 6.6. It appears that the motion of a particle along the infinite axis x can be separated,

$$\Psi_\alpha(x) = \Psi(x)\chi_\alpha ,$$

into free motion along the new dimension of the discrete-valued variable α [with quasimoment γ/Δ, compare with θ/Δ (1.2.12a)]

$$-(\chi_{\alpha+1,\lambda} - 2\chi_{\alpha\lambda} + \chi_{\alpha-1,\lambda})/\Delta^2 = \lambda\chi_{\alpha\lambda} ; \quad \chi_{\alpha\lambda} = \exp(\pm i\gamma\alpha)$$

and vibrations inside the finite continuous interval $0 \leqslant x \leqslant X$,

$$-d^2/dx^2\,\Psi_n(x) + V(x)\Psi_n(x) = (\varepsilon_n - \lambda)\Psi_n(x) ,$$

$$\Psi_n(0) = \exp(i\gamma)\Psi_n(X) , \quad \Psi_n'(0) = \exp(i\gamma)\Psi_n'(X) ,$$

where the energy λ of α-motion has a continuous spectrum bounded from above due to the discreteness of the α-variable [compare (1.3.15a)]:

$$0 \leqslant \lambda \leqslant 2/\Delta^2 .$$

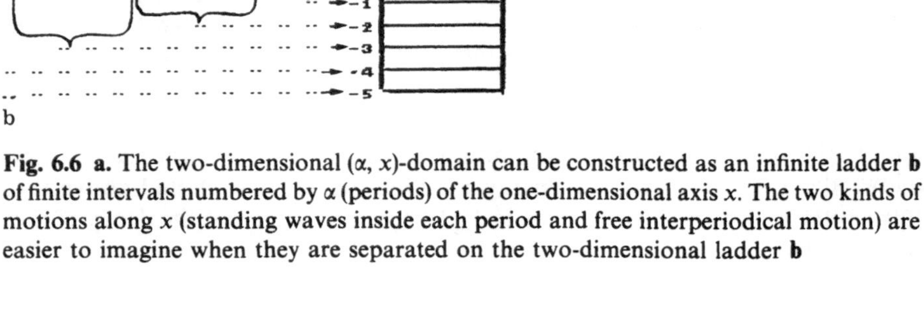

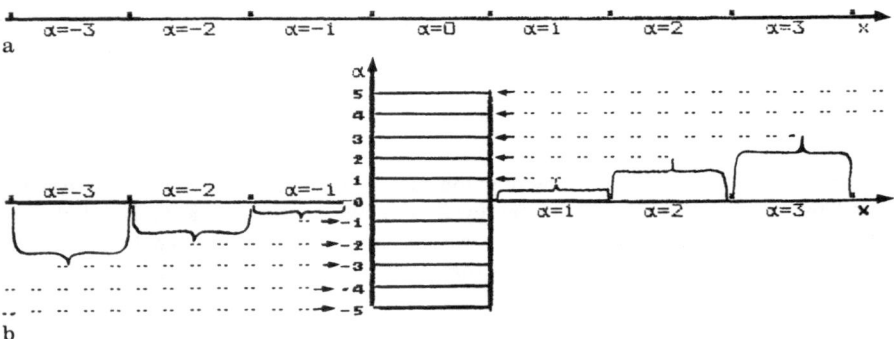

Fig. 6.6 a. The two-dimensional (α, x)-domain can be constructed as an infinite ladder **b** of finite intervals numbered by α (periods) of the one-dimensional axis x. The two kinds of motions along x (standing waves inside each period and free interperiodical motion) are easier to imagine when they are separated on the two-dimensional ladder **b**

The complete (stripped) spectrum of a particle in the periodic potential is the superposition of discrete energy levels ε_n (Sturm–Liouville eigenvalue problem on a finite interval) and finite continuous λ-band: $E = \varepsilon_n + \lambda$.

The ordinary momentum of free motion γ for the finite-difference Schrödinger equation appears to be a quasi momentum of motion in the periodic potential. All peculiarities of quasi-momentum are caused by the discreteness of the α-variable and should be familiar from Chap. 1.

Like the correspondence between the Bessel functions and sines, Neiman and Hankel functions and cosines, and exponentials we can speak about "Bloch-sines", "Bloch-cosines" and "Bloch-exponentials" (as well as about "Coulomb sines, cosines, exponents" etc.; Zakhariev 1990). In the forbidden zones the arguments of these functions become complex so that the solutions are exponentially increasing (or decreasing) as for the subbarrier motion. The general solutions on arbitrary pieces of the axis with a periodical field can be constructed as linear combinations with constant coefficients to be determined from the boundary conditions.

In particular, we can insert such an interval into the axis with another periodical field, shift the zones of oscillating (or exponentially increasing/decaying) solutions on separate intervals by additional potential barriers/wells or change the period there. Thus, the section of the forbidden zone can be inserted into the zone of allowed motion and serve as an effective "potential barrier". Systems of such barriers with resonance tunneling of Bloch waves can be combined (Chap. 1), to find the time evolution of Bloch wave packets (Olchovsky et al 1989, 1990; surprisingly high subbarrier speeds of wave packets and their slowing down at the end of the barrier were observed). In contrast, the interval of oscillating solutions (allowed zone) confined inside the forbidden zone can give bound states inserted into lacuna.

Bloch-type solutions can be used as building material in the different approaches to the inverse problem (on the whole axis, on the half axis, on the finite interval; Chap. 1) for construction of different quadratically integrable perturbations of the periodical potentials. Thus new classes of Bargmann-type potentials corresponding to the periodic reference Bloch solutions (particularly with bound states inside the conductive or forbidden zones) can be obtained.

The above ideas can be generalized to the many-channel and many-body systems. If at some point there is a wave only in one channel, we expect that at the neighboring points waves will appear in all other channels due to the perturbation of the periodic field. But there are N (according to the number of channels) eigenchannel states, for which the relative weights of the channel wave functions remain unchanged when the coordinate x is shifted by a period.

The motion in the external potential which is periodic in time can be treated in the same manner with introduction of quasi-energy.

6.7 Notes on the Literature

The multi-dimensional scattering problem for the difference Schrödinger operator was considered by *Eskina* (1970). Thus, the possibility of reconstruction of the potential from scattering data within an arbitrarily small solid angle was demonstrated.

It has been shown that scattering on two or more *fixed* centers can be reduced to the solution of simple algebraic equations where the scattering parameters on the individual centers serve as the initial data (Revai 1968).

The multi-dimensional finite-difference inverse problem with a potential depending on the velocity $V(\hat{p}^2, n, m)$ was solved by *Suzko* (1978).

The existence of a difference analogue of the formalism of *Faddeev* (1976) and *Newton* (1982) of reconstructing spherically nonsymmetric potentials has not yet been established. Probably, a discrete model of the theory would be useful in the multi-dimensional inverse problem from pedagogical point of view. As M. G. Krein and I. S. Kats have mentioned in the preface to the Russian edition of the book by Atkinson, 1964: "The development of the theory of discrete and continuous problems often proceeded independently and out of step; sometimes decades passed before certain links of the continuous chain of analogies were fully realized. . . The search for analogies between these problems still remains the source of new studies."

A generalization of the Darboux (Crum–Krein) method to multi-dimensional problems has been proposed by *Andrianov* et al (1984). Here, from a known solution of a single n-dimensional partial differential equation, the exact solution $\{\psi_i\}$ of a set of partial differential equations coupled by some matrix differential operator H_{ij} is constructed in an explicit form.

Bagrov et al (1982) has considered the exact solutions to the relativistic Klein–Gordon and Dirac equations for a charge in an electromagnetic field when the variables in the equations of motion are separable (Hundreds of such solutions already exist.)

Nonstationary inverse problems for hyperbolic equations have been studied by *Nizhnik* (1973). Multi-dimensional inverse problems for various equations are dealt with in the book by *Buchgeim* (1983).

The exact solution of the Schrödinger equation with a noncentral parabolic potential was considered by *Guha* and *Mukherjee* (1987). The generalized Levinson theorem for the case of a noncentral potential was demonstrated by *Newton* (1977).

The inverse problem in atmospheric optics was dealt with by *Naatz* (1986). Solution of the inverse problem in the WKB-approximation for noncentral potentials permitting separation of the variables in spheroidal coordinates was considered by *Funke* and *Zakhariev* (1987).

The multi-dimensional inverse problem in the adiabatic representation was formulated by *Vinitsky* and *Suzko* (1990). They proposed a procedure of constructing a wide class of exactly solvable models.

Chapter 7
Multi-particle Systems

7.1 General Remarks

During the last twenty-five years significant progress in the three-body problem has been made which was stimulated by the appearance of the Faddeev integral equations (Belyaev 1990; Merkuryev, Faddeev 1985; Schmid, Ziegelman 1974). A similar formalism of integral equations was also being developed for more complicated systems (Komarov et al 1985; Faddeev, Merkuryev 1985). In parallel with the development of methods for solving integral equations, exploration of another approach to the problem for which the conventional differential form of the Schrödinger equation serves as a starting point continues. Here Faddeev's idea concerning the separation of components of the wave function corresponding to different particle arrangements in fragments is utilized, or the Faddeev equations in the coordinate representation are solved directly (Merkuryev 1970–1980; Faddeev, Merkuryev 1985). The integral and differential approaches are to a large extent complementary, since the approximations applied in these approaches differ from each other. The Faddeev approach is not considered here. We refer the reader to *Belyaev* (1990) and *Faddeev* and *Merkuryev* (1985).

We shall limit ourselves to the reduction of multi-particle problems to multichannel (one-dimensional) problems, considered in Chap. 5. For the simplest but physically significant example, the phenomenon of tunneling by complex particles through potential barriers is chosen (Sect. 7.3). Another example is the method of K-harmonics (Simonov, Badalyan 1967), i.e., by expansion in hyperspherical functions (Sect. 7.5). This method was used in solving the nuclear inverse problem with the aid of the direct one, that is, for determination of nucleon–nucleon potentials from the characteristics of nuclei. Significant progress was made in this problem owing to the technique of K-harmonics, although in the multi-particle inverse problem now only the first steps are being made (Bursak et al 1982; Gorbatov et al 1979–1988).

In Sect. 7.4, with the aid of the multi-channel formalism we consider the scattering of particles by a complex target, taking into account the excitation of rotations and oscillations within the framework of the model of collective motions. It is possible, in principle, to reconstruct a nonspherical field of a target which rotates and oscillates. The Levinson theorem also exhibits certain unique features in the multi-particle case (Sect. 7.6).

In the case of complex multi-particle processes involving rearrangement of particles the direct reduction to multi-channel equations (coupled ordinary differential Schrödinger equations with a local interaction matrix) encounters significant difficulties. The possibility of overcoming them (Fano, Rau 1986; Zakhariev 1974–1988; Zhigunov, Zakhariev 1974; Vinitsky, Suzko 1990; Dubovik et al 1989) in the direct and inverse problems is discussed in Sect. 7.7.

7.2 Asymptotic Hamiltonians and Boundary Conditions

As the most simple example we shall consider three interacting particles. In the center-of-mass system the Schrödinger equation is

$$H\psi \equiv \left\{ -\frac{\hbar^2}{2M_k}\Delta_{\mathbf{R}_k} - \frac{\hbar^2}{2\mu_{ij}}\Delta_{\rho_{ij}} + V_{12} + V_{13} + V_{23} \right\} \times \psi(\mathbf{R}_k, \boldsymbol{\rho}_{ij}) = E\psi(\mathbf{R}_k, \boldsymbol{\rho}_{ij}) , \tag{7.2.1}$$

where

$$1/M_k = 1/(m_i + m_j) + 1/m_k ; \quad 1/\mu_{ij} = 1/m_i + 1/m_j ;$$
$$i \neq j \neq k = 1, 2, 3 ,$$

while $(\mathbf{R}_k, \boldsymbol{\rho}_{ij})$ represents one of the three sets of the Jacobi coordinates (Fig. 7.1):

$$\boldsymbol{\rho}_{ij} = \mathbf{r}_i - \mathbf{r}_j ; \quad \mathbf{R}_k = \mathbf{r}_k - (m_i\mathbf{r}_i + m_j\mathbf{r}_j)/(m_i + m_j) .$$

For fixing the unique solution ψ the equation of motion must be supplemented with boundary conditions. They are formulated in various ways depending on the arrangement of the particles in the asymptotic region: 1 + (2, 3); 2 + (1, 3); (1, 2) + 3; 1 + 2 + 3, where the brackets unite the pairs forming bound states, while the particles outside the brackets are free to leave the other ones. To each such partition of the system there corresponds an asymptotic Hamiltonian H_{ij} or H_0 which is obtained from H at large distances between the groups when the interaction between them vanishes (Fig. 7.2):

$$H = \begin{cases} H_{ij} + V_{ik} + V_{jk} \xrightarrow[R_k \to \infty]{} H_{ij} \equiv -\dfrac{\hbar^2}{2M_k}\Delta_{\mathbf{R}_k} - \dfrac{\hbar^2}{2\mu_{ij}}\Delta_{\rho_{ij}} + V_{ij} ; \\ H_0 + V_{12} + V_{13} + V_{23} \xrightarrow[R_k \to \infty;\ \rho_{ij} \to \infty]{} H_0 \\ \equiv -\dfrac{\hbar^2}{2M_k}\Delta_{\mathbf{R}_k} - \dfrac{\hbar^2}{2\mu_j}\Delta_{\rho_{ij}} . \end{cases} \tag{7.2.2}$$

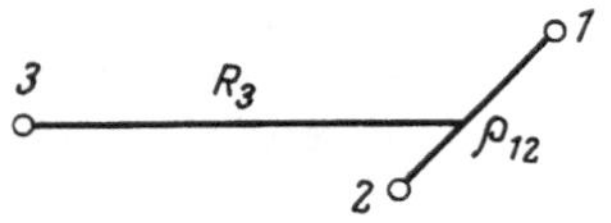

Fig. 7.1. The Jacobi coordinates for a three-body system, showing one of the possible arrangements

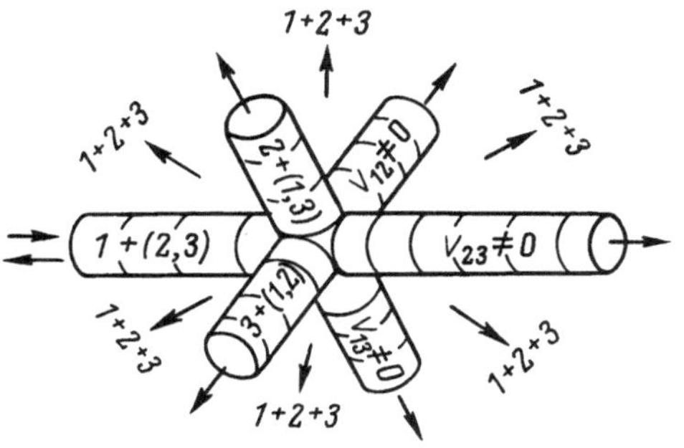

Fig. 7.2. The interaction region of three-body system in six-dimensional configuration space (conditional three-dimensional design). *Hypercylinders*, along which the waves of two-fragment channels with definite groups of particles propagate, diverge from a single center, where mixing of the channels takes place, in different directions and do not overlap asymptotically. The *arrows* indicate the directions of the incident and outgoing waves of the relative motion of the fragments. An incident flux arriving from a certain direction, upon reaching the center, is distributed over all the open channels

For different directions of the six-dimensional configuration space of three particles (7.2.1) transforms into

$$H_{ij}\psi = E\psi \quad \text{for} \quad R_k \to \infty; \quad H_0\psi = E\psi \quad \text{for} \quad R_k, \quad \rho_{ij} \to \infty\,, \tag{7.2.3}$$

in which separation of the variables $\mathbf{R}_k$, $\boldsymbol{\rho}_{ij}$ is already possible. The asymptotic behavior of the wave function ψ in channels differing in particle arrangement is conveniently represented by different systems of Jacobi coordinates. Thus, for processes involving particle rearrangement and with E lower than the disintegration threshold (into three particles) of the system ($E < 0$), we have:

$$\psi \underset{R_k \to \infty}{\sim} \sum_{\alpha}{}' \Psi_\alpha(\mathbf{R}_k)\chi_\alpha(\boldsymbol{\rho}_{ij})\,, \tag{7.2.4}$$

where $\chi_\alpha(\boldsymbol{\rho}_{ij})$ are functions of the bound states of the pair (i, j):

$$\left[-\frac{\hbar^2}{2\mu_{ij}}\Delta_{\rho_{ij}} + V_{ij}(\boldsymbol{\rho}_{ij})\right]\chi_\alpha(\boldsymbol{\rho}_{ij}) = \varepsilon_\alpha\chi_\alpha(\boldsymbol{\rho}_{ij})\,; \tag{7.2.5}$$

$\Psi_\alpha(\mathbf{R}_k)$ are functions of free propagating waves of the relative motion of the particle k and the center of mass of the pair (i, j); the prime on the summation indicates that summation is performed only over those states of the (i, j) pair to excite which the total energy of $E > \varepsilon_\alpha$ is sufficient.

For $E > 0$, disintegration of the system into three particles becomes possible. The asymptotic behavior of such states is conveniently represented in hyperspherical coordinates (Simonov, Badalyan 1967; Cox 1966). Instead of two ordinary three-dimensional vectors $\mathbf{R}_k$ and $\boldsymbol{\rho}_{ij}$, a single six-dimensional one $\boldsymbol{\rho}_6$ is introduced. It has the absolute value

$$\rho_6 = \sqrt{(M_k/\mathcal{M})R_k^2 + (\mu_{ij}/\mathcal{M})\rho_{ij}^2} \tag{7.2.6}$$

($\mathscr{M}$ is a certain auxiliary mass) and a direction given by five angles $\boldsymbol{\Omega}_5$. These may be the three Euler angles defining the orientation in space of a triangle with particles 1, 2, and 3 at its vertices and two angles determining its shape. Often so-called bispherical coordinates $\boldsymbol{\rho}_6$, $\boldsymbol{\Omega}_5 \equiv \{\alpha, \theta, \phi, \delta, \varphi\}$ are used where α is the "angle" determining the ratio of R_k to ρ_{ij} and the distribution of the total energy between the motions along $\mathbf{R}_k$ and $\boldsymbol{\rho}_{ij}$:

$$\tan\alpha = (m_{ij}/M_k)^{\frac{1}{2}}\rho_{ij}/R_k \, , \tag{7.2.7}$$

θ, ϕ and δ, φ are the angular variables of the vectors $\mathbf{R}_k$ and $\boldsymbol{\rho}_{ij}$.

The solution of the equation $H_0\psi = E\psi$ assumes the form of converging and diverging hyperspherical waves as $\rho_6 \to \infty$. When the initial channel is the one with two fragments, the absence of converging hyperspherical waves in ψ must be required:

$$\Psi_{\rho_6 \to \infty} \sim q(\boldsymbol{\Omega}_5)\exp(\mathrm{i}k_6\rho_6)/\rho_6^{5/6} \, , \tag{7.2.8}$$

where q is the amplitude of disintegration into three fragments.

The boundary conditions for ψ in two-particle channels are chosen by defining the flux of incident waves in ψ_α, that is we choose the constant coefficients of the converging spherical waves in the expansion of $\Psi_\alpha(\mathbf{R}_k)$ in the partial L-components. For example, in the case of an incident plane wave in the channel α_0:

$$\psi(\mathbf{R}_k, \boldsymbol{\rho}_{ij}) \xrightarrow[R_k \to \infty]{} \sum_{\alpha LM}' \mathrm{i}^L \frac{\sqrt{4\pi(2L+1)}}{R_k} \{\exp(-\mathrm{i}k_\alpha R_k + L\pi/2)\delta_{\alpha\alpha_0}$$

$$- S_{\alpha LM\alpha_0}\exp[\mathrm{i}(k_\alpha R_k - L\pi/2)]\} Y_{LM}(\boldsymbol{\Omega}_k)\chi_\alpha(\boldsymbol{\rho}_{ij}) \, . \tag{7.2.9}$$

The summation Σ' is performed only over the open channels, and the amplitudes S of the outgoing waves in Ψ_α are determined from the solution of the Schrödinger (7.2.1). The asymptotic behavior of Ψ_α can be chosen in the form of standing waves

$$\psi(\mathbf{R}_k, \boldsymbol{\rho}_{ij}) \xrightarrow[R_k \to \infty]{} \sum_{\alpha < M}' \sqrt{4\pi(2L+1)}\, \mathrm{i}^L \frac{1}{R_k} [a_{\alpha LM}\sin(k_\alpha R_k - L\pi/2)$$

$$+ b_{\alpha LM}\cos(k_\alpha R_k - L\pi/2)] Y_{LM}(\boldsymbol{\Omega}_k)\chi_\alpha(\boldsymbol{\rho}_{ij}) \, , \tag{7.2.9'}$$

often used in numerical calculations owing to the convenience in calculating purely real $\Psi_{\alpha\mathbf{R}_k}$. In (7.2.9′) the vector $\mathbf{a}$ with the components $a_{\alpha LM}$ is defined arbitrarily, while the vector $\mathbf{b}$ related to the reaction matrix $\mathbf{K}(\mathbf{b} \equiv \mathbf{Ka})$ is obtained from the equation of motion. If the total charge of the (i, j) pair differs from zero and the particle is also charged, then for the asymptotic behavior of ψ one should use the known Coulomb functions instead of the free-motion functions of particle k relative to (i, j).

Reactions with three free particles incident from different directions are feasible in gases. In formulating the respective boundary conditions it is necessary to take into account that waves decreasing less rapidly than $1/\rho_6^{5/2}$ may contribute to the asymptotic behavior of ψ.

Reduction of the multi-particle Schrödinger equation to a set of one-dimensional multi-channel equations. This is one of the most effective ways of solving the few-body problem when reactions involving the rearrangement of particles are forbidden. Now we shall consider a one-dimensional three-body model which has much in common with the motion of a particle in an external field (Sect. 6.3). Assume particle 3 to be incident on the pair (1, 2) bound by an infinitely deep well V_{12}. In this case rearrangement is impossible and as the basis for expanding the total wave function Ψ it is convenient to use the states of the internal motion of the pair $\chi_\alpha(\rho) \xrightarrow[\rho\to\infty]{} 0$:

$$-\frac{\hbar^2}{2\mu}\chi''_\alpha(\rho) + V_{12}(\rho)\chi_\alpha(\rho) = \varepsilon_\alpha\chi_\alpha(\rho);$$

$$\rho = x_2 - x_1; \quad \frac{1}{\mu} = \frac{1}{m_1} + \frac{1}{m_2}; \tag{7.2.10}$$

$$\psi(x, \rho) = \sum_\alpha \Psi_\alpha(x)\chi_\alpha(\rho); \quad x = x_3 - (m_1x_1 + m_2x_2)/(m_1 + m_2). \tag{7.2.11}$$

The boundary conditions for ψ with respect to the variable ρ are then satisfied automatically because χ_α tends towards zero at large ρ. For the functions $\Psi_\alpha(x)$ of the relative motion of particle 3 and the (1, 2) pair in the states α we obtain the multi-channel equations (5.2.1) by following the Bubnov–Galerkin principle (the index β corresponding to the asymptotic conditions as $x \to \infty$ is omitted):

$$-\frac{\hbar^2}{2M}\Psi''_\alpha(x) + \sum_{\alpha'} V_{\alpha\alpha'}(x)\Psi_{\alpha'}(x) = (E - \varepsilon_\alpha)\Psi_\alpha(x),$$

$$\frac{1}{M} = \frac{1}{m_1 + m_2} + \frac{1}{m_3}, \tag{7.2.12}$$

where

$$V_{\alpha\alpha'}(x) = \int \chi_\alpha(\rho)\left[V_{13}\left(x_3 - x_1 \equiv x + \rho\frac{m_2}{m_1 + m_2}\right)\right.$$
$$\left. + V_{23}\left(x_3 - x_2 \equiv x - \rho\frac{m_1}{m_1 + m_2}\right)\right]\chi_{\alpha'}(\rho)\,d\rho. \tag{7.2.13}$$

In accordance with the conservation of the total energy E, the motion of particle 3 relative to the (1, 2) pair takes place in the channel α with an energy of $E - \varepsilon_\alpha$, since the energy ε_α is spent for the internal motion of the quasideuteron. Particle 3 and the (1, 2) pair can be imagined to be swinging on an "energy swing" and exchanging portions of energy $\Delta\varepsilon = \varepsilon_\alpha - \varepsilon_{\alpha'}$ when they are close enough to interact with each other: the higher the excitation of the pair, the lower the energy to which particle 3 drops.

Problems involving a target with both discrete and continuous excitation spectra also reduce approximately to equations such as (7.2.12). To describe elastic scattering with account of virtual excitations of the target to states belonging to the continuous spectrum, one can use a basis of Sturm functions similar to (6.3.19,20). Even if the wave function of the system is expanded in an arbitrary, purely discrete, basis of L_2 functions with respect to the variable of internal motion of the target, it is not difficult to formulate the necessary boundary conditions for equations such as (7.2.12) (Zhigunov, Zakhariev 1974). More surprising is the possibility of providing for the correct behavior in the open scattering channels when Ψ is desired in the form of an expansion in the L_2-basis for all the variables. In a system of algebraic equations for the constant coefficients of the expansion, the part of space variables is assumed by the indices of the quantum numbers (Filippov et al 1984; 1985). [For the one-dimensional case see (Heller, Yamani 1974).]

The best way to understand the physical meaning of (7.2.12) is to apply them to a concrete phenomenon. This will be done in the next section.

7.3 Tunnelling Through Potential Barriers by Several Particles

As a first example where the specific features of multi-particle are revealed, we shall consider quantum tunnelling. The case of a single particle penetrating barriers was discussed in Sect. 1.3.

Two particles in an external field. In the quasiclassical approximation, the penetration of a barrier by a single particle is determined by the exponential factor

$$\exp\left[-\frac{1}{\hbar}\int_a^b\sqrt{2m(V(x)-E)}dx\right]. \tag{7.3.1}$$

It turns out that for identical particles in the two limiting cases of a pair of particles combined in a single one with doubled mass and energy and a pair of independent particles the quasiclassical approximation yields the same exponential factor. Indeed, in both cases the exponent is twice that of the one-particle expression (7.3.1). In the first case this is due to m, $V(x)$, and E being multiplied by 2 inside the square root and in the second case the exponents are added up in the product of the probabilities of independent tunnelling processes for the individual particles.

In the intermediate case where the pair of particles forms a quasideuteron of finite dimension, in the approximation that neglects excitations of the composite particle *enhancement* of the tunnelling effect takes place (Sokolov, Zakhariev 1964).

Let us clarify the above. The Schrödinger equation has the form:

$$-\frac{\hbar^2}{2M}\frac{\partial^2\psi(x,\rho)}{\partial x^2}-\frac{\hbar^2}{m}\frac{\partial^2\psi(x,\rho)}{\partial\rho^2}+[V_{12}(\rho)+V_1(x_1)$$

$$+V_2(x_2)]\psi(x,\rho)=\mathscr{E}\psi(x,\rho)\,, \tag{7.3.2}$$

where $x=(x_1+x_2)/2$; $M=2m$; $\rho=x_2-x_1$, and V_{12}, V_1, and V_2 are the interaction potentials between the particles and between each particle and the barrier. In the expansion (7.2.11) of the wave function ψ in the states χ_α of the internal motion of the pair [solutions of (7.2.10)] we shall retain a single term corresponding to the ground state χ_0:

$$\psi(x,\rho)=\sum_\alpha \Psi_\alpha(x)\chi_\alpha(\rho)\approx\Psi_0(x)\chi_0(\rho)\,. \tag{7.3.3}$$

The function $\Psi_0(x)$ describes the motion of the center of mass of the quasideuteron. Instead of the set of equations (7.2.12), in the approximation (7.3.3) for this function we obtain from (7.3.2) the sole equation

$$-\frac{\hbar^2}{2M}\Psi_0''(x)+V_{00}(x)\Psi_0(x)=E_0\Psi_0(x)\,;\quad E_0=\mathscr{E}-\varepsilon_0\,. \tag{7.3.4}$$

Here

$$V_{00}(x)=\int\chi_0^2(\rho)(V_1+V_2)d\rho$$

is the potential barrier acting on the center of mass of the pair. Owing to averaging over ρ, the form of $V_{00}(x)$ is smoothed out compared to the double barrier $2V(x)=V_1(x)+V_2(x)$ for the pair of particles merged into a single point (Fig. 7.3). For the energy $E_0=2E$ indicated in Fig. 7.3 the integral in the exponent in (7.3.1) over the interval between the turning points is significantly greater for $2V$ than for V_{00} and, correspondingly, the transmission probability of the extended quasideuteron is higher. Truly, at low energies the elongation of the sub-barrier path may weaken the penetrability of V_{00} in comparison with $2V$, in spite of the fact that the height of V_{00} decreases because V_{00} simultaneously becomes wider. The "spreading out" of the barrier in the approximation (7.3.3) can be explained as follows. The particles which form the pair oscillating in the ground state, when moving away from the center of mass start to feel the barrier *before* the center of mass enters the barrier. This means that for an extended composite particle the barrier is switched on in advance, although not fully. When the center of mass is already within the range of action of the barrier, the particles still spend part of the time beyond it (being on different sides of the center-of-mass point, one of them may be beyond the barrier, while the other one may be within its range).

We will not discuss the obvious fact that the penetrability is enhanced due to the addition of the binding energy of the quasideuteron to the translational energy of its center of mass.

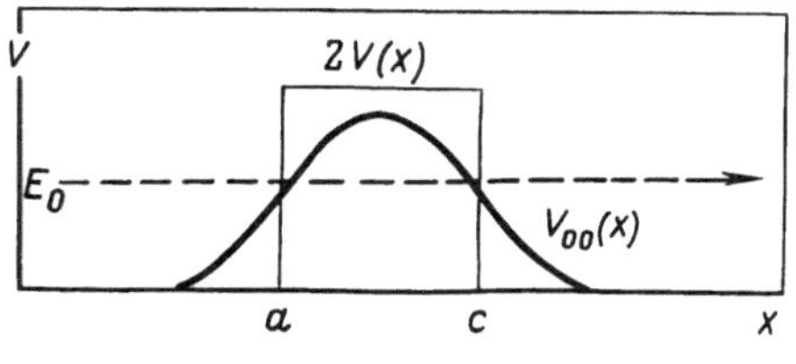

Fig. 7.3. Smoothing out of the potential barrier $2V(x)\to V_{00}(x)$, acting on the center of mass of the quasideuteron, with account of the dimension of the complex particle in the one-level approximation; the pair is "frozen" in the ground state of the internal motion

Neglecting in (7.3.3) all the channels except the one with $\alpha = 0$ means that the internal motion of the pair is frozen in the state $\chi_0(\rho)$. What is interesting is that the same enhancement of the penetrability is achieved in the adiabatic approximation also, when, on the contrary, the quasideuteron adapts itself to the external field as strongly as possible. In this case, taking into account a single level yields an equation similar to (7.3.4) except that $V_{00}(x)$ is replaced by $\mathscr{E}_0(x) - \varepsilon_0$ where $\mathscr{E}_0(x)$ is the position of the level of the pair at a fixed x in the field $V_{12} + V_1 + V_2$ [compare (6.3.13)]

$$\psi(x,\rho) \approx \Psi_0(x)\chi_0(\rho x);$$

$$-\frac{\hbar^2}{m}\frac{\partial^2}{\partial\rho^2}\chi_\alpha(\rho,x) + (V_{12} + V_1 + V_2)\chi_\alpha(\rho,x) = \mathscr{E}_\alpha(x)\chi_\alpha(\rho,x)\,, \tag{7.3.5}$$

where $x = \text{const}$;

$$-\frac{\hbar^2}{2M}\Psi_0''(x) + \mathscr{E}_0(x)\Psi_0(x) = \mathscr{E}\Psi_0(x)\,.$$

The function $\mathscr{E}_0(x)$ describes the gradual rise of the energy level with insertion of the barrier $V_1 + V_2$ into the well V_{12} and the lowering of this level as the latter is removed, when the fixed center of mass position x shifts from the left to the right in the region with $V_1 + V_2 \neq 0$. Therefore the form of $\mathscr{E}_0(x) - \varepsilon_0$ also represents a diffuse barrier $2V(x)$, although smoothed out to a somewhat lesser extent than $V_{00}(x)$.

We can describe this phenomenon through a classical analogue. Let two wheels, with an initial velocity insufficient for reaching the top, be rolling up a hill. They can move independently (Fig. 7.4a) or on a common axis (Fig. 7.4b). The latter is the analogue of a merged pair. In the approximation (7.3.3) the quasideuteron can be likened to two wheels connected one behind the other as in a bicycle. In this case the exchange of energy (via the frame of the bicycle) permits the wheels to pass through barriers which would be completely reflecting if they were independent or merged. When the first wheel reaches the height at which it should have stopped having spent all its initial kinetic energy, it is pushed ahead by the second wheel which is at a lower level and still has a reserve of velocity. Having passed the top, the leading wheel starts supplying

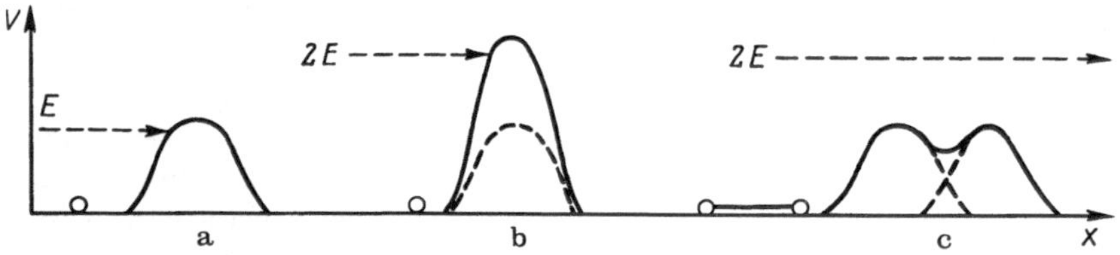

Fig. 7.4a–c. Classical interaction with a barrier. **a** barrier $V(x)$ for a single classical particle; **b** barrier for two particles merged into one; **c** smoothed out barrier for a composite particle of finite dimension

energy to the following wheel. The spreading of the quantum barrier $2V(x) \to V_{00}(x)$ (Fig. 7.3) in this case may be compared to the potential for the center of mass of the bicycle which is equal to the sum of the potentials for each of the two wheels shifted by the length of the frame, as shown in Fig. 7.4c.

Influence of excitations of the internal states of the quasideuteron on barrier penetration. Now let us discuss what happens if other channels of the system are taken into account. Owing to conservation of the total energy, excitation of the quasideuteron leads to a decrease in the center-of-mass energy. As a result, the motion of an excited composite particle as a whole occurs at a large depth under the barrier, so the penetrability decreases. If the dimension of the quasideuteron is increased by making the well $V_{12}(\rho)$ wider, then, on the one hand, $V_{00}(x)$ will smooth out more markedly and on the other hand the spectrum of internal states of the pair will become more dense and can be easily excited. The first will favor the penetrability, while the second will hinder it. Competition between these influences is what determines the change of the penetrability of a barrier in passing from merged to independent particles: first the transparency increases and then it falls. However, taking into account excitations not only influences quantitative characteristics of a phenomenon, but also yields qualitatively new effects, as will be shown in the next paragraphs.

Symmetry violation for penetrability. For a single quantum particle the probabilities of passage through a barrier in opposite directions are precisely identical, regardless of the form of the potential. The same holds for two particles in the limiting cases of independent and merged motions, as well as for the quasideuteron when excitations are neglected or when the excitation channels are closed and take part in the process only virtually.

Let us clarify the difference arising in the mechanisms of passage through a nonsymmetric barrier of a complex particle incident from the right and from the left when the real excitation of the channels is taken into account. Consider the potential $V_j(x_j)(j = 1, 2)$ depicted in Fig. 7.5. Let the quasideuteron be incident from the left in its ground state. Owing to the sharp change in the potential at its left-hand edge, strong excitation of the pair occurs, which consumes the center-of-mass kinetic energy and hinders tunnelling the barrier. When the particles are incident from the right, nearly all the barrier is passed through by the quasideuteron in the ground state, owing to the weak excitation in the field of the slowly

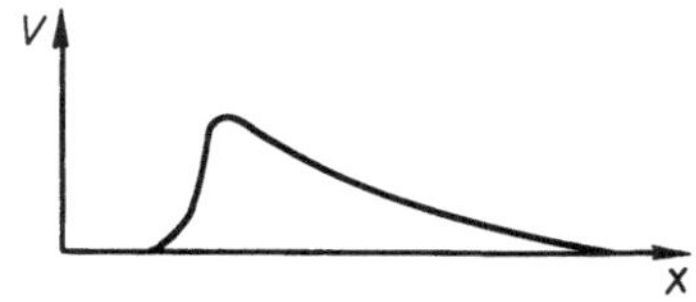

Fig. 7.5. Asymmetric barrier, equally penetrable from both sides for a single particle, for independent particles, and for a quasideuteron below the excitation threshold of states of its internal motion. The symmetry of the penetrability is violated when the real excitations of the composite particle are taken into account

changing $V_j(x)$ in the right-hand part. On the other hand, excitation at the very end of the barrier, when it has practically been passed, has little influence on the penetrability. (In a similar manner asymmetry is obtained when a single particle passes through a potential barrier of a target capable of being excited. Such phenomena must manifest themselves when nucleons, α, and other particles pass through a Coulomb barrier into and out of a nucleus. It could be that the described phenomenon serves as one of the mechanisms also responsible for the directional penetrability of biomembranes.)

The opposite is observed when quasideuterons are scattered in an excited state. In this case the sharp perturbation by the external field aids transition to the ground state and the barrier is penetrated more readily from the left. All these qualitative arguments have been confirmed by numerical calculations (Amirkhanov, Zakhariev 1965).

A "bicycle" with a spring connecting the wheels instead of a frame serves as a classical analogue of a composite excited particle. The total energy of the bicycle is distributed between the translational motion of the device as a whole and the vibration of the wheels connected by the spring. A bicycle with an ordinary frame passes over a nonsymmetric hill, such as the one depicted in Fig. 7.5, identically in both directions with an energy $E > V_{max}$, while with the spring it may be "reflected" as it arrives from the left, if compression of the spring takes up an energy larger than $E - V_{max}$ on the steep slope. In contrast, on the smooth rise from the right the "internal" motion consumes practically no energy.

Resonance tunnelling. This effect, discussed in Sect. 1.4, can also be observed in multi-particle systems. We shall consider a one-dimensional problem of the passage of a composite particle through the two barriers shown in Fig. 1.3a. If the energy of the particle's quasistable state in between the barriers lies below the excitation threshold of the internal states (only the elastic scattering channel is open), then all the above statements concerning a single particle also hold.

The situation changes in the case of two or more open channels. In general one cannot expect total penetrability if the incident particle flux is not present in all the open channels. Otherwise, more conditions would have to be imposed on the asymptotic behavior of the wave function in the open channels than can be satisfied. Consider the example of two open channels ($\alpha = 0$; 1) when the incident particles are not excited. Then, for the barriers to be totally transparent, it would be necessary to require the absence in Ψ_0 and Ψ_1 of waves travelling towards the incident wave (four conditions) on both sides of the barrier, and in addition, the coefficient of the incident wave in ψ_1 must be zero (the fifth condition). Thus, we obtain an overdetermined problem since even when four homogeneous boundary conditions are imposed the solution of the two-channel problem may exist only for certain discrete energies. Only if the choice of the form of potential is arbitrary can one expect the indicated conditions to be satisfied for certain special types of barriers.

If the conditions of resonance tunnelling are satisfied when incident waves are present in both channels, the ratio between their amplitudes is determined by the concrete form of the external field. No freedom in defining the ratio of amplitudes exists in this case.

Tunnelling of Multi-particle Systems

When the number of interacting particles increases, enhancement of the barrier penetrability may become significantly stronger. This occurs if the large dimensions of complex systems (for smoothing out the barriers) combine with the suppression of excitations of internal motion. It is then necessary to overcome the tendency of the effect to weaken due to the density of excited states increasing with the number of particles in the system and its size. Such complicated *super-penetrating* complexes (Zakhariev 1974–1988) could be systems involving pairing of particles (as in superconductors, "superfluid" nuclei), nuclei with closed shells and other complexes with interparticle correlations of long range. Excitations in such complexes are hindered owing to the existence of a gap $\Delta\varepsilon$ between the ground state and higher states. It is only necessary that the energy gap be sufficiently large and the external field to not destroy the pairing. (In classical physics a slowly moving long train going up a high hill will exhibit super-penetrability owing to the energy transfer from each wagon situated far from the summit to those wagons that are close to the top.) Strongly coupled clusters not only overcome obstacles more easily, but are also more transparent to other particles.

The transparency of barriers should also be enhanced by partial suppression of excitations in Fermi systems owing to the Pauli principle. An interesting example of this is provided by model calculations of the multi-particle penetrability performed by applying the time-dependent Hartree–Fock method (TDHF). *Bonche* et al (1976) considered the passage of a plane homogeneous nuclear layer through a plane barrier of finite width, $\sim$10 fm. The height of the barrier was chosen to be 10 or 20 MeV, the energy of the translational motion of the layer was about 10 MeV per nucleon, and the density was equal to the average density for real nuclei (Fig. 7.6a).

When the height of the barrier was 10 MeV, less than 1% of the total flux was reflected. This is significantly less than for a single particle (50% in the quasiclassical approximation). When the height of the barrier was 20 MeV, 71% of the particles were reflected, which is also much more than the 1.3% for an individual nucleon. If one compares this with the penetrability of a system of N independent particles, the difference between the passage and reflection turns out to be even more significant: 1.3% raised to the power N.

From Fig. 7.6b it can be seen that the system which has passed through the barrier is less excited than the reflected one. This is shown by the difference between their velocities. Such behavior is in accordance with the predictions that follow from the above arguments concerning complex Fermi systems.

Madler et al (1984; 1987), *Zakhariev* and *Plekhanov* (1987) have investigated the symmetry violation of the penetrability of a nuclear layer through a nonsymmetric Coulomb barrier in the approximation of the TDHF method. We

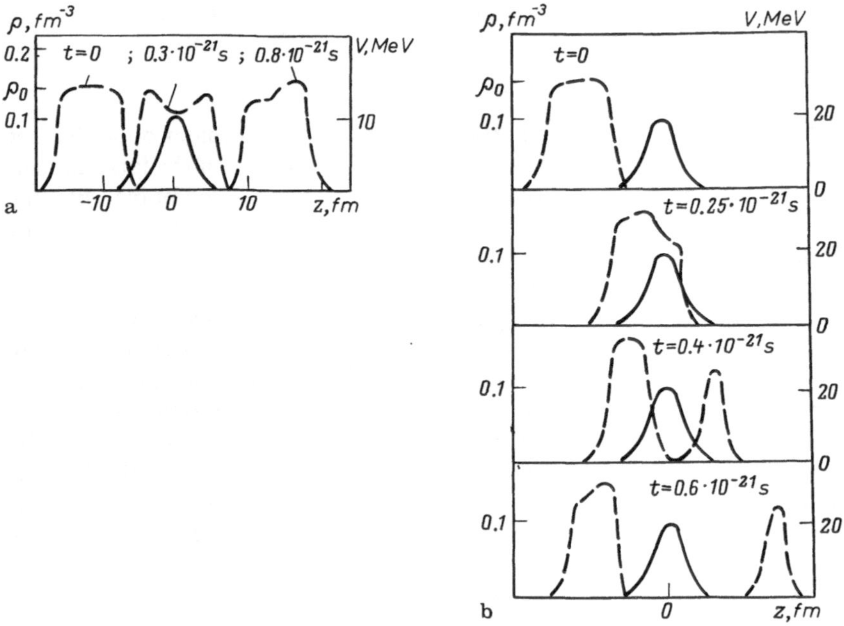

Fig. 7.6a,b. Passage of a layer of nuclear matter (the *dashed line* represents the density profile) through a potential barrier (*solid line*) calculated in the Hartree–Fock approximation for an initial center-of-mass kinetic energy equal to 10 MeV per nucleon. The height of the barrier was 10 MeV (**a**) and 20 MeV (**b**). In case **a** the penetrability is 50 times and in case **b** 20 times greater than for an individual nucleon with $E = 10$ MeV. From (Bonche et al 1976)

shall now consider one more typical example of reduction of a multi-particle problem to a multi-channel one.

7.4 Excitation of the Collective Degrees of Freedom of Multi-particle Systems

Instead of following the individual behavior of each particle in large complexes, often the more economical description of a multi-particle system is the approximation of a small number of collective variables. At low energies this approach turns out to be sufficiently effective: the properties of a number of states of nuclei and molecules are well explained within the framework of the model of collective rotational and vibrational motions.

The collective base functions χ_α are also used as the "building material" for the scattering wave functions. This solution of the multi-channel equations is the most widely applied.

The technique of strong channel coupling also provides the possibility of reconstructing the "optical" potential of the nucleus (the angular and radial dependence) in spite of its rotations and vibrations, which seem to point out the original interaction between the incident particles and the spherically asymmetrical target (Zakhariev, Suzko 1975).

Now we shall consider the features peculiar to such an inverse problem utilizing the example of particle scattering on a rigid axially symmetric rotator in the adiabatic approximation when the field of the target has no time to change, that is, undergo perceptible rotation, during the interaction time (Drosdov 1955). In this case the problem reduces to the reconstruction, considered in Chap. 6, of a motionless multi-dimensional potential $V(\vec{r})$. To this end, data on the scattering of particle beams incident from various sides of the target are necessary. In our case the rotator exposes different sides to the beam, while the beam itself does not alter its position. One must only transform the scattering data from the laboratory frame to represent them in the reference frame of the target.

The wave function of the rotator has the form

$$\chi_\alpha(\omega) \equiv \sqrt{(2I+1)/8\pi^2}\, D^I_{MK}(\omega)\,, \tag{7.4.1}$$

where $D^I_{MK}(\omega)$ is the Wigner function, ω are the Euler angles determining the orientation of the target in space, I is the total angular momentum of the target, K is its projection onto the symmetry axis of the rotator; M is the projection of I onto the z axis.

Assume that the particles are incident on a nucleus (rotator) in the state $\alpha_0 \equiv \{I_0 M_0 K\}$. The wave function of the whole system has, in the adiabatic approximation, the factorized form

$$\psi_{I_0 M_0 K}(\mathbf{k}, \mathbf{r}, \omega) = \psi(\mathbf{k}, \mathbf{r}, \omega)\, D^{I_0}_{M_0 K}(\omega)\,, \tag{7.4.2}$$

where $\psi(\mathbf{k}, \mathbf{r}, \omega)$ is the wave function of particles scattered by the field $V(\mathbf{r}, \omega)$ of the target with a fixed orientation ω:

$$\left[-\frac{\hbar^2}{2m}\nabla^2_{\mathbf{r}} + V(\mathbf{r}, \omega) - E\right]\psi(\mathbf{k}, \mathbf{r}, \omega) = 0\,. \tag{7.4.3}$$

The asymptotic behavior of ψ at large distances r is

$$\psi_{I_0 M_0 K}(\mathbf{k}, \mathbf{r}, \omega) \underset{r\to\infty}{\sim} \left\{\mathrm{e}^{ikz} + \frac{1}{r} q(k, \Omega, \omega)\mathrm{e}^{ikr}\right\} D^{I_0}_{M_0 K}(\omega)\,, \tag{7.4.4}$$

where $q(k, \Omega, \omega)$ is the scattering amplitude for particles of energy $E = k^2\hbar^2/2m$ in the field $V(\mathbf{r}, \omega)$ in the direction $\Omega \equiv \{\theta, \varphi\}$.

From experimental data, however, as a result of averaging over the orientations of nuclei one obtains the amplitudes $q_{I_0 M_0 K,\, IMK}(k, \Omega)$, instead of $q(k, \Omega, \omega)$. Expanding in the total set of functions χ_{IMK} of (7.4.1), the product

$$q(k, \Omega, \omega)\chi_{I_0 M_0 K}(\omega) = \sum_{IM} q_{I_0 M_0 K,\, IMK}(k, \Omega)\chi_{IMK}(\omega)\,. \tag{7.4.5}$$

and then multiplying both sides by $\chi^*_{I_0M_0K}$ and summing over K, one can obtain $q(k, \Omega, \omega)$.

The amplitudes $q(k, \Omega, \omega)$ determine the scattering matrices in the reference system of the nucleus itself, precisely in which the inverse problem with a motionless field $V(r)$, discussed in Chap. 6, is solved. Thus, for reconstruction of $V(r)$, of the entire matrix $q_{I'M'K', IMK}(k, \Omega)$ only those amplitudes that correspond to transitions of the nucleus from the ground state are needed, in accordance with (7.4.5). This solution is simpler than that of the nonadiabatic case.

If the conditions of adiabaticity are violated, the scattering process is described by a system such as (5.2.1) with the interaction matrix $\| V_{\alpha lm, \alpha' l' m'}(r) \|$, also coupling the equations with different quantum numbers $\alpha \equiv \{I, M\}$. It is convenient to use the multi-channel equations in the representation of the total momentum of the particle and the target J (for the functions $\psi_{\alpha \equiv \{J\mathcal{M}LI\}}$, instead of $\psi_{\alpha \equiv \{I\mathcal{M}lm\}}$), in which sets of equations corresponding to different J, $\mathcal{M} = M + m$ are separated.

Transformation of amplitudes under transition to the reference frame of the target. In the frame of the nucleus the polar axis coincides with the symmetry axis of the nucleus, $V(r) \equiv V(r, \omega = 0)$. The equations for the partial waves have the form

$$-\frac{\hbar^2}{2m}\Psi''_{LKL'}(r) + \frac{\hbar^2}{2m}\frac{L(L+1)}{r^2}\Psi_{LKL'}(r)$$

$$+ \sum_{L''} V_{LKL''}(r)\Psi_{L''KL'}(r) = E\Psi_{LKL'}(r) , \tag{7.4.6}$$

where the projection K for the axially symmetric target is a good quantum number:

$$V_{LKL'K'}(r) = \int d\Omega Y^*_{LK}(\Omega)V(r)Y_{L'K'}(\Omega) = V_{LKL'}(r)\delta_{KK'} . \tag{7.4.7}$$

To relate $S^K_{LL'}$ with $q(k, \Omega, \omega)$, we expand the right-hand side of (7.4.4) in the spherical harmonics Y_{LM} and then pass over to the reference frame connected rigidly with the nucleus, using the formula $Y_{LM}(\Omega) = \sum_K D^L_{MK} Y_{LK}(\Omega')$ where Ω' and K are in the new reference frame. As a result we have

$$\psi(k, r, \omega)_{\substack{\sim \\ r \to \infty}} \sum_{LK} [\mathrm{i}^L\sqrt{4\pi/(2L+1)}\, j_L(k, r)D^L_{0K}(\omega)$$

$$+ q_{LK}(k, \omega)h^{(1)}_L(k, r)]\, Y_{LK}(\Omega') , \tag{7.4.8}$$

where

$$q_{LK}(k, \omega) = \sum_M D^L_{MK}(\omega)q_{LM}(k, \omega)$$

is the amplitude of the partial scattered wave when incident waves are present in all channels. Clearly, it is composed of the amplitudes corresponding to the single input channel (l, K) and weighted in accordance with (7.4.8):

$$q_{LK}(k, \omega) = \sum_{L'K'} \mathrm{i}^{L'} \sqrt{4\pi(2L' + 1)}\, D^{L'}_{0K'}(\omega) q_{LK, L'K'}(k) \,. \tag{7.4.9}$$

Taking into account the orthogonality of D-functions we obtain the desired relation:

$$q_{LK, L'K'=K}(k) = \frac{\sqrt{2L' + 1}}{\mathrm{i}^{L'} 16\pi^{5/2}} \int D^{L'}_{0K}(\omega) q_{LK}(k, \omega)\, \mathrm{d}\omega \,. \tag{7.4.10}$$

7.5 The Method of Hyperspherical Functions (K-harmonics)

In solving the Schrödinger equation by the expansion of the wave function ψ perhaps the most critical stage is the choice of the basis $\{\chi_\alpha\}$. The set of functions χ_α represents a sort of language in which the investigated phenomenon or object is supposed to be described. Thus, the goal is to find such a language in which as few words as possible (i.e., individual functions χ_α) are required for transmitting the properties of the investigated object in a sufficiently clear manner. From this perspective hyperspherical functions are interesting. Although the range of their application has not yet been fully revealed, there is already no doubt that they have become a valuable additional instrument for the study of multi-particle quantum problems (see Avery 1989).

To reduce the multi-particle Schrödinger equation to a set of one-dimensional equations in the total wave function ψ it is necessary to factorize into known basis elements χ_α the dependence upon all the space coordinates, except for a single coordinate. Then the coefficients of the expansion Ψ_α will be functions of this single variable.

Which of the $3n - 3$ coordinates of n bodies in the center-of-mass system must be chosen as such an argument? For identical particles the approximate function ψ^N is usually desired in the form of the sum

$$\psi^N = \sum_{\alpha}^{N} \Psi^N_\alpha \chi_\alpha \,,$$

in which each term possesses the same symmetry with respect to particle permutations as the whole function ψ. Then adding or discarding individual terms should not violate the required symmetry of ψ^N. This occurs if the chosen variable is invariant relative to any arbitrary transposition of the particles.

For simplicity we shall first consider a two-body system. The absolute value of the vector of relative distance remains invariant under transposition of the two particles, $|\boldsymbol{\rho}_{12}| = |\mathbf{r}_1 - \mathbf{r}_2| = |\mathbf{r}_2 - \mathbf{r}_1|$, while the conventional spherical functions $Y_{LM}(\omega)$ serve as the basis $\{\chi_\alpha\}$ (we do not yet consider the spin and isospin dependences):

$$\psi(\boldsymbol{\rho}_{12}) = \sum_{lm} \Psi_{lm}(\rho_{12}) Y_{lm}(\omega)/\rho_{12} \,. \tag{7.5.1}$$

The harmonics with even l are symmetric, and the ones with odd l are antisymmetric relative to particle transpositions. Therefore only the terms of required symmetry are retained in the sum, which makes these calculations simpler than for the case of non-identical particles.

Now let us go to the three-body system. The relative distribution of the three particles is determined by six coordinates. These may be the coordinates $\boldsymbol{\rho}_{12}$ and $\mathbf{R}_3$ (Fig. 7.1). But neither $\boldsymbol{\rho}_{12}$, nor $\mathbf{R}_3$ represent the required invariants. Thus, although $\boldsymbol{\rho}_{12}$ does not alter under the transposition of particles 1 and 2, it is not invariant with respect to the substitution of the third particle for any one of the first two. If, however, instead of $\boldsymbol{\rho}_{12}$ and $\mathbf{R}_3$, one introduces hyperspherical coordinates as in (7.2.6–8) and adopts the single six-dimensional vector $\boldsymbol{\rho}_6$, then its modulus (7.2.6) is the desired invariant. For the remaining five coordinates one takes the angles $\boldsymbol{\Omega}_5$ defining the direction of the vector $\boldsymbol{\rho}_6$. (They may be introduced in various ways.) Rotation in the six-dimensional space corresponds to rotation with simultaneous particle transposition in the space of usual variables $\{\boldsymbol{\rho}_{12}, \mathbf{R}_3\}$, as well as to the change of $|\mathbf{R}_3|$ and $|\boldsymbol{\rho}_{12}|$ for a constant $|\boldsymbol{\rho}_6|$.

Like the harmonics $Y_{lm}(\omega)$ one defines the hyperspherical functions $Y_K(\boldsymbol{\Omega}_5)$, generalizing the Y_{lm} to a larger number of dimensions: they represent eigenfunctions of the angular part of the Laplacian Δ_{ρ_6}. The K is meant to stand for the set of five quantum numbers, instead of the two, l and m, in the two-body problem. One can choose $\mathbf{K} \equiv \{K, L, M, \nu, \omega\}$, where K is the quantum number of the "global momentum"; L and M represent the total angular momentum and its projection, respectively; ν characterizes transpositional symmetry; ω removes the remaining degeneracy. Like Y_{lm}, the functions $Y_{\mathbf{K}}$ are orthonormal:

$$\int Y^*_{\mathbf{K}}(\boldsymbol{\Omega}_5)\, Y_{\mathbf{K}'}(\boldsymbol{\Omega}_5)\, d\Omega_5 = \delta_{\mathbf{KK}'} , \tag{7.5.2}$$

Bound States

Each individual harmonic $Y_K(\boldsymbol{\Omega}_5)$ possesses a certain symmetry with respect to transpositions. This symmetry may be even or odd under the transposition of some or other particles. In a product with the spin–isospin function of adjoint symmetry they form totally antisymmetric base functions $\chi_K(\boldsymbol{\Omega}_5, s, T)$, that is, K-harmonics convenient for describing multi-particle Fermi systems (s and T represent the spin and isotopic spin of the system).

Like (7.5.1), we seek the total wave function in the form

$$\psi(\boldsymbol{\rho}_6, s, T) = \sum_{\mathbf{K}} \Psi_{\mathbf{K}}(\rho_6)\chi_{\mathbf{K}}(\boldsymbol{\Omega}_5, s, T)/\rho_6^{5/2} , \tag{7.5.3}$$

and for the coefficients of the expansion $\Psi_{\mathbf{K}}$ we obtain, as usual, the coupled equations

$$\left[-\frac{\hbar^2}{2M}\frac{\partial^2}{\partial\rho_6^2} + \frac{\mathscr{K}(\mathscr{K}+1)}{\rho_6^2}\right]\Psi_{\mathbf{K}}(\rho_6) + \sum_{\mathbf{K}'} V_{\mathbf{KK}'}(\rho_6)\,\Psi_{\mathbf{K}'}(\rho_6) = E\Psi_{\mathbf{K}}(\rho_6) , \tag{7.5.4}$$

where $\mathscr{K} = K + 3/2$. Unlike the interaction matrix $\| V_{lml'm'} \|$ in the two-body problem (6.1.2) which falls off as $r \to \infty$ more rapidly with faster decrease of

$V_{12}(r)$, the matrix $\| V_{\mathbf{KK}'}(\rho_6) \|$, even in the case of two-body forces of finite range decreases with the increase of ρ_6 not faster than ρ_6^{-3}. This is due to the fact that the interaction region in configuration space in the case of a system with three particles is not limited: the hypercylinders ("pipes" in Fig. 7.2), inside of which the individual two-particle potentials V_{ij} differ from zero tend toward infinity, as shown in Fig. 7.2. In addition integration in the expression for $V_{\mathbf{KK}'}$ is performed over a hypersphere of radius ρ_6 in this space. Thus for any ρ_6 the potentials V_{ij} in the expression for $V_{\mathbf{KK}'}$ occurring in the integrand, do not decrease in certain directions $\boldsymbol{\Omega}_5$ of the six-dimensional space. The decrease of $V_{\mathbf{KK}'}$ with $\rho_6 \to \infty$, however, occurs because of the contribution to the integral of successively smaller intervals of the angles $\boldsymbol{\Omega}_5$. The intersections of the hypercylinders with the hypersphere of radius ρ_6 when ρ_6 increases are seen at these $\boldsymbol{\Omega}_5$:

$$V_{\mathbf{KK}'}(\rho_6) = \left\langle \chi_{\mathbf{K}} \middle| \sum_{i \neq j} v_{ij} \middle| \chi_{\mathbf{K}'} \right\rangle \xrightarrow[\rho_6 \to \infty]{} 0 \, . \tag{7.5.5}$$

The system (7.5.4) must be supplemented with boundary conditions imposed on $\Psi_{\mathbf{K}}(\rho_6)$. For ψ to be limited, it is necessary to choose $\Psi_{\mathbf{K}}(0) = 0$, and in the case of bound states $\Psi_{\mathbf{K}}(\rho_6)$ must also be required to vanish at infinity. It is very complicated to provide for the proper asymptotic behavior of ψ in the form of (7.5.3) in scattering problems. We shall deal with this issue later.

As the quantum number K increases, the "centrifugal" potential $\mathscr{K}(\mathscr{K}+1)/\rho_6^2$ in (7.5.4) also increases. As a result, the motion in states with K higher than a certain value turns out to be totally *sub-barrier* motion. It is natural to expect the contribution of such states to be significantly suppressed, and to expect that this suppression to be stronger with higher values of K. The smoothness of the $\boldsymbol{\Omega}_5$ angular dependence of ψ also influences the convergence of the expansion (7.5.3). The smoother this dependence, the fewer terms one has to take into account in (7.5.3). For two-particle potentials with a hard core the convergence becomes noticeably worse since it is necessary to make ψ become zero on the surface of the repulsive core, which in the six-dimensional configuration space is a bundle of "hypercylinders" (Fig. 7.2) piercing into the region where the bound-state function ψ is concentrated. A rigorous proof of the convergence of approximations within the method of K-harmonics has not yet been provided.

In the general case of $n > 3$ particles the vector $\boldsymbol{\rho}_{3n-3}$ is introduced in the $3n-3$-dimensional space. The modulus is, $\rho_{3n-3} = \sqrt{\sum_{i=1}^{n-1} \xi_j^2}$, where ξ_j are the Jacobi coordinates of the n particles. The corresponding hyperspherical functions $Y(\boldsymbol{\Omega}_{3n-4})$ depend on $3n-4$ angular variables $\boldsymbol{\Omega}_{3n-4}$, and $\mathbf{K}$ is a set of $3n-4$ quantum numbers. The set of equations for the coefficients of the expansion $\Psi_{\mathbf{K}}(\rho_{3n-3})$ of the wave function Ψ coincides in form with (7.5.4) where only $\mathscr{K} = K + (n-2)3/2$.

As K increases, a rapid growth of the number of harmonics with the given value of K is characteristic of a base of hyperspherical functions. The higher the number of particles n in the system, the sharper this growth. For practical computations one needs a rule for classifying the basis functions depending on their degree of significance. In 1969 *Fabre* pointed out the special role of *potential harmonics.* He made the natural assumption that it is especially important to take into account those harmonics $\chi_{\mathbf{K}}$ for which the matrix elements of the forces acting within the system differ from zero $V_{\mathbf{KK}_{\min}}(\rho_6) \neq 0$, where $K_{\min}$ is the minimal possible value of K for the given n. Such potential harmonics were constructed by *Efros* (1972; 1983) for $n \leqslant 4$, and eventually by *Gorbatov* et al (1979–88) for arbitrary n and differing components of realistic N–N interactions, for example, tensor forces.

As n increases, the problem is additionally complicated by the weak convergence of the approximations $\Psi = \sum_{\mathbf{K}} \Psi_{\mathbf{K}} \chi_{\mathbf{K}}$ in the case of two-particle potentials with a repulsive core. This difficulty can be overcome by introducing pair-correlation operators $\hat{O}$, the action of which is similar to that of the Jastrow factors equating $\Psi = \hat{O}\tilde{\psi}$ to zero in the region of strong repulsion (where $\tilde{\psi}$ is a smooth auxiliary function) (Gorbatov 1983). However, unlike the known Jastrow method of correlation factors, here one can use the operator $\hat{O}^{-1}$, the inverse of $\hat{O}$, to write the equation $\tilde{H}\tilde{\psi} = E\tilde{\psi}$ for a function whose expansion in K-harmonics converges well. Now, unlike the initial Hamiltonian the effective Hamiltonian $\tilde{H}$ no longer contains any dangerous terms with strong repulsion.

Gorbatov (1983) proposed a method to simplify the set of coupled equations for Ψ_K such as (7.5.4). In those equations where the centrifugal potential $\mathscr{K}(\mathscr{K}+1)/\rho_{3n-3}^2$ is large, the terms involving second derivatives with respect to ρ_{3n-3} were discarded under the assumption that the kinetic energy was relatively small. As a result, part of the differential equations in the system (7.5.4) becomes algebraic. It turns out that the uncertainty introduced by such simplification into the energy of a bound multi-body state is quite readily estimated. It is possible to strictly determine the upper and lower boundaries for exact eigenvalues. This is done as follows.

Let $H^{(0)}$ be the operator of the system with discarded derivatives, and let $\psi^{(0)}$ and $E^{(0)}$ be the respective eigenfunction of the ground state and the energy eigenvalue, $E^{(0)} = \langle \psi^{(0)} | H^{(0)} | \psi^{(0)} \rangle$. We shall show that $E^{(0)}$ represents the lower limit for the exact energy E_{exact}. If in the given expression for $E^{(0)}$ the function $\psi^{(0)}$ is replaced by some other function ψ which may be the exact one, then the functional $\langle \psi | H^{(0)} | \psi \rangle$ increases, since by definition $E^{(0)}$ represents its minimum value (for bound states the sets (7.5.4) are obtained from the Ritz variational principle). If, on the other hand, the positive definite operator of the discarded kinetic energies is also added to $H^{(0)}$, then there is even more reason for $E_{\text{exact}} = \langle \psi | H | \psi \rangle \geqslant \langle \psi^{(0)} | H^{(0)} | \psi^{(0)} \rangle$; and, vice versa, if in the functional $\langle \psi | H | \psi \rangle$ the function $\psi^{(0)}$ is substituted for the exact ψ, we then obtain the upper boundary for E_{exact}. This is because having violated the condition of the functional being minimal, we shift its value upwards. Thus, with the aid of a known approximate function $\psi^{(0)}$ strict limits are given for the unknown exact quantity E_{exact}:

$$\langle \psi^{(0)} | H | \psi^{(0)} \rangle \geqslant \langle \psi | H | \psi \rangle = E_{\text{exact}} \geqslant \langle \psi^{(0)} | H^{(0)} | \psi^{(0)} \rangle \, .$$

In the preceding chapters various versions of the formulation of inverse problems for the special case of a two-particle system were discussed. Multi-body studies reveal new possibilities of reconstructing interactions from observed particle data when the number of particles varies. Thus, the method of K-harmonics was applied for solution of a special inverse problem: the determination of nucleon–nucleon forces from the properties of nuclei with a mass number in the range $2 < A \leqslant 16$.

In a series of works (see the bibliographies in *Gorbatov* et al (1979–1988)) the N–N-potential was chosen by varying the forms of its individual components (central, tensor, and others). The authors arrived at the conclusion that, although the forces that they found correctly describe the energy characteristics, they lead to values for the characteristics deduced from experiments involving electromagnetic fields such as radii of nuclei and the quadrupole momentum of the deuteron that are too low.

Fabre de la Ripelle (1988) criticized the Independent Particle Model in nuclear physics by using the method of hyperspherical harmonics.

A fast convergent K-harmonic decomposition was proposed by *Haftel* and *Mandelzweig* (1989).

States in the Continuous Spectrum

The harmonics $Y_{\mathbf{K}}$ do not serve as universal building material for construction of multi-particle wave functions ψ. Therefore to apply them successfully it is worth knowing which properties of quantum multi-particle systems are only described with difficulty by using $Y_{\mathbf{K}}$. In scattering problems the angular size of the composite particles traveling towards infinity tends to zero. For reconstruction of their internal structure with the aid of the angular variables $\boldsymbol{\Omega}_{3n-4}$ harmonics of infinitely high rank would be required. When such asymptotic components are present in ψ, its expansion in $Y_{\mathbf{K}}$ diverges. The method of overcoming this difficulty will be discussed below.

It is instructive to bear in mind another limiting case, most favorable for the method of K-harmonics, when ψ consists of a single $Y_{\mathbf{K}}$ function. Consider several particles interacting between themselves only by means of a hyperspherically symmetric potential $V(\rho_{3n-3})$ independent of the angles $\boldsymbol{\Omega}_{3n-4}$. In this case the variables in the Schrödinger equation separate in the hyperspherical reference frame. The matrix $V_{\mathbf{KK}'}$ in the set (7.5.4) becomes diagonal, $V_{\mathbf{KK}'}(\rho_{3n-3}) = V(\rho_{3n-3})\delta_{\mathbf{KK}'}$, and for each partial hyperradial component $\Psi_{\mathbf{K}}(\rho_{3n-3})$ a simple one-dimensional equation is obtained:

$$-\frac{\hbar^2}{2M}\Psi''_{\mathbf{K}}(\rho_{3n-3}) + \frac{\mathcal{K}(\mathcal{K}+1)}{\rho^2_{3n-3}}\Psi_{\mathbf{K}}(\rho_{3n-3}) + V_{\mathbf{KK}}(\rho_{3n-3})\Psi_{\mathbf{K}}(\rho_{3n-3}) = E\Psi_{\mathbf{K}}(\rho_{3n-3}) \,. \quad (7.5.6)$$

The one-dimensional one-channel theory of the direct and inverse problems considered in Part 1, is totally applicable in this case. Multi-particle bound states with $k = \mathrm{i}\kappa_j(\mathrm{Im}\{\kappa_j\} = 0,\ \mathrm{Re}\ \{\kappa_j\} > 0)$ may be present in the potential

$V(\rho_{3n-3})$. When $\rho_{3n-3} \to \infty$, $V(\rho_{3n-3}) \to 0$ and the energy is higher than the three-fragment scattering threshold ($K^2 > 0$), the asymptotic behavior of the function $\Psi_{\mathbf{K}}$ may be represented in the form of a combination of incident and outgoing waves:

$$\Psi_{\mathbf{K}}(\rho_{3n-3}) \xrightarrow[\rho_{3n-3} \to \infty]{} \mathrm{e}^{-ik\rho_{3n-3}} - s_{\mathbf{K}}(|k|)\mathrm{e}^{ik\rho_{3n-3}} .$$

For short-range multi-particle forces $V(\rho_{3n-3} \geqslant a) = 0$ similar behavior is observed immediately when $\rho_{3n-3} \geqslant a$ except that instead of exponentials the corresponding Hankel functions are used. Such short-range interaction means that even if only one of the particles is removed from the others at a distance greater than a, all the remaining n particles become free.

The total energy $E = k^2\hbar^2/2M$ may be distributed over the free particles in various ways. The harmonic $Y_{\mathbf{K}}$ in the wave function

$$\psi(\rho_{3n-3}) = \Psi_{\mathbf{K}}(\rho_{3n-3}) Y_{\mathbf{K}}(\mathbf{\Omega}_{3n-4})/\rho_{3n-3}$$

fixes for each set K definite angular and energy distributions for particles incident from various directions and with various momenta in the interaction region and going out after scattering with phase shifts $\delta_{\mathbf{K}}$.

Owing to the completeness of the system $Y_{\mathbf{K}}$ it is possible with the discrete (and actually even finite) set $Y_{\mathbf{K}}$ to describe any asymptotic relative motion of many free fragments of the systems (just as the continuous angular distribution of the two-fragment reaction products is determined by the number of spherical functions Y_{lm}). This is still another important advantage of the method of K-harmonics (Delves 1959–1962).

It can be expected that at moderate energies a notable contribution to the scattering cross section will be provided only by harmonics with small values of the global momentum $\mathcal{K}$, for which the centrifugal barrier $\mathcal{K}(\mathcal{K} + 1)/\rho_{3n-3}^2$ is not too high. The increase of this barrier with increasing K prevents the penetration of waves from outside into the interaction region, as well as the exit of reaction products from it.

7.6 The Levinson Theorem

The peculiarity of the formulation of the theorem for multi-particle systems already manifests itself in the approximate description of the interaction of complex composite clusters with the aid of effective two-particle potentials.

The Pauli Principle

If two colliding particles A and B are bound states of several particles, then for a consistent description of their scattering it is necessary to solve the Schrödinger equation corresponding to the motion of all the elementary components of the system. However, in treating experimental data at an energy lower than the threshold of inelastic scattering, E_{thresh}, the interaction of A and B is often reduced approximately to the effective pair potential $V_{AB}(r_{AB})$. Naturally, in the case of such a formulation the relationship between the number of bound states $A + B$ and the difference between the phase shifts $\delta(0) - \delta(\infty)$ obtained from the conventional two-particle Levinson theorem for potential V_{AB} scattering of A on B, cannot be considered a rigorous result. Thus, at an energy higher than the threshold of inelastic scattering where new channels open, the concept of the elastic scattering phase shift makes no sense. Nevertheless, it is possible to write an analogue of the Levinson theorem for the chosen phenomenological potential V_{AB} (Swan 1955). The more rigorous formulation of the theorem will be mentioned below.

Multi-channel Equations with Different Threshold Energies

The multi-channel Levinson theorem considered in Chap. 5 can be generalized to the case of several interacting particles if their motion is described by the system (7.2.12). An essentially new point in this case is the change of the number of phase eigenstates when the energy of the system surpasses the threshold. The number is conserved only in the gap between two adjacent threshold values. Consequently, the Levinson relation (5.2.14) for bound channels, which assumes N to be constant for all $0 \leqslant E < \infty$, cannot be applied.

Wright (1965) has shown for the case of a three-body system that the difference between the sums of phase shifts of the eigenstates in (5.2.14) must be replaced by the sum of the corresponding differences at the threshold energy values E_j:

$$\sum_j \sum_m [\delta_m(E_j) - \delta_m(E_{j+1})] = n\pi \, . \tag{7.6.1}$$

7.7 Three-particle Potentials

Two-particle forces must be deduced from two-particle problems, with the exception of such cases as neutron–neutron scattering which can be done experimentally only with great difficulty. The question concerning the existence of three-particle potentials and the form they have, must, naturally, be answered by studies of three-body systems under the assumption that the pair interactions are already known.

A Model Without Particle Rearrangement

To start, we shall demonstrate the possibility of finding three-particle forces using the simple example of scattering of a particle on a pair bound by an infinite potential well V_2. This is similar to the model considered in (7.2.10–13). Let one of the target particles be fixed at the origin of the reference frame and other particles move in one dimension along the semi-axis $x \geqslant 0$. We shall assume that the pair potentials $V_1(x_1)$ and $V_{12}(x_1 - x_2)$ and the three-particle potential $V_{123}(x_1, x_2)$ are limited and of finite range. In this case the incident particle is capable only of exciting the target, while knock-out, pickup, and breakup reactions are forbidden. Expanding the wave function of the system $\psi(x_1, x_2)$ in the purely discrete set of states of the internal motion of the quasideuteron we obtain, as in (6.3.7–11), multi-channel equations with the interaction matrix $V_{\alpha\beta}(x_1)$. Having reconstructed this matrix from the inelastic scattering data, we express through its matrix elements $V_{\alpha\beta}(x_1)$, by the formula (6.3.27), the sum of the desired three-particle and the known pair potentials $V_{123}(x_1, x_2) + V_1(x_1) + V_{12}(x_1 - x_2)$.

Since the solution of one-dimensional inverse problems on the basis of the lower part of the spectrum worked successfully, one may expect that here also it will be sufficient for the approximate reconstruction of $V_{123}(x_1, x_2)$ to use only the scattering submatrix for open channels at not too high energies. This is in line with the Kotel'nikov theorem, that is, that data of the lower part of the spectrum, if measured with sufficient precision, contain information on all the important details.

Reactions Involving Particle Rearrangement

In the general case of the direct three-particle problem the standard practice is to apply equations of the Faddeev type. In the inverse problem this is inconvenient, because coupling of channels (of partial equations) occurs through integro-differential terms, i.e., by an effective *nonlocal* interaction. In contrast, utilization of the Gelfand–Levitan–Marchenko formalism requires ordinary differential equations with *local* potential terms. Therefore another approach to the three-body direct problem was suggested (Zhigunov, Zakhariev 1974) as a basis for the inverse one.

Consider the same three-body model as in the preceding section, only with a finite potential $V_{12}(x)$ binding the quasideuteron. Although the particles move along a single "string" (Fig. 7.7a), the whole configuration space of the system is two-dimensional. It is depicted in Fig. 7.7b, so as to represent the regions of particle interaction in a clearer manner. The finite-range two-body potentials V_1, V_2, and V_{12} differ from zero along the strips that depart from the area near the origin in the (x_1, x_2) plane. If the energy axis is perpendicular to the plane of the paper then the strips represent the potential channels. The cross sections of these channels are of the same shape as those of the potentials V_1, V_2, V_{12}. The wavy line in Fig. 7.7b indicates where the finite-range three-body potential differs from zero. If at least one of the particles is removed from the others, $V_{123} = 0$. Elastic and inelastic waves of the first particle scattered by the target (2, 3) are spread out along the horizontal strip $V_1 \neq 0$. Other rearrangement reactions are also permitted in this model system: i) the *knock-out* of the second particle accompanied by capture of the first particle to a bound state (1, 3), i.e., particle 2 travels along the strip $V_2 \neq 0$, while the first particle vibrates across it; ii) the *pickup* reaction when the (1, 2) pair travels away along the inclined strip; iii) breakup waves spread out in between the strips if the energy is above the threshold of three-body free motion.

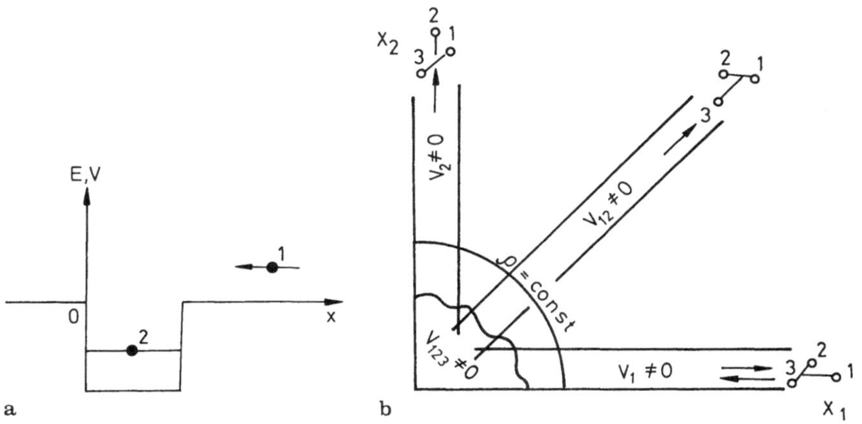

Fig. 7.7. a Particles 1 and 2 moving along x in the external short-range finite potential well (the third particle is fixed at the origin). **b** Two-dimensional configurational space for the system of two particles shown in (a). The two-body potentials are nonzero inside the strips starting from the origin. The *wavy line* marks the region of action of three-body forces. For fixed hyperradius the particles are moving along the arc (ρ = const). Compare with the four-body picture in Fig. 7.9

Internal motion of the (2, 3), (1, 2), and (1, 3) pairs takes place across the corresponding strips.

The asymptotic behavior of Ψ along the two-fragment strips assumes the form (7.2.4), where α_0 denotes the incident channel:

$$\Psi(x_1, x_2) \rightarrow \begin{cases} \xrightarrow[x_2 \text{ finite}]{x_1 \to \infty} & \Phi^{\mathrm{I}} = \sum\limits_{\alpha} (\delta_{\alpha\alpha_0} \exp(-ik_\alpha x_1) - S_{\alpha\alpha_0} \exp(ik_\alpha x_1)) \Phi_\alpha(x_2)\,, \\ \xrightarrow[x_1 \text{ finite}]{x_2 \to \infty} & \Phi^{\mathrm{III}} = -\sum\limits_{\alpha} S_{\gamma\alpha_0} \exp(ik_\gamma x_2) \cdot \Phi_\gamma(x_1)\,, \\ \xrightarrow[x_1 - x_2 \text{ finite}]{x_1 + x_2 \to \infty} & \Phi^{\mathrm{II}} = -\sum\limits_{\alpha} S_{\beta\alpha_0} \exp(ik_\beta (x_1 + x_2)/2) \Phi_\beta(|x_1 - x_2|) \end{cases} \tag{7.7.1}$$

where Φ_α, Φ_β, Φ_γ are the bound states of the (2, 3), (1, 3), and (1, 2) pairs (7.2.5):

$$\begin{aligned} &-\hbar^2 \Phi_\alpha''(x_2) + 2m_2(V_2 - \varepsilon_\alpha)\Phi_\alpha(x_2) = 0\,, \\ &-\hbar^2 \Phi_\beta''(|x_1 - x_2|) + 2(m_1 + m_2)(V_{12} - \varepsilon_\beta)\Phi_\beta(|x_1 - x_2|) = 0\,, \\ &-\hbar^2 \Phi_\gamma''(x_1) + 2m_1(V_1 - \varepsilon_\gamma)\Phi_\gamma(x_1) = 0 \end{aligned} \tag{7.7.2}$$

and $k_\alpha^2(E - \varepsilon_\alpha)2m_1/\hbar^2$, $k_\gamma^2 = (E - \varepsilon_\gamma)2m_2/\hbar^2$, $k_\beta^2 = (E - \varepsilon_\beta)2(m_1 + m_2)/\hbar^2$.

The breakup asymptotic behavior can be written in accordance with (7.2.8).

Two suitable methods have been proposed independently and simultaneously that reduce the three-particle Schrödinger equation to systems of one-dimensional differential equations (Macek 1968; Zhigunov, Zakhariev 1974; see also Fano, Rau 1986; Matveenko 1985; Vinitsky, Soloviev 1986).

Following *Macek* (1968) the motion of the whole system is divided into hyperradial (and hyperangular) parts as in (7.5.3):

$$\Psi(\boldsymbol{\rho}, \alpha) = \sum_s F_s(\rho)\Phi_s(\alpha, \rho)/\rho^{1/2} . \tag{7.7.3}$$

Unlike the free hyperangular harmonics $Y_{\mathbf{K}}(\boldsymbol{\rho})$ in (7.5.3), the basis functions $\Phi_s(\alpha, \rho)$ with fixed hyperradii ρ are the eigenfunctions of the hyperangular part of the Hamiltonian containing all the two-body potentials:

$$-(\hbar^2/2\mu\rho^2)\, d^2\Phi_s(\alpha, \rho)/d\alpha^2 + [V_{12}(\alpha, \rho) + V_1(\alpha, \rho) + V_2(\alpha, \rho)]\Phi_s(\alpha, \rho) = E_s(\rho)\Phi_s(\alpha, \rho) . \tag{7.7.4}$$

To make the meaning of $\Phi_s(\alpha, \rho)$ more clear, we have shown in Fig. 7.8 the shape of the two-body potential cross section along the arc ρ = const. The angular motion corresponds to oscillations in an infinitely deep potential well of the width $\rho\pi/2$ with three additional potential miniwells at its bottom representing the pair potentials. The length of the arc increases with the ρ-value, the miniwells move apart so that the mutual influence of bound states in different miniwells is weakened and they become exact pair bound states in the limit $\rho \to \infty$. The levels $E_s(\rho)$ above $E > 0$ correspond to the three-fragment decay states. They are the ordinary centrifugal barriers in (7.5.4) of hyperradial motion for the coefficients $\Psi_{\mathbf{K}}(\rho)$ of the expansion (7.5.3):

$$-F_s''(\rho) + \sum_{s'} \hat{\Lambda}_{ss'}(\rho)F_{s'}(\rho) + \sum_{s'} V_{ss'}^{(3)}(\rho)F_{s'}(\rho)2\mu/\hbar^2[E_s(\rho) + 1/4\rho^2] \times F_s(\rho) = 2\mu/\hbar^2 E F_s(\rho) \tag{7.7.5}$$

where

$$\hat{\Lambda}_{ss'}(\rho) \equiv \Lambda_{ss'}^{(1)}(\rho) + \Lambda_{ss'}^{(2)}(\rho)\partial/\partial\rho = -2\mu/\hbar^2 \int \Phi_s(\alpha, \rho)\partial^2/\partial\rho^2\Phi_{s'}(\alpha, \rho)\partial\alpha - 2\mu/\hbar^2 \int \Phi_s(\alpha, \rho)\partial/\partial\rho\,\Phi_{s'}(\alpha, \rho)\partial\alpha\partial/\partial\rho,$$

$$V_{ss'}^{(3)}(\rho) = 2\mu/\hbar^2 \int \Phi_s(\alpha, \rho)V_{123}(\alpha, \rho)\Phi_{s'}(\alpha, \rho)\partial\alpha . \tag{7.7.6}$$

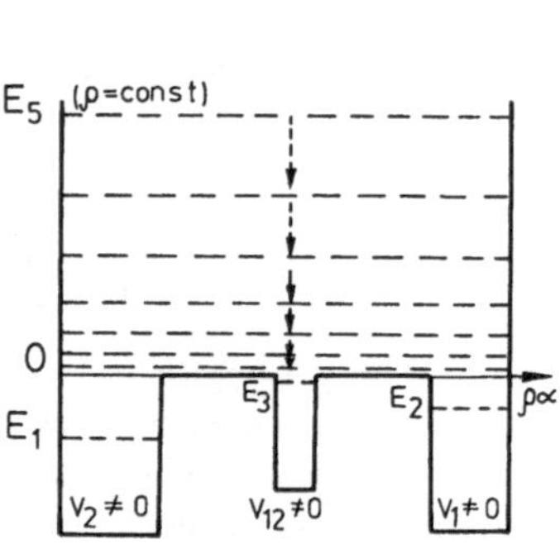

Fig. 7.8. The potential relief for the three-body hyperangular motion along the arc (ρ = const, see Fig. 7.7b). When ρ increases the two-body potential wells move away from one another and their energy levels E_1, E_2, E_3 become the bound-state energies of pairs (2, 3), (1, 3), and (1, 2). The ($E > 0$)-levels in the "maxi-well" correspond to virtual (break-up) hyperangular standing waves. These levels come closer to $E = 0$ with increasing ρ and form the hypercentrifugal potentials $E(\rho)$ for hyperradial motion

Although the physical asymptotic conditions are formulated with different sets of Jacobi coordinates whose schemes are symbolically represented at the corresponding strips in Fig. 7.7, the hyperspherical coordinates (α, ρ) along various directions of the configuration space transform into the suitable Jacobi coordinates (Zakhariev 1987). For large ρ, $F_s(\rho)$ represent the wave functions of the relative motion of fragments in various channels, so that ordinary differential equations describe all the reactions involving particle rearrangement.

The inverse problem for the multi-channel equations (7.7.5) may be solved by ordinary methods after transformation removing first derivatives in (7.7.5), as was proposed by *Vinitsky* et al (1989, 1990) and *Dubovik* et al (1989).

The other method is to single out the asymptotic tails that have a known analytic form (7.7.1) (Zhigunov, Zakhariev 1974):

$$\begin{aligned}&\Psi(x_1, x_2) - \Phi^{\mathrm{I}}[1 - \exp(-ax_1)] - \Phi^{\mathrm{III}}[1 - \exp(-ax_2)] \\ &- \Phi^{\mathrm{II}}[1 - \exp(-ax_1)][1 - \exp(-ax_2)] \equiv X(x_1, x_2)\,. \end{aligned} \tag{7.7.7}$$

The factors in square brackets are introduced to make the functions vanish at $x_1 = 0$ and $x_2 = 0$.

Expanding the remaining part $X(x_1, x_2) \equiv X(\rho, \alpha)$ in free hyperspherical harmonics $Y_{\mathbf{K}}(\alpha)$

$$X(\rho, \alpha) = \sum_{\mathbf{K}} F_{\mathbf{K}}(\rho)\, Y_{\mathbf{K}}(\alpha)/\rho^{1/2} \tag{7.7.8}$$

we obtain for the coefficients $F_{\mathbf{K}}(\rho)$ a set of equations

$$\begin{aligned}&-\frac{\hbar^2}{2\mu} F''_{\mathbf{K}}(\rho) + \frac{\mathscr{K}(\mathscr{K}+1)}{\rho^2} F_{\mathbf{K}}(\rho) + \sum_{\mathbf{K}'} V_{\mathbf{KK}'}(\rho) F_{\mathbf{K}'}(\rho) - EF_{\mathbf{K}}(\rho) \\ &= J_{\mathbf{K}\alpha_0}(\rho)\,; \quad \mathscr{K} \equiv (K - \tfrac{1}{2})\end{aligned} \tag{7.7.9}$$

with the sources

$$J_{\mathbf{K}\alpha_0}(\rho) = \int Y^*_{\mathbf{K}}(\alpha)(E - H)(\Psi - X)\,d\alpha \tag{7.7.10}$$

and interaction matrix

$$V_{\mathbf{KK}'}(\rho) = \int Y^*_{\mathbf{K}}(\alpha)[V_1 + V_2 + V_{12} + V_{123}]\, Y_{\mathbf{K}}(\alpha)\,d\alpha\,. \tag{7.7.11}$$

The inhomogeneous terms J in (7.7.9) prevent using the Gelfand–Levitan–Marchenko equations which take into account particle rearrangement. It is possible, however, to introduce a Bargmann approximation $V^{\mathrm{B}}_{\mathbf{KK}'}(\rho)$, $f^{\mathrm{B}(1)}(\rho)$, $f^{\mathrm{B}(2)}(\rho)$ for the interaction matrices and the linearly independent matrix solutions of the homogeneous part of (7.7.9). With their aid one can write J and the approximate general solution of the multi-channel system (7.7.9):

$$F^{\mathrm{B}}_{\mathbf{K}\alpha_0}(\rho) = \sum_{\mathbf{K}'} A_{\mathbf{K}'\alpha_0} f^{\mathrm{B}(1)}_{\mathbf{KK}'}(\rho) + \sum_{\mathbf{K}'} B_{\mathbf{K}'\alpha_0} f^{B(2)}(\rho) + F^{\mathrm{B}(J)}_{\mathbf{K}\alpha_0}(\rho) \tag{7.7.12}$$

where $F^{\mathrm{B}(J)}_{\mathbf{K}\alpha_0}(\rho)$ represents a partial solution of (7.7.9):

$$F^{\mathrm{B}(J)}_{\mathbf{K}\alpha_0}(\rho) = \sum_{\mathbf{K}'} \int \mathring{G}^{\mathrm{B}}_{\mathbf{KK}'}(\rho, \rho')\, J_{\mathbf{K}'\alpha_0}(\rho')\,d\rho' \tag{7.7.13}$$

and $\mathring{G}^{B}_{KK'}(\rho, \rho')$ is the matrix Green function for the homogeneous part of (7.7.9).

The coefficients $A_{KK'}$, $B_{KK'}$ and the Bargmann parameters are fixed by the boundary conditions (for $E < 0$ the function X must vanish at large distances) $\rho \geqslant \rho_0$ by fitting $F_{\alpha\alpha_0}(\rho) = 0$, $F'_{\alpha\alpha_0}(\rho) = 0$ at different energies. The same parameters determine the potential matrix $V_{\alpha\alpha'}(\rho)$. The scattering data enter into the approximate solutions through the source terms $J_{\alpha\alpha_0}(\rho)$. If it is desirable to include the information for $E > 0$, $F_{\alpha\alpha}(\rho)$ must be required to have only outgoing breakup waves. The three-particle potential is determined from the matrix $V^{B}_{KK'}(\rho)$, similarly to (6.3.25,27).

Interaction region for four particles. In Fig. 7.9 we show the generalization of Fig. 7.7 for the problem of one-dimensional motion along the x-axis of three particles 1, 2, 3 interacting with a fourth one fixed at the origin. All the two-body, three-body, and four-body potentials V_{ij}, V_{ijk}, V_{1234} are supposed to be short-range. In the center-of-mass coordinate system the configuration space is three-dimensional. The domain where the two-body potential V_{ij} is nonzero appears to be a planar layer with a width equal to the range of the potential. Inside this layer two chosen particles are close to one another and other particles are arbitrarily disposed. The intersection of two orthogonal layers is a tube inside which the particles are grouped into two pairs of interacting particles $(i, j) + (k, l)$, and along the tube the distance between the pairs varies. The cross section of three such tubes in Fig. 7.9b is rectangular.

The intersection of three plane layers gives a tube corresponding to a three-body compact cluster (i, j, k) and a fourth particle l, with varying distance between l and (i, j, k) along the tube. The cross section of four such tubes in Fig. 7.9b is a star. Three-body interactions can be nonzero inside these tubes. Four-body forces of finite range are nonzero near the origin.

7.8 Notes on the Literature

The scattering theory for many-body systems, developed by generalization of the technique of configuration mixing in the shell model to single-particle states in the continuous spectrum has been presented by *Machaux, Weidenmüller*

a
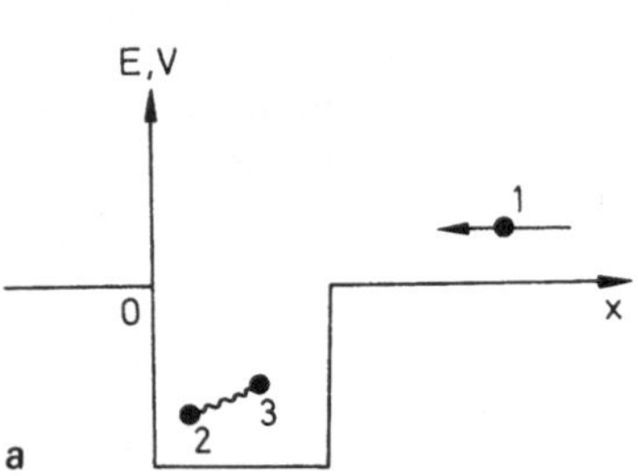

Fig. 7.9 (*continued*)

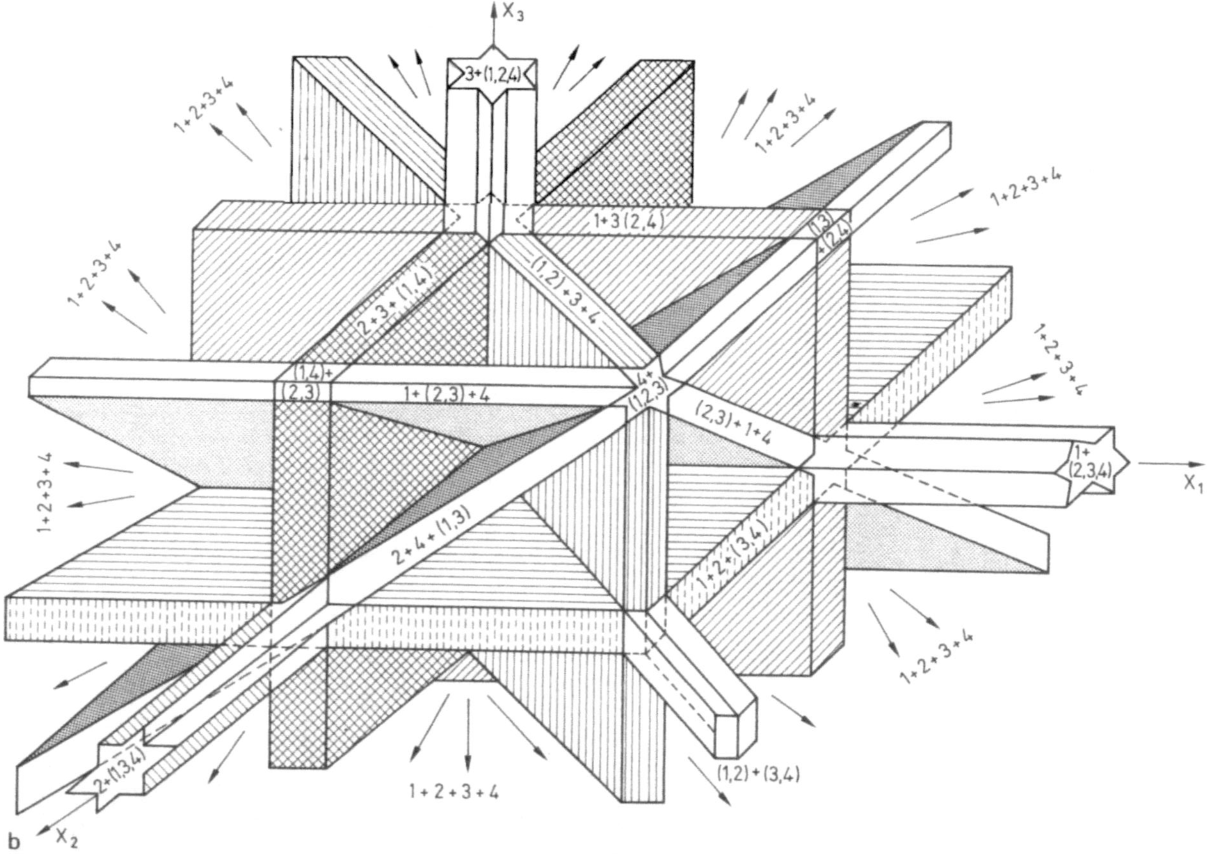

Fig. 7.9. **a** A system of three particles moving along the x-axis (one dimension). Their interaction with the fourth particle is shown as a finite potential well. **b** The interaction domains of the four-particle system. Six planar layers correspond to the domains where short-range two-body potentials V_{ij} are nonzero. Seven cylindrical tubes formed by intersections of these layers correspond to different cluster arrangements of particles: $(i, j, k)+l$; $(i, j)+(k, l)$. The domains outside the layers correspond to the total break-up of the system: $1+2+3+4$.

(1969) and *Rotter* (1984). Within this approach the coordinate dependence of the wave function ψ is described by auxiliary base functions in which ψ is expanded, while for the coefficients of the expansion integral equations are obtained. In contrast with *Machaux, Weidenmüller* (1969) *Zhigunov* and *Zakhariev* (1974) discussed the methods of the so-called unified reaction theory, in which one of the space coordinates remains in the argument of the expansion coefficients.

Variational methods in scattering problems. The coefficients Ψ_α of the expansion in basis sets can be sought not only by direct methods (Bubnov–Galerkin) from the condition that the individual projections of equation discrepancies be equal to zero, but also by a variational procedure. It is remarkable that the equations that are obtained for the Ψ_α are often the same in both approaches. We then have an additional justification for applying direct methods. Under certain conditions the new approach yields more accurate results.

In variational methods solution of the Schrödinger equation is replaced by the equivalent problem of seeking the function ψ by providing for a certain functional $T(\psi)$ a stationary value, that is, with respect to the variation of ψ:

$$\delta T(\psi) = 0\ . \tag{7.8.1}$$

A small deviation $\Delta\psi$ from the exact solution ψ does not alter $T(\psi)$ in the first (linear) approximation; the first variation of $T(\psi)$ equals zero. If the functional T is constructed so that it has the meaning of the desired physical quantity, for instance, the scattering amplitude, then the sensitivity to uncertainties in the trial function of amplitude computations becomes weaker since $T(\psi)$ is stable. We shall consider an extremely simplified example.

Let the trial functions ψ depend on a countable set of parameters Ψ_α which at certain values of $\Psi_\alpha = \Psi_\alpha^{\text{exact}}$ make $T(\psi)$ exactly equal to the physical quantity $T(\psi^{\text{exact}})$. (In the general case the Ψ_α are *functions* of part of the space variables, while here we assume that Ψ_α to be constants in the configuration space.) We shall fix these values for all the parameters except a certain Ψ_{α_0}. Then the condition (7.8.1) will correspond to the derivative of $T(\Psi_{\alpha_0})$ a t the point where $\Psi_{\alpha_0}^{\text{exact}}$ is equal to zero:

$$\left.\frac{\partial T(\Psi_{\alpha_0}\Psi}{\partial \Psi_{\alpha_0}}\right|_{\Psi_{\alpha_0} = \Psi_{\alpha_0}^{\text{exact}}} = 0\ .$$

In Fig. 7.10 the comparatively weak change in T is demonstrated for a relatively significant $\Delta\Psi_{\alpha_0}$(ΔT differs from zero only in the second order of $\Delta\Psi_{\alpha_0}$; $\Delta T = (d^2T/d^2\Psi_{\alpha_0})\,(\Delta\Psi_{\alpha_0})^2 + \ldots$).

One must, however, bear in mind that when the functional is stationary small errors are ensured only for ψ sufficiently close to ψ^{exact}. Thus, in the example just considered, the parameter Ψ_{α_0} leaving the zone where the curve $T(\Psi_{\alpha_0})$ is nearly horizontal (Fig. 7.10) leads to a sharp increase of the uncertainty ΔT.

Especially favorable is the case when the variation of the functional $T(\psi)$ exhibits the property of *definite sign*. This occurs in the eigenvalue problem (Ritz method) as well as in the computation of the scattering length ($k \to 0$). Then the

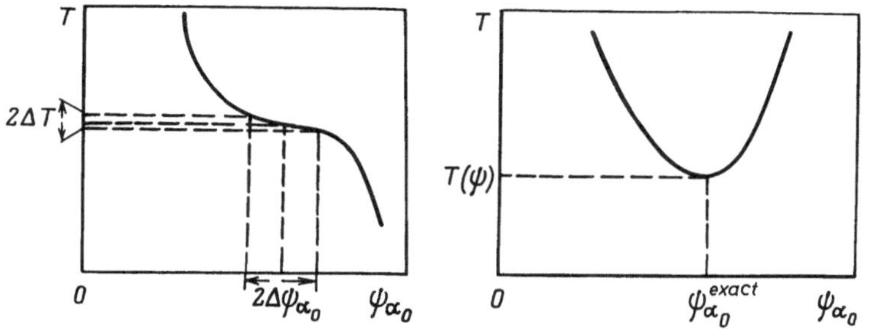

Fig. 7.10. a, b. Dependence of the variational functional T on the parameter Ψ_{α_0}: **a** when T is stationary; **b** when T is minimal

condition that $T(\psi)$ be stationary transforms into the condition that it be minimized (Fig. 7.10b). In this case the approximate value of $T(\Psi_{\alpha 0})$ gives the one-sided limit for the sought T: its exact value $T(\psi)$ is sure to lie below $T(\Psi_{\alpha 0})$.

From the beginning the variational approach was applied in quantum mechanics only in problems dealing with bound states. The variational principle for scattering was formulated by *Hulten* in 1944 and, in a somewhat different form, by *Kohn* in 1948. At present the Hulten–Kohn principle has been extended to arbitrary reactions (Mentkovsky 1982; Zhigunov, Zakhariev 1974).

The solutions of multi-particle equations of the Faddeev type have been discussed by *Belyaev* (1988), *Merkuryev* (1970–1980), and *Merkuryev* and *Faddeev* (1985). Here we shall only point out the method proposed by *Amirkhanov* (1971; 1973) which is suitable both for solution of Faddeev equations in the coordinate representation and for multi-channel equations within the unified reaction theory.

We shall consider this method by using the example of scattering in an external field of a pair of particles (1, 2) possessing a single bound state at an energy below the disintegration threshold. We expand $\psi(R, \rho_{12})$ in the states $\chi(\rho_{12})$ of the relative motion of the particles composing a quasideuteron, including virtual states of the continuum $\chi(k, \rho_{12})$:

$$\psi(R, \rho) = \Psi_0(R)\chi_0(\kappa, \rho_{12}) + \int dk \Psi(k, R)\chi(k, \rho_{12}) \,. \tag{7.8.2}$$

For Ψ_0, $\Psi(k, R)$ we have the set of equations ($2M = \hbar^2 = 1$):

$$\begin{aligned} \Psi_0''(R) + k_0^2 \Psi_0(R) &= V_{00}(R)\Psi_0(R) + \int dk \, V_{0k}(R)\Psi(k, R) \,; \\ \Psi''(k, R) + k^2 \Psi(k, R) &= \int dk' \, V_{kk'}(R)\Psi(k', R) + V_{k0}(R)\Psi_0(R) \,, \end{aligned} \tag{7.8.3}$$

where V_{ij} are the interaction matrix elements in the functions χ. The expressions in the right-hand sides of (7.8.3) are quadratically integrable functions of R. Therefore they can be expanded in the basis L_2 by limiting the consideration to the approximation of a finite number of terms:

$$\Psi_0''(R) + k_0^2 \Psi_0(R) = \sum_n^N A_{0n} \chi_n^{L_2}(R) ;$$

$$\Psi''(k, R) + k^2 \Psi(k, R) = \sum_m^N A_{km} \chi_n^{L_2}(R) . \tag{7.8.4}$$

With the aid of the Green functions $\mathring{G}_k(R, R')$ of free motion in the channels we write the solutions $\Psi_0(R)$ and $\psi(k, R)$ in the form of linear combinations of known expressions with the coefficients A_{0n} and A_{km} which still have to be determined:

$$\Psi_0(R) = e^{-ik_0R} + \sum_n^N A_{0n} \int dR' \mathring{G}_0(R, R') \chi_n^{L_2}(R') ;$$

$$\Psi(k, R) = \sum_m A_{km} \int dR' \mathring{G}_k(R, R') \chi_m^{L_2}(R') . \tag{7.8.5}$$

Substituting these functions into the right-hand sides of (7.8.3), equating them to the right-hand sides of (7.8.4), and projecting the obtained equalities onto $\chi_n^{L_2}$, we obtain a set of equations for the coefficients A_{0n} and A_{km}:

$$\begin{aligned} A_{0n} = {} & \int dR \chi_n^{L_2} V_{00}(R) e^{-ikR} + \int dk \sum_m^M A_{km} \int V_{0k}(R) \chi_n^{L_2}(R) \\ & \times \left\{ \int dR' \mathring{G}_k(R, R') \chi_m^{L_2}(R') \right\} dR + \sum_{n'}^k A_{0n'} \int \chi_n^{L_2}(R) V_{00}(R) \\ & \times \left\{ \int dR' \mathring{G}_0(R, R') \chi_{n'}^{L_2}(R') \right\} dR ; \\ A_{km} = {} & \int dk' \sum_{m'}^M A_{k'm'} \int \chi_m^{L_2}(R) \left\{ \int dR' \mathring{G}_k(R, R') \chi_{m'}^{L_2}(R') \right\} dR \\ & + \int V_{k0}(R) e^{-ik'_0R} \chi_m^{L_2}(R) dR + \sum_n^N A_{0n} \int \chi_m^{L_2}(R) V_{k0}(R) \\ & \times \left\{ \int dR' \mathring{G}_k(R, R') \chi_n^{L_2}(R') \right\} dR . \end{aligned} \tag{7.8.6}$$

This set of integro-algebraic equations can be reduced to a purely algebraic set by replacing, by some approximations the integrals over k by the corresponding summations, or by expanding these integrals in the basis $\chi_a^{L_2}(\mathbf{k})$, taking advantage of the decrease of A_{km} as $\sim 1/k^2$ at large k. A similar solution scheme can be also applied to more complicated problems involving particle rearrangement. For example, in the Faddeev equations

$$\left[\frac{\hbar^2}{2M} \frac{\partial^2}{\partial R_i^2} + \frac{\hbar^2}{2m_i} \frac{\partial^2}{\partial \rho_i} - V_{jk} + E \right] \psi_i = V_{jk} \psi_j + V_{jk} \psi_k , \tag{7.8.7}$$

where V_{ij} is the short-range potential of the particles i and j ($i \neq j \neq k = 1, 2, 3$),

ψ_j is the part of the wave function $\psi = \sum_{j=1}^{3} \psi_j$ which transforms into the eigenfunction of the asymptotic Hamiltonian as $R_j \to \infty$. In the Amirkhanov method the right-hand side of (7.8.7) is expanded in functions from L_2. This is possible because the *infinite* hypercylinders of the six-dimensional configuration space of three particles (where V and ψ_j are not zero, Fig. 7.2) depart in different directions and exhibit significant overlapping in a *limited* region.

Effective forces. To simplify the description of a scattering problem, especially of multi-particle systems, one often uses an effective potential, the nature of which may have a physical explanation outside the framework of the formalism being considered. The accumulation of information about various examples of connections between the original and effective forces is quite instructive. As an example, we shall point out the distortion by a medium of a field acting between two particles in vacuum, and the popular presentation of the idea of "Grand Unification" of strong, weak and electromagnetic interactions.

It is interesting that the external electron which is in a bound state of an impurity atom in a semiconductor moves at distances from its nucleus which are by two orders of magnitude (!) larger than the inter-atomic distance (Efros 1972; 1983). The dimensions of the impurity atoms are so large that they can overlap even if their density is a million times smaller than that of the atoms in the semiconductor. This results in an impurity conductivity.

If the atom is in vacuum, then the last electron is in a field nearly like the one in the hydrogen atom (the remainder of the electron shell and the nucleus have a common charge equal to $+e$). Such an enormous difference is due to polarization of the medium which strongly screens the electron charge, so that the effective potential is ε times smaller, $e^2/\varepsilon r$, instead of e^2/r. In addition, the effective mass of the electron m^* also decreases (in Germanium $m^* \approx 0.1\ m$; $\varepsilon = 16$ and, as a result the orbit radius $a^* \approx 160\ a_B$ where $a_B = \hbar^2/me^2$).

Our second example concerns the unity of fundamental interactions. Modern field theory has been successful in unifying electromagnetic and weak forces. The inclusion of the strong interaction within the unification scheme is still under development. The sameness of the nature of such unlike fields can be understood with the aid of the following nonrigorous arguments given by *Kirzhnitz* and *Linde* (1983).

The three types of interaction potentials which have essentially different dependences on the distance between the particles are shown in Fig. 7.11. To electromagnetic forces where massless photons are exchanged there correspond $V_{\text{em}}(r)$ that decrease with r as $\sim 1/r$. Thus opposite charges or currents flowing in the same direction are attracted. The exchange of massive bosons yields the Yukawa potential $V_{\text{weak}} \sim e_{\text{wk}}^2 \mathrm{e}^{-\mu r}/r$. (The W bosons have been only recently observed.) The strong interaction of quarks increases linearly with r; the forces acting between the nucleons are a result of mutual "screening" of the quarks.

These potentials can be represented as a unique "Coulomb" interaction distorted by various vacuua, where vacuum is defined as a medium with an infinite number of particles. Thus, the exponentially decreasing V_{weak} is obtained

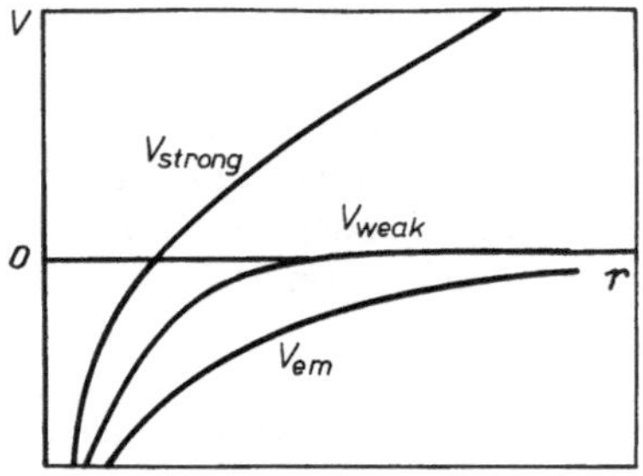

Fig. 7.11. The form of the potentials for the strong, weak, and electromagnetic interactions. From (Kirzhnitz, Linde 1983)

in a way similar to how Coulomb attraction of charged particles in vacuum transforms into Yukawa attraction in a conducting medium. The Yukawa interaction is also obtained for currents in a superconducting vacuum, owing to the development of non-decaying induced currents. Now, the potential $V_{\text{strong}} \approx r$ can be considered to be a consequence of the action of the vacuum which compresses the forces lines of the charges into a "string" connecting them (like a superconductor that pushes out magnetic fields), which requires an energy $\sim r$.

Progress in the solution of the three-body problem has also led to success in taking into account three-particle collisions in systems with an infinite number of particles described by statistical methods (Merkuryev 1970–1980). (Statistical physics is formulated at present on the basis of a unique approach for quantum and classical systems, for stationary states and non-equilibrium processes (Balesku 1975). This scheme also includes relativistic effects.)

Relativistic quantum mechanics of three or more particles was developed by *Sokolov* (1984). He also proposed a solution of the inverse multi-particle problem (reconstruction of the nonlocal interaction).

Optical potentials for relativistic equations may be found in (Arnold 1982). The nuclear multi-body problem has been dealt with by *Ring* and *Shuck* (1980).

Gordov and *Tvorogov* (1984) have described particle systems, part of which obey classical and the other part obey quantum laws of motion. Exactly solvable quantum multi-particle models are dealt with in the monograph written by *Baxter* (1982), in chapter 4 of book by *Mahan* (1981), and classical ones by *Perkus* (1987).

Quasiclassical methods were applied by *Popov* (1986) for solution of three-body scattering problems.

Exactly solvable models for systems of three nonidentical particles with a single two-body and separable three-body potential have been considered by *Fuda* (1973).

The theory of strong coupling between channels in the case of disintegration processes has been presented in detail by *Kamimura* (1986).

The high effectiveness of Monte-Carlo Green functions for describing multi-particle systems was shown by *Efros* (1988).

The relation of the nodal structure and phase shifts of zeroincident-energy wave functions for multi-particle scattering was discussed by *Iwinski* et al (1985; 1986).

The exact solution of some many-body problems by transformation to higher-dimensional spaces was considered by *Perelomov* (1979) [see also Gipson (1982)].

The inverse problem involving both the processes with rearrangement and break-up was formulated by *Dubovik* et al (1989), *Vinitsky* and *Suzko* (1990).

Bibliography

Ablowitz MJ, Segur H: *Solitons and the Inverse Scattering Transform* (Soc. Ind. Appl. Math, Philadelphia 1981)

Abraham PB, Moses HF: Phys. Rev. A22, 1333 (1980)

Abramov DI: Sov. J. Theor. Math. Phys. **58**, 244 (1984); ibid. **63**, 32 (1985); Sov. Dokl. Akad. Nauk, **298**, 585 (1988)

Adamyan MN: Sov. J. Theor. Math. Phys. **22**, 236 (1975); ibid. **48**, 70 (1981); Ukr. Phys. J. **27**, 972 (1982)

Agranovich ZS, Marchenko VA: *The Inverse Problem of Scattering Theory* (Gordon and Breach, New York 1963)

Aguilera VC, Peltier SM, Plastino-Navarro A, Ley-Koo E, de Llano M: J. Math. Phys. **23**, 2439 (1982)

Albeverio S: Ann. Phys. **71**, 167 (1972)

Albeverio S, Gestezy F, Hoegh-Krohn R, Holden H: *Solvable Models in Quantum Mechanics* (Springer, Berlin-Heidelberg 1988)

Alfaro V, Regge T: *Potential Scattering* (Wiley, New York 1965)

Alhassid Y, Gursey F, Iachello F: Ann. Phys. **148**, 346 (1983); ibid. **167**, 181 (1986); ibid. **173**, 68 (1987); J. Phys. A**22**, L947 (1989)

Amirkhanov IV, Mirkasimov RM, Grusha GV: Sov. Vestn. MGU. ser. Phys. astr. **N5**, 579 (1971); ibid. **N2**, 149 (1973); Sov. Nucl. Phys. **26**, 207 (1977); Sov. Part. Phys. **12**, 651 (1981)

Amirkhanov IV, Puzynin IV, Puzynina TP, Zakhariev BN: in *Theory of Quantum Systems with Strong Interactions* (Kalinin University Press, Kalinin 1989) p. 55

Amirkhanov IV, Zakhariev BN: Sov. J. Exp. Theor. Phys. **49**, 1097 (1965)

Andrianov AA, Borisov NV, Ioffe MV: Sov. Theor. Math. Phys. **61**, 17; 183 (1984); ibid. **72**, 97 (1987); Sov. Vestn. LGU, **4**, N4 (1988); Phys. Lett. **105A**, 2; **109A**, 143 (1984)

Aneva BL: J. Phys. **A22**, 129 (1989)

Arnold LD: Phys. Rev. **C25**, 936 (1982)

Asthana P, Kamal AN: Z. Phys. **C19**, 37 (1983)

Atkinson FV: *Discrete and Continuous Boundary Problems* (Academic, New York 1964)

Avery J: *Hyperspherical Harmonics Applications in Quantum Theory* (Kluwer Acad. Reidel, Dordrecht 1989)

Babenko VA, Petrov NM: Sov. Theor. Math. Phys. **75**, 61 (1988)

Babikov VV: *Method of Phase Functions in Quantum Mechanics* (Nauka, Moscow 1988)

Bagrov VG, Gitman DM, Ternov IM: *Exact Solutions of Relativistic Wave Equations* (Nauka, Novosibirsk 1982)

Baldock RA, Robson BA, Barret RF: Nucl. Phys. **A366**, 270 (1981); **A381**, 138 (1982)

Balesku R: *Equilibrium and Nonequilibrium Statistical Mechanics* (J. Wiley, New York 1975)

Bang E, Ershov SN, Gareev FA, Ivanova SP: Sov. Part. Phys. **9**, 287 (1978); ibid. **11**, 813 (1980); Lett. Nuovo Cim. **32**, 420 (1981); Nucl. Phys. **A339**, 89 (1980); Phys. Scr. **18**, 289; 298 (1978)

Barcilon V: J. Math. Phys. **15**, 429 (1974)
Bargmann V: Rev. Mod. Phys. **21**, 488 (1949)
Barthelemy M-C: Ann. Inst. Henri Poincaré, **7**, 115 (1967)
Barut AO, Únal N: J. Math. Phys. **27**, 3055 (1986)
Barz BI, Bolotin YuL, Inopin EV, Gonchar VYu: *Hartree–Fock Method in Nuclear Theory* (Naukova Dumka, Kiev 1982)
Baxter FRS: *Exactly Solved Models in Statistical Mechanics* (Academic, London 1982)
Baz AI: Sov. J. Expt. Theor. Phys. **70**, 397 (1976)
Belyaev VB: *Lectures on Few-Body Systems* (Energoatomizdat, Moscow 1986); (English translation, Springer, Berlin-Heidelberg 1990)
Ben-Israel A, Greville TNE: *Generalized Inverses: Theory and Applications* (Wiley, New York 1974)
Benn J, Scharf G: Nucl. Phys. **A134**, 481 (1969); Helv. Phys. Acta. **40**, 271 (1967)
Beregy, P, Zakhariev BN, Nijazgulov SA: Sov. Part. Nucl. **4**, 512 (1973)
Berezansky YuM: *The Expansion into the Set of Eigenfunctions of Selfadjoint Operators* (Naukova Dumka, Kiev 1965)
Berezin FA, Shubin MA: *The Schrödinger Equation* (Moscow University Press, Moscow 1983)
Berezovoy VP, Pashnev AI: Sov. J. Theor. Math. Phys. **70**, 146 (1987)
Berryman JG, Green RR: Phys. Lett. **A65**, 13 (1978)
Binnig G, Rohrer H: Rev. Mod. Phys. **59**, 515 (1987)
Bogdanov IV: Sov. Theor. Math. Phys. **65**, 35 (1985)
Bogdanov IV, Demkov YuN: Sov. J. Exp. Theor. Phys. **82**, 1798 (1982); Sov. J. Theor. Math. Phys. **93**, 3 (1987)
Bogolyubov NN Jr, Brankov YG, Zagrebnov VA, Kurbatov AM, Tonchev NS: *The Aproximating Hamiltonian Method in Statistical Physics* (Acad. Sci. Bulg., Sofia 1981)
Bolle D, Osborn TA: J. Math. Phys. **20**, 1121 (1979); ibid. **22**, 883 (1981); Nucl. Phys. **A351**, 377 (1981); Ann. Phys. **101**, 119 (1976); Phys. Rev. **A22**, 101 (1980); ibid. **A26**, 3062 (1982)
Bolsterly M: Phys. Rev. **177**, 1443 (1969)
Bonche P, Koonin S, Negele JW: Phys. Rev. **C13**, 1229 (1976)
Brandt S, Dahmen HD: *The Picture Book of Quantum Mechanics* (Wiley, New York 1985)
Brink DM: in *Frontiers in Nuclear Dynamics.* Proc. Int. School. 171, Erice. (Plenum, New York 1985)
Bruckstein AM, Kailath T: J. Math. Phys. **28**, 2914 (1987)
Buchgeim AL: *Volterra Equations and Inverse Problems* (Nauka, Novosibirsk 1983)
Burdet G, Giffon M, Predazzi E: Nuovo Cim. **36**, 1337 (1965); ibid. **A44**, 138 (1966)
Bursak AV, Gorbatov AM, Krylov YuN, Rudyak BV: Sov. Nucl. Phys. **36**, 1138 (1982)
Buslaev VS, Faddeev LD: Sov. Math. Dokl. **1**, 451 (1960); Sov. DAN USSR, **143**, 1067 (1962): in Problems in Mathematical Physics Vol. 82, (LGU, Leningrad 1966)
Butera P, Girardello L: Nuovo Cim. **54A**, 141 (1967)
Calogero F: *Variable Space Approach to Potential Scattering* (Academic, New York 1967)
Calogero F, Degasperis A: Lett. Nuovo Cim. **23**, Ser. 2, 143 (1978)
Calogero F, Simonov YuA, Surkov EL: Nuovo Cim. **47**, 178 (1967); Phys. Rev. **C5**, 1493 (1972)
Campi M, Harrison M: Am. J. Phys. **34**, 260; 1122 (1966); **35**, 133 (1967)
Case KM, Chui SC, Lau CW: J. Math. Phys. **14**, 916; 1643 (1973); ibid. **15**, 143; 974; 2166 (1974); ibid. **16**, 1435 (1975)
Case KM, Kac M: J. Math. Phys. **14**, 594 (1973)
Chadan K: Compt. Rend. **299**, ser2, 271 (1984)

Chadan K, Grosse H: Compt. Rend. **299**, ser2, 1305 (1984)
Chadan K, Kobauashi R: Compt. Rend. **303**, ser2, 329 (1986)
Chadan K, Musette M: Inv. Probl. **5**, 257 (1989)
Chadan K, Sabatier PC: *Inverse Problems in Quantum Scattering Theory* (Springer, Berlin-Heidelberg 1989)
Christov ECh: *Nonlinear Evolution Equations for Approximate Solution of Inverse Problems.* Ph.D Thesis, JINR, Dubna (1981)
Churchill JN: Am. J. Phys. **55**, 372 (1987)
Ciaffaloni M: Nuovo Cim. **29**, 420 (1963)
Clemence DP: Inv. Probl. **5**, 269 (1989)
Coen S: J. Math. Phys. **22**, 1127 (1981)
Cohen BL: Am. J. Phys. **33**, 97 (1965)
Cooper CG, Mackintosh RS: Inv. Probl. **5**, 707 (1989)
Cooper F, Ginocchio JN, Khare A: Phys. Rev. **D36**, 2458 (1987)
Coudray C: These ORSAY, N2160 (1979)
Coudray C: Lett. Nuovo Cim. **19**, 319 (1977)
Coudray C, Coz M: Ann. Phys. NY. **61**, 488 (1970); J. Math. Phys. **12**, 1166 (1972); **14**, 1574 (1973)
Cox J: Ann. Phys. **39**, 216 (1966); J. Math. Phys. **5**, 1065 (1964); **8**, 2327 (1967)
Cox J, Garcia HK: J. Math. Phys. **16**, 1402 (1975)
Coz M, Rochus R: J. Math. Phys. **18**, 2223, 2232 (1977); ibid. **22**, 1596 (1981)
Coz M, Geramb HV, Kuberczyk J, Lumpe JD: Z. Phys. **A326**, 345 (1987); **A328**, 259 (1987)
Crandal RE, Litt BR: Ann. Phys. **146**, 458 (1983); J. Phys. **A16**, 3005 (1983)
Crum MM: Quart. J. Math. **6**, 121 (1955)
Dammert O: J. Math. Phys. **24**, 2163 (1983); **27**, 461 (1985)
Delves LM: Nucl. Phys. **9**, 391 (1959); **20**, 275 (1960); **29**, 268 (1962)
Demkov YuN: *Variational Principles in Scattering Theory* (Fizmatgiz, Moscow 1958)
Demkov YuN, Ostrovsky VN: *Method of Zero-Range Potentials in Atomic Physics* (Plenum, New York 1988)
Denisov AM: Sov. J. Comp. Math. **17**, 753 (1977); **22**, 858 (1982)
Deo BB, Swain S, Swain S: Phys. Rev. **C32**, 1247 (1985)
Djibuti VI, Krupennikova NB: *Method of Hyperspherical Functions in the Quantum Mechanics of Few-Body Systems* (Mecniereba, Tbilisi 1983)
Draper JE: Am. J. Phys. **47**, 525 (1979); **49**, 749 (1980)
Dreyfus T: J. Math. Phys. **9**, 187 (1976); J. Math. Anal. Appl. **64**, 114 (1978)
Drosdov SI: Sov. J. Exp. Theor. Phys. **28**, 734 (1955)
Dubovik VM, Markovsky BL, Suzko AA, Vinitsky SI: Phys. Lett. A **142**, 133 (1989)
Dubrovin BA: Sov. Usp. Mat. Nauk **36**, 11 (1981)
Dubrovin BA, Malanyuk TM, Krichever IM, Makhankov VG: Sov. Part. Nucl. **19**, 579 (1988)
Dubrovin BA, Krichever IM: Sov. Funct. Anal. **9**, 41 (1975); Sov. DAN USSR, **229**, 15 (1976)
Eastham, MSP, Kalf H: *Schrodinger-type Operators with Continuous Spectra* (Pitman, London 1982)
Efimenko TG, Zakhariev BN, Zhigunov VP: Ann. Phys. NY. **47**, 275 (1968)
Efros AL: *Physics and Geometry of Disorder* (Nauka, Moscow 1982)
Efros VD: Sov. Nucl. Phys. **15**, 226 (1972); **38**, 874 (1983); Dissertation, Institute of Atomic Energy, Moscow (1988)

Elphik C: J. Math. Phys. **28**, 1243 (1987)
Eskina MS: in *Trans. Seminar on Functional Analysis*. Vol. 2, 207. Math. Inst. Acad. Sci. Ukr. SSR. Kiev (1970)
Euler L: M. S. Academiac Exhibuit die 13, Jan. 1780; Instutiones Calculi Integralis, **4**, 533–543 (1794)
Exner P, Sheba P, Dittrich J: J. Math. Phys. **26**, 2000 (1985); ibid. **28**, 386 (1987); Preprint JINR, Dubna. E2-87-213; 214 (1987)
Fabre de la Ripelle M: Rev. Roum. Phys. **14**, 1215 (1969); Ann. Phys. **123**, 185 (1979); **147**, 281 (1983); J. Math. Phys. **24**, 1992 (1983); in *Microscopical Methods in the Theory of Few-Body Systems* (Kalinin University Press, Kalinin 1988); Few-Body Syst. **6**, 157 (1989)
De Facio B, Moses HE: J. Math. Phys. **21**, 1716 (1980)
Faddeev DK, Faddeeva VN: *Computational Methods of Linear Algebra* (Nauka, Moscow 1969); Trans. Steklov Math. Inst. **53**, 387 (1959)
Faddeev LD: J. Math. Phys. **4**, 72 (1963); Sov. Probl. Math. **3**, 93 (1974); J. Sov. Math. **5**, 334 (1976); Trans. Steklov Math. Inst. CLXXVI, **4**, 4 (1987);
Fano U, Rau ARP: *Atomic Collisions and Spectra* (Academic, Orlando 1986)
Fernandez DJ: Lett. Math. Phys. **8**, 337 (1984)
Fiedeldey H, Lipperheide R, Naidoo K, Sofianos S: Phys. Rev. **C33**, 1581 (1986)
Filippov GF, Vasilevsky VS, Chopovsky LL: Sov. Part. Nucl. **15**, 1338 (1984); ibid. **16**, 349 (1985)
Firsova NE: Zapisky Nauchn. Sem. LOMI, **51**, 183 (1975); [trans. J. Sov. Math. **11**, 487 (1979)]; ibid. **60**, 196 (1985); Sov. Math. Sborn. **130**, 349 (1986)
Flugge S: *Practical Quantum Mechanics* (Springer, Berlin-Heidelberg 1971)
Frank WM, Land DJ, Spector RM: Rev. Mod. Phys. **43**, 36 (1971)
Fuda MG: Phys. Rev. **C7**, 1365 (1973)
Funke H, Zakhariev BN: Commun. JINR, Dubna, N 12–85, 35 (1985); Phys. Lett. **B185**, 265 (1987); *Proc. Conf. Inverse Problems* Montpellier, 1986 (Academic, New York 1987)
Fushchich WI, Shtelen WM, Serov NI: *Symmetry Analysis and Exact Solutions of Nonlinear Equations of Mathematical Physics* (Naukova Dumka, Kiev 1989)
Gagnon L, Winternitz P: J. Phys. **A22**, 469 (1989)
Gasymov MG, Guseinov GSh: Sov. Diff. Uravnen. **25**, 588 (1989)
Gavurin MK: Sov. J. Izv. Vuz. Ser. Math, N5/6, 18 (1958)
Gazdy B: Phys. Lett. **A61**, 89 (1977)
Gelfand IM, Levitan BM: Sov. J. Dokl. Acad. Nauk USSR, **77**, 557 (1951); Am. Math. Soc. Trans. **1**, 253 (1951)
Gelfand IM, Vilenkin NYa: *Generalized Functions* Vol. 4 (Fizmatgiz, Moscow 1961)
Gendenstein LE, Krive IV: Sov. Usp. Phys. Nauk. **146**, 553 (1985)
Gerasimenko NI: Sov. Theor. Math. Phys. **75**, 187 (1988)
Gibson WG: Phys. Rev. **A36**, 564 (1987)
Gipson JM: Phys. Rev. Lett. **48**, 1511 (1982)
Girard R, Kroger H, Labelle P: Phys. Rev. **A37**, 3195 (1988)
Good RH: Amer. J. Phys. **40**, 343 (1972)
Goldberg A: Am. J. Phys. **35**, 177 (1967)
Goncharsky AV, Cherepashuk AM, Yagola AG: *Inverse Astrophysical Problems* (Znanie, Moscow 1987)
Goncharsky AV, Bakushinsky AB: *Iterative Methods of Solution of Ill-Posed Problems* (Nauka, Moscow 1989); Ill-Posed Problems (Moscow Univ. Press, Moscow 1989)

Gorbatov AM: Thesis, JINR Dubna (1983)

Gorbatov AM, Bursak AV, Krylov YuN, Nikishov PYu, Kolganova EA: in *Microscopic Calculations for Light Nuclei* (Kalinin University Press, Kalinin 1979–1989)

Gordov EP, Tvorogov SD: *Method of Semiclassical Representation in Quantum Theory* (Nauka, Novosibirsk 1984)

Gorelik GK: *Why is Space Three-Dimensional*? (Nauka, Moscow 1982); *Space Dimension* (MGU, Moscow 1983)

Gostev VB, Mineev VS, Frenkin AR: Sov. DAN, **262**, 1364 (1982); Sov. J. Theor. Math. Phys. **56**, 74 (1983); ibid. **59**, 432 (1984); ibid. **62**, 472 (1985); ibid. **68**, 45 (1986)

Gradstein IS, Rizhik IM: *Tables of Integrals*, Nauka, Moscow (1971)

Grinevich PG: Sov. Theor. Math. Phys. **69**, 307 (1986)

Groetsch CW: *The Theory of Tikhonov Regularization for Fredholm Equations of the First Kind*. Research Notes in Math. Vol. 105 (Pitman, London 1984)

Grosse H, Martin A: Nucl. Phys. **B132**, 125 (1978); ibid. **B148**, 413 (1979); Lect. Notes Phys. **116**, 68 (1979); Phys. Rep. **C60**, 341 (1980)

Guha A, Mukherjee S: J. Math. Phys. **28**, 840 (1987)

Guseinov GS: Sov. J. Math. Notes. **23**, 709 (1978); ibid. **32**, 737 (1982); Sov. Math. Dokl. **17**, 1684 (1976); Sov. DAN, **231**, 1045

Hadamard J: *Le probleme de Cauchy et les êquations aux derives partiells linêares hyperboliques* (Hermann, Paris 1932)

Haftel MI, Mandelzweig VB: Ann. Phys. **189**, 29 (1989)

Haymaker RH, Rau ARP: Am. J. Phys. **54**, 928 (1986)

Hefter EF: Phys. Rev. A32, 1205 (1985); **C34**, 1588 (1986)

Hefter EF, Kartavenko VG: JINR Rapid Comm. N3. Dubna (1987)

Heller EJ, Yamani HA: Phys. Rev. **A9**, 1201 (1974)

Henkin GM, Novikov RG: Inv. Probl. **4**, 103 (1988)

Hoffman-Ostenhof M, Hoffman-Ostenhof T: J. Math. Phys. **25**, 2490 (1984)

Holevo AS: *Probable and Statistical Aspects of Quantum Theory* (North-Holland, Amsterdam 1982)

Hooshyar MA, Razavy M: Can. J. Phys. **59**, 1627 (1981); Phys. Rev. **C25**, 1187 (1982); ibid. **29**, 20 (1984)

Hron M, Razavy M: Can. J. Phys. **55**, 1434 (1977), J. Phys. **A14**, 2215 (1981)

Hulten L: Kungl. Phys. Sáll. Lund Fórh. **14**, 1 (1944)

Humi M: J. Phys. **A18**, 1085 (1985); J. Math. Phys. **27**, (1986)

Hwang JS: J. Math. Phys. **28**, 1287 (1987)

Imshenecky VG: Zapis. Imper. Akad. Nauk. Peterburg, 1–21 (1882)

Infeld L, Hull TE: Rev. Mod. Phys. **23**, 21 (1951)

Ivanov VK: *Theory of Linear Ill-Posed Problems and its Applications* (Nauka, Moscow 1978)

Iwinski ZR, Rosenberg L, Spruch L: Phys. Rev. **A31**, 1229 (1985); **A33**, 946 (1986)

Jauho AP, Nieto MM: Superlattices and Microstructures **2**, 407 (1986);

Jauho AP: Lecture at "Advanced Summer School in Microelectronics", Espoo, Finland (H.C.O. Tryk, Copenhagen 1987)

Jaulent M, Jean C: Ann. Inst. Henri Poincaré, Sect. A, **25**, 105; 119 (1976); ibid. **17**, 363 (1972); Commun. Math. Phys. **28**, 177 (1972); J. Math. Phys. **17**, 1351 (1976); ibid. **23**, 258 (1982)

Jost R, Kohn W: Phys. Rev. **88**, 382 (1952)

Jost R, Pais A: Phys. Rev. **82**, 840 (1951)

Kac M: Rocky Mount. J. Math. **4**, 511 (1974)

Kamimura M: Progr. Theor. Phys. Suppl. **89**, 1 (1986)

Kamuntavichus GP: Few-Body Systems, **1**, 91 (1986); Sov. Izv. Acad. Nauk. ser. phys. **51**, 8 (1987); Sov. Part. Nucl. **20**, 261 (1989)

Kay I, Moses HE: Nuovo Cim. **22**, 689 (1961); Comm. Pure Appl. Math. **14**, 435 (1961)

Kiselev AA, Lyapcev AV: *Quantum–Mechanical Perturbation Theory (Diagram Method)*, (Leningrad University Press, Leningrad 1989)

Kirzhnitz DA, Linde AD: in *Science and Humanity* (Znanie, Moscow 1983)

Klein JR: Am. J. Phys. **44**, 754 (1976)

Kohn W: Phys. Rev. **74**, 1763 (1948)

Kojushner MA: *Tunnel Effects* (Znanie, Moscow 1983)

Kolkunov VA: Sov. Nucl. Phys. **10**, 1296, N6 (1969)

Komarov VV, Popova AM, Shablov VL: Sov. Part. Nucl. **16**, 407 (1985), and references therein

Korn G, Korn T: *Mathematical Handbook* (McGraw-Hill, New York 1968)

Kovalsky JM, Fry JL: J. Math. Phys. **28**, 2407 (1987)

Krasnoselsky MA, Vainikko GM, Zabreiko PP, Rutitsky YaB, Stetsenko VYa: *Approximate Solution of Operator Equations* (Nauka, Moscow 1969)

Kravitsky AO: Sov. Diff. Equat. **4**, 165 (1968)

Krein MG: Sov. DAN. USSR. **76**, 21; 345 (1951)

Krichever IM: Sov. Usp. Math. Nauk **44**, 121 (1989)

Kristensson G: J. Math. Phys. **27**, 804 (1986)

Kukulin VI, Krasnopolsky VM, Horacek J: *Theory of Resonances. Principles and Applications* (Academia, Prague 1989)

Kuperin YuA, Makarov LA, Merkuryev SP, Motovilov SP, Popov BS: Sov. J. Theor. Math. Phys. **69**, 100 (1986); ibid. **75**, 431; ibid. **76**, 242 (1988)

Kvitko AN, Pivovarchik VN: Sov. Ukr. Phys. J. **32**, 152 (1987)

Kwong W, Rosner JL: Progr. Theor. Phys. Suppl. **86**, 366 (1986)

Kwong W, Rosner JL, Schonefeld JF, Quigg C, Thacker HB: Am. J. Phys. **48**, 926 (1980)

Lambert F, Corbella O, Thome ZD: Nucl. Phys. **B90**, 267 (1975)

Lapidus R: Am. J. Phys. **50**, 663 (1982)

Leeb H, Fideldey H, Lipperheide R: Phys. Rev. **C32**, 1223 (1985)

Leeb H, Schmizer WA, Fiedeldey H, Sofianos SA, Lipperheide R: Inv. Probl. **5**, 817 (1989)

Lend'el VI, Salak K: *Nonrelativistic Quantum Scattering Theory* (Vishya Shkola, Lvov 1983)

Leon JJ: J. Math. Phys. **21**, 2572 (1980); ibid. **22**, 965 (1981)

Lessie D, Spadaro J: Am. J. Phys. **54**, 909 (1986)

Lev FM: Forschr. Phys. **31**, 75 (1983)

Levi D: Inverse Probl. **4**, 165 (1988)

Levinson N: Kgl. dansk. vid. sels. Mayh.-fis. medd. **24**, 1 (1949)

Levitan BM: *Inverse Sturm–Liouville Problems* (Nauka, Moscow 1984); in *Problems of Mechanics and Mathematical Physics* (Nauka, Moscow 1976) p. 166

Levitan BM, Gasymov MG: Sov. Usp. Math. Nauk, **19**, 1 (1964)

Levitan BM, Sargsyan IS: *Introduction to Spectral Theory* (Nauka, Moscow 1970)

Levy-Leblond JM: Am. J. Phys. **47**, 1045 (1979)

Lipperheide R, Fiedeldey H: Z. Phys. **A286**, 45 (1978); **A301**, 81 (1981); Phys. Rev. **A32**, 3095 (1985); **C26**, 770 (1982)

Lopes CA, Massida V: Nuovo Cim. **A58**, 160 (1968); Nucl. Phys. **A149**, 33 (1970)

Lundin DB, Kozel VA: Theor. Funct., Funct. Anal. Appl. 11–17 (Kharkov University Press, Kharkov 1973)

Lyance VE: Sov. Mat. Sbornik, **72** (114), 537 (1967)

Ma Z-Q, Ni G-J, Liang Y-G: J. Math. Phys., **26**, 1995 (1985); Phys. Rev. **D31**, 1482 (1985); ibid. **D34**, 565 (1986)

Macek JH: J. Phys. **B1**, 831 (1968)

Machaux C, Weidenmuller HA: Shell Model Approach to Nuclear Reactions (North-Holland, Amsterdam 1969); [Rotter I: Sov. Part. Nucl. **15**, 762 (1984)]

Madler P, Nikishov PYu, Zakhariev BN: Comm. JINR Dubna E4-84-487 (1984); Sov. Izv. AN USSR ser. Phys. **51**, 2073 (1987)

Mahan GD: *Many-Particle Physics* (Plenum, New York 1981)

Malyarov VV, Poplavsky IV, Popushoy MN: Sov. Nucl. Phys. **16**, 873 (1972); ibid. **18**, 1140 (1973); ibid. **22**, 987; 860 (1975); ibid. **25**, 72 (1977); ibid. **37**, 1128 (1978); ibid. **30**, 1126 (1979)

Manakov SV: Sov. J. Exp. Theor. Phys. **67**, 543 (1974); J. Math. Phys. **21**, 2749 (1980); Sov. Usp. Math. Nauk. **31**, 245 (1985)

Marchenko VA: Sov. Dokl. Acad. Nauk. USSR. **104**, 695 (1955)

Marchenko VA: *Spectral Theory of Sturm–Liouville Operators* (Naukova Dumka, Kiev 1972)

Marchenko VA: *Sturm–Liouville Operators and Their Applications* (Naukova Dumka, Kiev 1977)

Martin PA: Lectures in Acta Phys. Austriaca Suppl. **23**, 157 (1981)

Matveenko AV: J. Phys. **B18**, L645 (1985)

May K-E: *Das Inverse Streuproblem in der nuclearen Schwerionenphysik*, Dissertation, Universitat Giessen, (1988)

May R, Noye J: in *Computational Techniques for Differential Equations* (North-Holland, Amsterdam 1984)

McGuire JB, Hurst CA: J. Math. Phys. **29**, 155 (1988)

McVoy KW: in *Fundamentals in Nuclear Theory* (IAEA, Vienna 1967) Chap. 8 p. 419

McWilliams B: Phys. Rev. **D20**, 1221 (1979)

Melnikov VN, Zakhariev BN: Rep. Math. Phys. **18**, 353 (1980); Phys. Lett. **65**, 13 (1978)

Mentkovsky YuL: *Particles in Nuclear–Coulomb Fields* (Energoatomizdat, Moscow 1982)

Merkuryev SP: Sov. Theor. Math. Phys. **5**, 372 (1970); ibid. **32**, 187 (1977); **38**, 201 (1978); **56**, 60 (1983); Sov. Nucl. Phys. **24**, 289 (1976); Nucl. Phys. **A233**, 395 (1974); Ann. Phys. **130**, 395 (1980)

Merkuryev SP, Faddeev LD: *Quantum Scattering Theory for Few-Body Systems* (Nauka, Moscow 1985)

Meyer-Vernet N: Am. J. Phys. **50**, 353 (1982)

Mikhailov AV, Shabat AB, Yamilov RI: Sov. Usp. Math. Nauk. **42**, 3 (1987)

Mikhlin SG: *Variational Methods of Mathematical Physics* (Nauka, Moscow 1970); *Numerical Realization of Variational Methods* (Nauka, Moscow 1966)

Mielnik B: J. Math. Phys. **25**, 3387 (1984)

Moses HE: Phys. Rev. **102**, 559 (1956); J. Math. Phys. **18**, 2243 (1977); **20**, 2047 (1979); Lect. Notes Phys. **130**, 260 (1979); J. Phys. **A16**, 303 (1983)

Moses HE, Prosser RT: J. Math. Phys. **24**, 2146 (1983)

Moses HE, Tuan FS: Nuovo Cim. **13**, 197 (1959)

Munchov M, Scheid W: Phys. Rev. Lett. **44**, 1299 (1980); Phys. Lett. **141B**, 1 (1984)

Muzafarov VM: Sov. J. Theor. Math. Phys. **64**, 208 (1985); **70**, 30; **71**, 3 (1987); Mod. Phys. Lett. **2**, 177 (1987) Inv. Probl. **4**, 185 (1988)

Naatz IE: *The Method of the Inverse Problem in Atmospheric Optics* (Nauka, Novosibirsk 1986)

Nachman AI, Ablowitz MJ: Stud. Appl. Math. **71**, 243 (1984)
Naidoo K, Fiedeldey H, Sofianos SA, Lipperheide R: Nucl. Phys. **A419**, 13–22 (1984)
Natanson GA: Sov. J. Theor. Math. Phys. **38**, 219 (1979)
Neudachin VG, Kukulin VI, Krasnopolsky VM, Obukhovsky IT: Sov. Nucl. Phys. **20**, 883 (1974); Phys. Rev. **C11**, 128 (1975)
Neimark MA: *Linear Differential Operators* (Nauka, Moscow 1969)
Neuman J, Wigner E: Phys. Z. **30**, 465 (1929)
Newton RG: J. Math. Phys. **18**, 1348 (1977); Inv. Probl. **1**, 127 (1985)
Newton RG: *Scattering Theory of Waves and Particles*. 2nd Ed. (Springer, New York 1982); *Inverse Schrödinger Scattering in Three Dimensions* (Springer, Berlin-Heidelberg 1990)
Newton RG: *Proc. Conf. Inverse Scattering* 1983 (Soc. Ind. Appl. Math. Philadelphia 1983)
Newton RG: J. Math. Phys. **21**, 493 (1980); **22**, 2192 (1981); **23**, 594 (1982); **24**, 2152 (1983); in *Scattering Theory in Mathematical Physics* Vol. 193 (Reidel, Dordrecht 1974); Phys. Rev. Lett. **62**, 1811 (1989)
Newton RG, Fulton T: Nuovo Cim. **3**, 677 (1956); Phys. Rev. **107**, 1103 (1957)
Nichitiu F: *Phase Analysis in Physics of Nuclear Interactions* (Mir, Moscow 1983)
Nieto MM: Phys. Rev. **A20**, 700 (1979); Phys. Lett. **145B**, 208 (1984)
Nizhnik AP: *Inverse Nonstationary Scattering Problem* (Naukova Dumka, Kiev 1989)
Novikov RG, Henkin GM: Sov. Usp. Math. Nauk. **42**, 93 (1987); Sov. DAN **292**, 814 (1987)
Novikov RG: Sov. J. Theor. Math. Phys. **61**, 199 (1984); Funct. Anal. Appl. **20**, 246 (1986); **22**, 11 (1988)
Olchovsky VS, Zakhariev BN, Shilov VS: Preprint JINR, P4-89-289, Dubna, 1989; *Proc. Int. Symp.* "Order, Disorder and Chaos in Quantum Systems" Dubna 1989 (World Scientific, Singapore 1990)
Oldham KB, Spanier J: *The Fractional Calculus* (Academic, New York 1974)
Parlett B: *The Symmetric Eigenvalue Problem* (Prentice-Hall, New York 1980)
Pashnev AI: Sov. J. Theor. Mat. Phys. **69**, 311 (1986)
Pavlov BS: *On Nonselfadjoint Schrödinger Operators* (Leningrad University Press 1966); Sov. Usp. Math. Nauk, **42**, 99 (1987)
Pechenick KR: J. Math. Phys. **22**, 1513 (1981); **24**, 406 (1983); **25**, 1900; 1924 (1984)
Perey CM, Perey FG: Atom. and Nucl. Data and Tables, **17** (1976)
Perkus JK: in *Simple Models of Equilibrium and Nonequilibrium Phenomena* (North-Holland, Amsterdam 1987)
Perelomov AM: Sov. Part. Nucl. **10**, 336 (1979)
Petrosyan VI, Slonov VN, Gulyaev YuV: Sov. DAN. USSR. **286**, 99 (1986)
Petrov NM, Tartakovskaya EV: Sov. Izv. Acad. Nauk, ser. Phys. **48**, 1978 (1984)
Pivovarchik VN: Sov. Ukr. Phys. J. **25**, 635 (1980); **26**, 1042 (1981); **27**, 1, 775, 1586; 1867 (1982); **32**, 152 (1987)
Pivovarchik VN, Suzko AA: in *Proc. Inst. Heat Mass. Trans.* 168, Minsk (1982)
Pivovarchik VN, Poplavsky IV, Suzko AA: Phys. Lett. **A101**, 72 (1984)
Pivovarchik VN, Suzko AA, Zakhariev BN: Phys. Scripta. **34**, 101 (1986); Sov. Izv. Acad. Nauk, ser. Phys. **49**, 2227 (1985)
Plekhanov EB, Suzko AA, Zakhariev BN: Ann. Phys. **39**, 313 (1982)
Ponomarev LI, Vinitsky SI: Sov. Part. Nucl. **13**, 1165 (1982)
Poplavsky IV: Sov. J. Theor. Math. Phys. **69**, 475 (1986); Sov. J. Nucl. Phys. **44**, 952 (1986); Ukr. Phys. J. **32**, 485 (1987); Sov. Izv. Acad. Nauk. **51**, 97 (1987)

Popov VS: *Quasiclassical Methods in Three-Body Scattering Theory* (MGU, Moscow 1986)
Popushoy MN: Sov. J. Theor. Math. Phys. **63**, 340 (1985); **69**, 466 (1986); Sov. Izv. Acad. Nauk. ser. Phys. **51**, 952 (1987); **52**, 150 (1988)
Poshel J, Trubovitz E: *Inverse Spectral Theory* (Academic, New York 1987)
Prakash M: J. Phys. A. Math. Gen. 9, 1847 (1976)
Proc. Conf. Inverse Problems Montpellier 1986 (Academic, New York 1987)
Prosser HT: J. Math. Phys. **21**, 2648 (1980)
Quigg C, Rosner JL: Phys. Lett. **B71**, 153 (1977); **B72**, 462 (1978); Phys. Rev. **D17**, 2364 (1978); Phys. Rep. **C56**, 167 (1979)
Rabotnov NS: *Quantum Tunnelling Effect. A Half-Century of Mysteries and Discoveries* (Energoatomizdat, Moscow 1983)
Reed M, Simon B: *Methods of Modern Mathematical Physics*, Vol. 3, *Scattering Theory* (Academic, New York 1979)
Regge T: Nuovo Cim. **9**, 491 (1958)
Rein D: Z. Phys. **220**, 394 (1969)
Revai J: Acta Phys. Acad. Sci. Hung. **25**, 307 (1968)
Ring P, Shuck PNY: The *Nuclear Many-Body Problem* (Springer, Berlin-Heidelberg 1980)
Rudyak BV, Suzko AA, Zakhariev BN: Phys. Scripta. **29**, 515 (1984)
Rudyak BV, Zakhariev BN: Inv. Probl. **3**, 125 (1987)
Sabatier PS: J. Math. Phys. **7**, 1515 (1966); **8**, 905 (1967); **9**, 1241 (1968)
Schrödinger E: Proc. Roy. Irish. Acad. **46A**, 183 (1940)
Schmid EW, Ziegelman H: *The Quantum Mechanical Three-Body Problem* (Pergamon, Braunschweig 1974)
Schonefeld JF, Kwong W, Rosner JL, Quigg C, Thacker HB: Ann. Phys. 128, 1 (1980)
Serebryakov VI: Sov. Trans. Mos. Math. Soc. **49**, 130 (1986)
Simonov YuA, Badalyan AM: Sov. Nucl. Phys. **3**, 630 (1967); **5**, 88 (1967)
Sitenko AG: *Scattering Theory* (Vishya Shkola, Kiev 1975)
Sitenko AG: *Theory of Nuclear Reactions* (Energoatomizdat, Moscow 1983)
Sklyanin EK: Zapisky Nauchn. Sem. LOMI **95**, 129 (1980); Funct. Anal. **21**, 86 (1987)
Skoblin YuA: Sov. Dokl. Acad. Nauk. **293**, 541 (1987)
Skriganov MM: in *Notes of Scientific Seminars*, N7, 149 (Steklov Math. Inst., Leningrad 1973)
Sofianos SA, Fiedeldey H, Allen LJ: Phys. Rev. **C31**, 2300 (1985)
Sokachev E, Stoyanov DT: Mod. Phys. Lett. **A1**, 577 (1987)
Sokolov SN: Rep. Math. Phys. **20**, 235 (1984); Dissertation, Serpukhov Institute of High-Energy Physics (1975)
Sokolov SN, Zakhariev BN: Ann. Phys. **14**, 229 (1964)
Solovtsev IL: Vest. Acad. Sci. BSSR, **N1**, 109 (1984); Izv. Vissh. Uch. Zaved. Fiz. **30**, 11–17 (1987)
Somersalo E: J. Math. Phys. **28**, 2416 (1987)
Soroko LM: Sov. J. Part. Nucl. **12**, 303 (1981)
Spector RM, Aly HH: J. Math. Phys. **5**, 1185 (1964); Nuovo Cim. **38**, 149 (1965)
Spruch L, Rosenberg L, Gerjuoy E, Rau A: Phys. Rev. **116**, 1034 (1959); Rev. Mod. Phys. **55**, 725 (1983)
Stefanescu IS: J. Math. Phys. **23**, 2190 (1982); Z. Phys. **C41**, 453 (1988)
Stein D, Green A: Amer. J. Phys. **50**, 1120 (1982)
Steinmann O: Helv. Phys. Acta. **30**, 515 (1957)

Steklov VA: Soobshcheniya Mat. Obshchestva Ser. 2, **10**, 97 (1907)
Stillinger FH, Herrik DR: Phys. Rev. **A11**, 446 (1975)
Strang G, Fix G: *An Analysis of the Finite Element Method* (Prentice-Hall, New York 1973)
Sukumar CV: J. Phys. **A18**, 2917; 2937 (1985); **A21**, L455 (1988)
Suslov SK: Sov. Usp. Math. Nauk **44**, 185 (1989)
Suzko AA: Dissertation, JINR, Dubna (1978); Phys. Scripta. **31**, 447 (1985); **34**, 5 (1986); Proc. Conf. Few-Body Systems 11, Japan (1986); Thesis JINR, Dubna (1987)
Suzko AA, Kolpashchikov VL: Sov. Ing. Phys. J. **50**, 316 (1986); Int. Symp. Electro-Magn. Theory, Budapest (1986)
Swan P: Proc. Roy. Soc. **228**, 10 (1955); Nucl. Phys. **46**, 669 (1963); Nucl. Phys. **A119**, 40 (1968)
Takhtajan LA, Faddeev LD: *The Hamilton Approach in Soliton Theory* (Nauka, Moscow 1986)
Tarantola A: *Inverse Problem Theory* (Elsevier, Amsterdam 1987)
Taylor JR: *Scattering Theory* (Wiley, New York 1972)
Temkin A: J. Math. Phys. **2**, 336 (1967)
Thaker HB, Quigg C, Rosner JL: Phys. Lett. **B74**, 350 (1978); Phys. Rev. **D18**, 274 (1978); **21**, 234 (1980)
Tikhonov AN, Arsenin VYa: *Methods of Solution of Ill-posed Problems* (Nauka, Moscow 1979)
Trubovitz E, Deift P: Comm. Pure Appl. Math. **30**, 321 (1977); **32**, 121 (1979); **37**, 647, 715 (1984)
Tsai TL: Amer. J. Phys. **34**, 260; 1122 (1966); **44**, 636 (1976)
Turbiner AV: Sov. Func. Anal. **22**, 92 (1988); Sov. JETP, **94**, 33 (1988); see also Leach PGL: Physica, **D17**, 331 (1985); J. Phys. **A20**, 5923 (1987)
Turchin VF, Kozlov VP, Malkevich MS: Sov. Usp. Phys. Nauk, **102**, 345 (1970)
Ulyanov VV: *Methods of Quantum Kinetics* (Kharkov University Press, Kharkov 1987)
Ushveridse AG: Sov. Short Commun. FIAN, N2, 3 (1988); Sov. Part. Nucl. **20**, 1185 (1989); Phys. Scripta **39**, 30 (1989)
Vasilevsky LS, Zhirnov NI: J. Phys. **B10**, 1425 (1977); Sov. J. Theor. Math. Phys. **13**, 108 (1972)
Venakides S: Comm. Pure Appl. Math. **41**, 3 (1988)
Vinitsky SI, Soloviev EA: J. Phys. **B18**, L557 (1985); ibid. **B19**, L765 (1986)
Vinitsky, SI, Suzko AA, Markovsky BL, Kadomtsev MB, Dubovik VM: Int. Seminar Geom. Aspects Quant. Theory, Dubna, IX-1988 in *Topological Phases in Quantum Theory*, 173 (World-Scientific, Singapore 1989)
Vinitsky SI, Kadomtsev MB, Suzko AA: Sov. Nucl. Phys. **51**, 952 (1990); Vinitsky SI, Suzko AA: ibid **52**, N9 (1990)
Wadati M, Kamija T: Progr. Theor. Phys. **52**, 397 (1974)
Wheeler JA: in *Studies in Mathematical Physics* (*Essays in Honor of V. Bargmann*) (Princeton University Press 1976); Phys. Rev. **89**, 1102 (1953)
Wiesner J, Zhidkov EP, Lelex V, Malyshev RV, Khoromsky BN, Khristov E, Ulehla I: Sov. Part. Nucl. **9**, 763 (1979)
Wilkinson JH: *Rounding Errors in Algebraic Processes* (Prentice-Hall, New York 1964)
Wong CW: Phys. Rep. **150**, 283 (1975); Workshop, *Numerical Treatment of Inverse Problems* (Springer, Heidelberg 1982)
Wright JA: Phys. Rev. **B139**, 137 (1965)
Wu T-Y, Omura T: *Quantum Theory of Scattering* (Prentice-Hall, New York 1962)

Yegikian RS, Zhidkov EP: Commun. JINR. 5-85-366; -858 (1985); P5-87-284 (1987)
Zakhariev BN: Preprint P4-90-46, JINR, Dubna, 1990
Zakhariev BN: Preprint JINR P4-7815, Dubna (1974); Sov. Izv. AN USSR. ser. Phys. **47**, 859 (1983); Proc. Conf. Inverse Problems p. 141, Montpellier 1986 (Academic, New York 1987) Few–Body syst. **4**, 25 (1988)
Zakhariev BN, Sokolov SN: Ann. d. Phys. **14**, 230 (1964)
Zakhariev BN, Pustovalov VV, Efros VD: Sov. Nucl. Phys. **8**, 406 (1968)
Zakhariev BN, Nijazgulov SA, Suzko AA: Sov. Nucl. Phys. **20**, 1273 (1974)
Zakhariev BN, Suzko AA: Sov. Nucl. Phys. **22**, 289 (1975)
Zakhariev BN, Melnikov VN, Rudyak BV, Suzko AA: Sov. Part. & Nucl. **8**, 290 (1977)
Zakhariev BN, Melnikov VN, Suzko AA: Sov. Izv. AN USSR, ser. Phys. **43**, 2206 (1979)
Zakhariev BN, Pivovarchik VN, Plekhanov EB, Suzko AA: Sov. Izv. Acad. Nauk. **44**, 949 (1980)
Zakhariev BN, Pivovarchik VN, Plekhanov EB, Suzko AA: in Microscopic Calculations for Light Nuclei, **30**, (Kalinin University Press, Kalinin, 1981)
Zakhariev BN, Pivovarchik VN, Plekhanov EB, Suzko AA: Sov. Part. & Nucl. 13, 1284 (1982)
Zakhariev BN, Nikishov PYu, Plekhanov EB: Sov. J. Nucl. Phys. 38, 95 (1983)
Zakhariev BN, Plekhanov EB: in *Theory of Quantum Systems with Strong Interactions* (Kalinin University Press, Kalinin, 1987) p. 34
Zakhariev BN, Suzko AA: in *Theory of Quantum Systems with Strong Interactions* (Kalinin University Press, Kalinin 1988) p. 20
Zakhariev BN, Zastavenko LG: Phys. Rev. A39, 5528 (1989)
Zakhariev BN, Kostov NA, Plekhanov EB: Sov. Part. Nucl. 8, 290 (1990)
Zakharov VE, Manakov SV, Novikov SP, Pitaevsky LP: *Theory of Solitons* (Nauka, Moscow 1980)
Zaslavsky OB, Ulyanov VV: Sov. Theor. Math. Phys. **71**, 260 (1987)
Zelenov EI: Sov. Theor. Math. Phys. **80**, 253 (1989)
Zhigunov VP, Korolyova TK: Preprint JINR P4-7815, Dubna (1974); Preprint IHEP 79–107, Serpukhov (1983); Nucl. Instr. Meth. **216**, 183 (1983); **A235**, 146 (1985)
Zhigunov VP, Zakhariev BN: *Methods of Close Coupling of Channels in Quantum Scattering Theory* (Atomizdat, Moscow 1974)
Zubarev AL: *Schwinger Variational Principle in Quantum Mechanics* (Energoizdat, Moscow 1981)

Subject Index